ISAAC NEWTON

ISAAC NEWTON

Alchemy, Prophecy, and the
Search for Lost Knowledge

JOHN CHAMBERS

Destiny Books
Rochester, Vermont • Toronto, Canada

Destiny Books
One Park Street
Rochester, Vermont 05767
www.DestinyBooks.com

Text stock is SFI certified

Destiny Books is a division of Inner Traditions International

Copyright © 2018 by the Chambers Family Trust

All rights reserved. No part of this book may be reproduced or utilized in any form or by any means, electronic or mechanical, including photocopying, recording, or by any information storage and retrieval system, without permission in writing from the publisher.

Library of Congress Cataloging-in-Publication Data
Names: Chambers, John, 1939–2017, author.
Title: The metaphysical world of Isaac Newton : alchemy, prophecy, and the search for lost knowledge / John Chambers.
Description: Rochester, VT : Destiny Books, 2018. | Includes bibliographical references and index.
Identifiers: LCCN 2017020306 (print) | LCCN 2017050113 (e-book) | ISBN 9781620552049 (hardcover) | ISBN 9781620552056 (e-book)
Subjects: LCSH: Newton, Issac, 1642–1727. | Philosophy, Medieval. | Philosophical theology. | Theology.
Classification: LCC B1299.N34 C43 2018 (print) | LCC B1299.N34 (ebook) | DDC 192—dc23
LC record available at https://lccn.loc.gov/2017020306

Printed and bound in the United States by Lake Book Manufacturing, Inc. The text stock is SFI certified. The Sustainable Forestry Initiative® program promotes sustainable forest management.

10 9 8 7 6 5 4 3 2 1

Text design and layout by Debbie Glogover
This book was typeset in Garamond Premier Pro with Gill Sans MT Pro and Arno Pro as display fonts

*To my sister-in-law, Bonnie Balas (1945–2012),
who made this book possible, and my wife, Judy,
who makes everything possible*

Contents

	Acknowledgments	ix
ONE	"A History of the Corruption of the Soul of Man"	1
TWO	The Newton Code	15
THREE	Newton's God	44
FOUR	Bloodbath in a Boghouse: Murder in the Fourth Century AD, Part 1	65
FIVE	The Severed Hand: Murder in the Fourth Century AD, Part 2	85
SIX	The Temptation of Saint Anthony	120
SEVEN	The Great Apostasy	130
EIGHT	Apocalypse 2060?	152
NINE	The Conversion of the Jews	172
TEN	With Noah on the Mountaintop	193
ELEVEN	In the Days of the Comet	218
TWELVE	Deconstructing Time	251

THIRTEEN	Chiron and the Star Globe	271
FOURTEEN	A Glitter of Atlantis	288
FIFTEEN	The Secret of Life	303
SIXTEEN	Masters of the *Prisca Sapientia,* Part 1: Aristarchus, Anaxagoras, Numa Pompilius	325
SEVENTEEN	Masters of the *Prisca Sapientia,* Part 2: Philolaus, Pythagoras, Moses, Pauli	346
EIGHTEEN	Son of Archimedes	364

APPENDIX A	Newton's Prophetic Hieroglyphs	390
APPENDIX B	Further Corruptions Found in the New Testament	392
APPENDIX C	Newton's Twenty-Three Queries Concerning the Word ὁμοούσιος	398
APPENDIX D	Newton on God's Anti-Trinitarian Introduction to John's Book of Revelation	402
APPENDIX E	Newton on Ancient Science	410
APPENDIX F	Newton's Translation of the Emerald Tablet	412
	Notes	414
	Bibliography	439
	Index	456

Acknowledgments

For their help in supplying me with valuable information and/or invaluable moral support, I would particularly like to thank the following: Arthur Davis, Roger Doyle, Martin Ebon, Terry Helbick, Brian Johnston, Cat Karell, Guyon Neutze, Henry Roper, Robert Ryan, Steven Sittenreich, and, of course, J.C.

CHAPTER ONE

"A History of the Corruption of the Soul of Man"

Epitaph
Intended for Sir Isaac Newton
In Westminster Abbey
ISAACUS NEWTONIUS:
Quem Immortalem
Testantur, Tempus, Natura, Coelum:
Mortalem
Hoc marmor fatetur.*

Nature and Nature's Laws lay hid in Night:
GOD said, Let Newton be! and all was light.
ALEXANDER POPE, 1730

Not long after the publication in 1687 of Isaac Newton's *Principia Mathematica* (Principles of Mathematics), the great work of science that completely changed the way we perceive the world around us, the Marquis de l'Hôpital, a prominent French mathematician, was shown a

---

*"Isaac Newton: whom Time, Nature, and the Heavens proclaim immortal, this marble reveals mortal."

copy. He read a little, then "cried out with admiration, 'Good god, what a fund of knowledge there is in that book!' He then asked the Doctor [who had shown him the book] every particular about Sir Isaac, even to the color of his hair, [and] said, 'Does he eat and drink and sleep? Is he like other men?'"[1]

Edmund Halley, the comet-chasing astronomer who edited the *Principia,* added to the book an ode that ended with the line "Nearer the Gods no mortal may approach." When he presented Queen Anne with a copy, Halley told her: "All science can be divided into two halves. The first is everything up to Newton. The second is Newton. And Newton had the better half."[2] The French philosopher Voltaire declared: "Before Kepler, all men were blind. Kepler had one eye; Newton had two."[3]

It has been almost three hundred years since Isaac Newton died, and he is still regarded by the world—Albert Einstein and Stephen Hawking affirmed this view—as the greatest scientist who ever lived. His achievements are astonishing, and, if we tend to forget this, it is only because they have become the background noise of the world in which we live.

His mathematical physics have, in the words of physicist Hermann Bondi, "entered the marrow of what we know without knowing how we know it."[4] The world from which Newton departed in 1727 was substantially different from the world into which he was born, and he himself accounted for much of the difference. Ninety percent of the physics and math we learn in high school comes from Isaac Newton. He invented calculus, discovered the binomial theorem, introduced polar coordinates, proved that white light was a mixture of colors, explained the rainbow, was the first to build a reflecting telescope, and demonstrated that one force, that of gravity, is responsible for pulling objects to the ground, guiding every celestial body in its orbit, and generating the tides.

These are only the main headings of Newton's work, which first broke upon a totally unprepared public in the *Principia Mathematica* of 1687. Innumerable individual discoveries and penetrating observations flow from these overarching categories. The *Principia* is universally considered to be the greatest work of science ever written. Newton published an adjunct work, *On Optics,* in 1702. It is considered to illustrate in exemplary fashion how to write a science textbook. But these epochal texts are only one-half of Newton's literary achievement. We know today that he spent almost his entire life writing a whole other book. This work consists of many separate but interlocked parts. If we were to give it a single title, it might be "History of the Corruption of the Soul of Man." It is five million words long—perhaps the minimum number of words required to cover the all-encompassing subject of mankind's troubled relationship to goodness and God.

This "History" consists of hundreds of glittering treatises and fragments of treatises on alchemy, chronology, mythology, the history of Christianity, the interpretation of biblical prophecy, and much more. We find in it none of the astonishingly brave and astute overturning of paradigms that characterize the *Principia Mathematica.* Newton was a pious Christian who believed the Bible was the word of God; that Noah and the Flood were real; and that the world began in 6000 BC—and he never goes beyond those boundaries. But his concept of Christ's relationship to God was heretical for his times. That is why, when Newton was alive, almost no one knew that he was writing this "second book." Newton made sure they didn't know.

The thousands of pages of this treasure trove of nonscientific writings were packed into two metal trunks that were hardly ever opened during the two centuries that they remained stored away on the estate of Newton's descendants. Suddenly, in 1936, the papers were auctioned off to the public at Sotheby's in London. Now we know that, however astonishingly diverse their contents seem, a single thread runs through them all: Everywhere, Newton is charting the descent of man's soul from perfection through constant falling-off and fretful renewals until, not far in our own future (Newton suggests 2060), everything ends with an apocalypse followed by a radically transformed world. Everywhere, Newton seems to be asking: What happened to us? We were once perfect; why are we now so far from that? How can we reclaim our birthright? What form did it take?

During the two centuries that these writings lay hidden away, a myth was fostered of Newton as a genius of incandescent brilliance when he formulated his equations during the day, and an exhausted, doddering fool when he scribbled away at his nonscientific observations during the night. This rumor may have originated in part with orthodox Christian theologians shaken to the core by the fear that Newton's unimpeachably towering intellect might have proved some of their basic assumptions to be wrong. Rivals and envious friends may have contributed to these rumors. And Newton himself my have found it advantageous to encourage them as much as he could.

The 1936 auction at Sotheby's was the first step in a lengthy process that would see the scholarly world only very gradually awaken to the realization that Newton had made a whole other statement to the world. Slowly, this new Summa Theologica made its way into some of the great libraries of the world; over the past eighty years, scholars have read its treatises with escalating attention. In 1998, Cambridge's Newton Project, currently based at Oxford University, began to make these documents available online; as of June 2016, 6.4 million words had been posted.[5]

Scholars now see that Newton deployed his genius just as powerfully over his nonscientific writings as he did over his scientific discoveries. In the words of Steven Snobelen, that "only now are scholars beginning to study Newton's manuscript corpus . . . to reconstruct a holistic view of the man in which his theology and natural philosophy are seen as equally important elements of the same grand unified project, the restoration of man's original pristine knowledge of God and the world."[6]

There is still reluctance to outright call this "second book" of Isaac Newton's a "Principia Theologica" or a "Principia Ethica," and to think of it as a companion, even a corrective, volume to the *Principia Mathematica*. But now that we know that Newton's untempered intellect animates every one of these treatises, it's possible to say, along with the Newton scholar James T. Force, that this kind of simplistic, "two-Newton" interpretation—one a young, brilliant scientist, the second a senile religious nut case—went by the boards when the Yahuda materials, and others, which were auctioned at Sotheby's in London in 1936 showed that Newton was working on these "unscientific" topics throughout his adult life and was, in fact, working on them at the very times that he did his greatest scientific writings.[7]

We can today see the possibility that one man, a product of many intellectual and religious currents in the seventeenth century, could write great scientific works, great works in church history, in biblical interpretation, and so on, as part of one great enterprise, that of understanding man and his place in the grand scheme of God's creation.

Just who was this astonishing man who invented modern science while writing a book of a thousand treatises on the state of the soul of mankind? Did he in fact eat, drink, and sleep like any other man? Let's begin with a brief biography of Sir Isaac Newton.

The prodigious and tormented life of Isaac Newton began on a small farm in the hamlet of Woolsthorpe, in Lincolnshire, England, on Christmas Day, 1642 (January 4, 1643, in most of continental Europe). His father, also Isaac, a competent if illiterate farmer, died three months before he was born. The infant Isaac, so tiny he could fit into a quart mug, wasn't expected to live, but he survived until the age of eighty-four, dying on March 20, 1727.

At the time of Isaac Newton's birth, it was believed that a "posthumous"—a child born after a parent's death—was blessed with healing powers and destined for greatness.[8] Certainly the child Isaac would have needed a degree of special inner power to get through the first decade of his life. When he was three, his mother, Hannah, married a sixty-two-year-old vicar and moved into his house a mile and a half away, leaving Isaac with his maternal grandparents. (She returned when he was ten, her second husband having died, and brought three stepsiblings with her.)

We know nothing of what Isaac did when he was in the care of his grandparents. Did he roam the woods, find solace in nature, and in his loneliness learn to inhabit a world of (as James Gleick writes) "forms, forces, and spirits, some real and some imagined"[9]—a world that helped him form the uncanny faculty of intuition that (as his colleague William Whiston claimed fifty years later) helped Newton arrive at truths before he'd worked out the mathematics that would lead him there?

Virtually motherless from the age of three to ten, Newton must have felt all the more keenly the lack of a father. Critic Frank Manuel wondered if Newton's all-consuming search to find proof of God's existence through his activities in the world did not stem in part from his yearning to find traces of his biological father. Scholar Gale Christiansen believes that Newton's twofold abandonment "played a significant if not decisive role in the development of his sensitive temperament and always enigmatic character" and explains why, in his years at Cambridge, "somber and secretive, he moved about his shuttered rooms like some gray disheveled ghost."[10] Other scholars view the effects as more devastating, one commenting that Newton "grew up isolated as much by a shamed sense of abandonment as by his overweening intellect, emerging as a psychopathic cornucopia of simmering rage and icy disdain."[11]

Hannah managed the estates of her two dead husbands well enough that by the end of the 1650s the Newton-Ascoughs were one of the thousand wealthiest landowning families in England (wealthy meant an income of £700 a year or more); she could afford to send Newton to the King's School in Grantham, seven miles away, where he took a room above the shop of the local apothecary. The school taught much Latin and some Greek, but Newton's real interest lay in making scale models of windmills (including one with a mouse on a treadmill turning the mill and grinding wheat);[12] of watermills that he placed in streams; of a four-foot-high water clock that still told the right time years after he went to Cambridge—of paper lanterns, kites with candles attached, sundials, and for his friends' sisters, "little tables, cupboards & other utensils" where they could "set their babies & trinkets."[13] Newton wrote poems; one was recited to William Stukeley by an eighty-two-year-old woman who told him she had been beloved of the young Newton.* He also drew, decorating "his whole room with pictures of his own making, [of king's heads,] birds, beasts, men, ships & mathematical schemes, & very well designed."[14]

Recently commentators have questioned whether Newton really was a "child engineer"; they believe Stukeley was too quick to accept as truth what was really a sort of village mythology vis-à-vis Newton. All agree that Newton was remote, arrogant, quick to take offense, and did well at school only when pushed—and then he did superbly. Early on Newton showed a certain steely toughness; when sufficiently aggravated by a school bully much bigger than himself, he beat him to a pulp. This toughness served him well when, some years later, he slid a jackknife under his eyeball to see how the light changed when he changed the shape of his eye; and it served him well when, a mere three years later, he wrote the *Principia* and changed the shape of the universe.

With one more year to go in school, Newton's mother called him back to Woolsthorpe; she wanted her eldest son to become a gentleman farmer. Newton tried; but assigned tasks always disintegrated around

---

*William Stukeley (1687–1765) was the first important antiquarian to declare that Stonehenge was a Druid temple. He grew up in Grantham and later came to know the much older Newton personally. Stukeley wrote a memoir of Newton based on interviews with the townspeople of Grantham; he included the old woman's poem, possibly written when Newton was fifteen.

> A secret art my soul prepares to try,
> If prayers can give me, what the wars deny.
> three crowns distinguishd here in order, doe
> present thir objects, to my doubtful view.
> Earths crown thats at my feet, I can disdain:
> which heavy is, & at the best but vain.
> but now a crown of thorns I gladly greet,
> sharp is this crown, but not so sharp as sweet.
> the radiant crown, which I above me see,
> is that of glory, & eternity.
>
> (STUKELEY, "MEMOIR OF NEWTON")

this young man lost in the world of books. Newton's schoolmaster had glimpsed his amazing potential; he begged his mother to send him back to school; Newton's uncle, a Cambridge-trained cleric, added his voice. Reluctantly, Hannah let her son return to King's School. He graduated head of his class and enrolled at Trinity College, Cambridge, in 1661.

An astounding mind, of unprecedented and seemingly unlimited capaciousness, incapable of superficiality and addicted to profundity, began to unfold in a tiny students' quarters next to Trinity Chapel. Newton was registered as a "sizar"; he performed menial tasks for students to keep his tuition down. Newton did this in body only; his mind was already beginning to roam through the universe of knowledge. He studied eighteen to twenty-two hours a day seven days a week. In his first year, the lean and hungry yeoman farmer's son chewed up, swallowed, digested, and rejected much of Plato and Aristotle. In his second year, Descartes's *Geometrie* turned him into a mathematician, though he would soon reject many aspects of Descartes's philosophy. Newton drank in Kepler and Galileo and other mathematicians near his own century, then circled restlessly back to the pre-Socratics of ancient Greece. In his third year, he began to create his own mathematics.

Instruction was rarely randomly distributed at Cambridge. The school was ruled by Aristotelianism. Enormously prestigious, it basically prepared young men for government or the ministry. Who you knew counted for a lot. Isaac Barrow, the first Lucasian Professor of Mathematics and a gifted mathematician in his own right, saw Newton's genius and took him under his wing. He coaxed him into sending a mathematical paper around and was the first to witness the odd mixture of pride and fear in Newton's eyes when mathematicians twice his age treated him with adulation. Newton was already on the road to becoming a tormented genius.

The Great Plague of 1665–66 closed Cambridge for eighteen months; Newton spent the time on his mother's farm at Woolsthorpe. He would call this period of his life "the prime of my age for invention."[15] Here Newton invented calculus, laid the foundations for the law of universal gravitation, and began to structure the laws of motion. Computing in the fields (perhaps lying beside the apple tree from which the famous apple fell), he became, at age twenty-four, the greatest mathematician in the world. Nobody knew it, except, perhaps, himself.

Newton had matriculated as a B.A. before the plague; back at Cambridge he took up a fellowship. At Woolsthorpe he had performed a revolutionary set of experiments that proved that white light was made up of the seven colors of the rainbow. The whole world had thought the opposite was true: that white light was primary, and the spectrum issued from it. His findings were published in the Royal Society's *Philosophical Transactions* and set up an awed furor that went on for four years. Newton's ability to ignore received opinion and make the world un-receive it was beginning to make its disruptive, astounding presence known.

Newton took his M.A. in 1668 and succeeded Isaac Barrow as Lucasian Professor of Mathematics in 1669, still only twenty-six. In 1671 he became the first person in the world to build a reflecting telescope. It did not receive light through a lens but reflected it from a concave mirror; Newton ground the lens himself and also made the tools to build it. He sent a copy to the Royal Society in London; it created a sensation, being much shorter than a refracting telescope and supplying many times the magnification.

Beginning in 1684, the astronomer Edmund Halley, having seen with astonishment some of Newton's unpublished work, coaxed, cajoled, and bullied Newton into writing the *Principia Mathematica,* which was completed in 1687. Snobelen writes that this masterpiece

> introduced not only a powerful, new mathematical physics with which natural philosophers could describe both terrestrial and celestial mechanics with unprecedented precision, but also demonstrated the law-like nature of the cosmos. . . . [Moreover, Newton] through his publications made lasting contributions to the inductive and experimental methods in science.[16]

Newton's fame spread. Soon he was the object of Europe-wide adulation. He moved from Cambridge to London in 1692. Not long after his arrival, he was pursuing two careers, the first as warden and then master of the London Mint, the second as president of the Royal Society of London for Improving Natural Knowledge, founded in 1660 and the oldest—and today still the most prestigious—organization of scientists in the world.

In both of these positions, Newton displayed an excellent grasp of worldly affairs and succeeded admirably. At the mint, he oversaw a complete makeover of Britain's coinage, designed a number of coins, and pursued counterfeiters with a vengeance. While at Cambridge he had in his clandestine writings on Christianity relentlessly tracked down fraud and corruption in the Catholic Church; now he relentlessly tracked these down in the real world.

Isaac Newton was a difficult person to get along with. In all his dealings, both private and public, he could be mean and vindictive. This served him well when it came to tracking down counterfeiters. It served him badly in his friendships, many of which were not friendships so much as sustained and difficult duels. He had a secret: he belonged to a heretical sect of Christianity—he was an Arian—and had many known, he might at the very least have lost his livelihood. This was another tension serving to keep him apart from people. Edward Dolnick writes—and all would concur—that "anyone dealing with Newton needed the delicate touch and elaborate caution of a man trying to disarm a bomb."[17] Victims of his roiled, suspicious, and unhappy mind included the Astronomer Royal John Flamsteed and the natural philosophers Robert Hooke and Gottfried Leibniz; all three came, and justifiably, to despise him.

Although Newton seems not to have been interested in women as such, and was a virgin all his life, in midlife he formed a friendship (that seemed more than a friendship) with the brilliant and charming Swiss mathematician Nicholas Fatio de Duillier, who was half his age. There was some talk between them of Fatio's moving to Cambridge to be near

Newton. Then Fatio abruptly disappeared (into the Continent), and, a month later, Newton had a nervous breakdown (regarding the causes of which, however, there is no easy answer). In sum, Newton was capable of reasonable friendship with those who were not too far below him in intelligence, such as John Locke and Edmund Halley. (This wasn't snobbery; Newton's almost unimaginable brilliance created a real gap between himself and other people.)

Newton died in 1728, aged 84, and was given a funeral at Westminster Abbey usually reserved for heads of state and peers of the realm. A huge monument was erected for him in the abbey. There his soul rested, presumably in peace, for one century, for two centuries— until, in early 1936, it awakened with a shock and looked around.

Isaac Newton was about to make a comeback. He was going to introduce new ideas to the world for a second time.

The man who would release the contents of Newton's two containers of nonscientific writings to the world was a corpulent redheaded British neo-Nazi named Gerard Wallop, who was in 1936 the direct descendant of Newton's stepniece Catherine Conduitt (née Barton) and therefore the custodian of the locked trunks. Also known as Lord Lymington, ninth Earl of Portsmouth (after the estate), Wallop admired Hitler and Mussolini, was stridently anti-Semitic, and advocated turning England into a federation of medieval fiefdoms run by dictatorial aristocrats just like himself.

It's ironic that the murky soul that was Lord Lymington's should have been the one to introduce to the world the bulk of Newton's brilliantly illuminating nonscientific writings. But Lymington needed money badly—not, as has been reported, to fund the British Fascist Party but rather to pay death duties on his late aunt's wealthy estate and for his own divorce. That's why, on July 13 and 14, 1936, 327 lots of manuscript pages covered with Newton's tiny, obsessively tidy handwriting went on the auction block at Sotheby's in London. When the gavel fell for the last time, the greater part of what were then called the Portsmouth Papers had been sold for £9,000 ($30,000 today).*

Most of the items were sold, but the knowledge they contained might not have gotten very far into the world if it hadn't been for two guardian angels, so to speak, who, though they didn't attend the auction, were well apprised of it and hastened to purchase large portions of the papers indirectly.

One was a heavyset Jew from Jerusalem who, avers Professor Richard H. Popkin, was once "one of the most hated people in Israel."[18] This was Abraham Yahuda (1877–1951), who received a Ph.D. in Semitic languages from the University of Nuremburg at the age of sixteen. Yahuda rapidly became perhaps the world's leading scholar of ancient Semitic languages; during his lifetime he was probably the only person on the planet who could speak ancient Assyrian.

Yahuda was a fervent Zionist as a young man, but for thirty years he quarreled ferociously with Zionist leaders as to what form the state of Israel should take. Yahuda didn't want it to be a modern technoscientific state but rather a resurrection of ancient Judaea and Israel; he even insisted he be its president! Eventually the great scholar renounced Zionism, leaving Palestine permanently and pursuing an academic career that took him to the greatest universities of the world and enabled him to befriend other exceptional figures. One of these was Albert Einstein, whom Yahuda helped escape from Nazi Germany in the late 1930s.

The other "guardian angel" who would ensure the world felt the presence of Newton's newly released manuscripts was an Englishman as slim as Yahuda was heavyset and possessed of an equally subtle and inquring mind. This was the Cambridge don and eminent economist John Maynard Keynes (1883–1946), who literally wrote the book on Keynesian economics and who served as an economic adviser and emissary for British prime ministers Winston Churchill and Clement Atlee.

---

*At a Sotheby's auction in 2001, a single manuscript page of Newton's "Paradoxical Questions Concerning the Morals and Actions of Athanasius and His Followers" went for £18,000 ($26,000 at the time). By late 2014, collectors were finding it hard to buy a single manuscript page of Newton's writings for less than £125,000 ($200,000).

Yahuda was somewhat retiring; Keynes was not and led a brilliant social life, lavishly entertaining artists and intellectuals at his country estate, and, despite the fact that he was gay (or perhaps because of it), marrying a ballerina from Diaghilev's Ballets Russes.

Yahuda acquired a considerable portion of the theology manuscripts made available at the Sotheby's auction. Keynes, probably just to keep them together, scooped up the bulk of the alchemical writings. Poring over the manuscripts, both men immediately saw how wrong Voltaire was when he declared that Newton worked at his nonscientific writings "to relieve the fatigue of severer studies."[19] They saw that Newton had focused the same vast piercing intelligence on alchemy and religion as he had on mathematics and physics; that he had explored these areas exhaustively and done so virtually from youth to old age; and that he had made new discoveries in both these fields. They both saw Newton's overall achievement from two new and different perspectives. Keynes declared in a public address in 1942 that Newton "was not the first scientist of the age of reason. He was the last of the magicians, the last of the Babylonians and Sumerians, the last great mind which looked out on the visible and intellectual world with the same eyes as those who began to build our intellectual inheritance rather less than 10,000 years ago."[20]

If Keynes believed Newton came close to being a magician, Yahuda believed he came close to being a Jew. The professor of Semitic languages was moved by Newton's assertion that the universe was created and is ruled by a single God of providence and dominion, to whom Jesus is subordinate in his role of the executor of God's will. These long-hidden writings on theology, which were essentially heretical, revealed Newton to be, wrote Yahuda, a "Judaic monotheist of the school of Maimonides" who believed that "Jehovah is the unique god."[21]

Keynes was intimately in touch with the highest heads of state and the most celebrated artists of his time; Yahuda swam in the deepest depths of academic endeavor; and neither hesitated to chat about the Newton papers and what they were discovering in them. Passing academics listened in. Keynes left his collection of Newton papers on alchemy to the Cambridge University Library in 1946, while Yahuda bequeathed his share to the Jewish National and University Library of Israel in Jerusalem in 1951. Since that time both sets of papers have been scrutinized carefully, with academics recognizing their importance. In the 1990s, a number of distinguished Newton scholars, including Professors Betsy Teeter Hobbs, Richard Westfall, and James Force, sought financial backing in Britain and America to publish a set of volumes of Newton's nonscientific writings. None was forthcoming; but suddenly in 1998 a windfall came in the form of a large grant from the British government to finance a website dedicated to reproducing as many as possible of the Newton papers auctioned at Sotheby's in 1936. As of 2017, the Newton Project has placed seven thousand works online and moving at the academic equivalent of ramming speed to put on many more.

But what exactly do these papers say? That, we will discover in the following seventeen chapters, as we try to plumb the deeper meanings of Isaac Newton's "History of the Corruption of the Soul of Man."

CHAPTER TWO

THE NEWTON CODE

"Sir Isaac Newton, Britain's greatest scientist, predicted the date of the end of the world—and it is only 57 years away."

Thus began the front-page story in London's *Daily Telegraph* for February 22, 2003. The end-of-the-world date was 2060. The *Telegraph* added some details: "Newton, who was also a theologian and alchemist, predicted that the Second Coming of Christ would follow plagues and war and would precede a 1,000-year reign by the saints on earth—of which he would be one."

So great is Isaac Newton's reputation, and so worried was humankind in 2003, that newspapers all around the globe immediately picked up the story that the creator of modern science had predicted the world would end in 2060. On February 23, *Maariv* and *Yediot Aharonot* newspapers in Israel ran the story on their front page. On February 24, major news media in Canada, the United States, and Europe broke the disconcerting news. Word of the prediction surfaced on Internet sites in Latin America, South Africa, India, China, Japan, and Vietnam. Some websites played the story for laughs, one running a photo of a mushroom cloud with the caption: "Party like it's 2060!"

The source of this alarming news was an unassuming Canadian academic named Stephen Snobelen. Assistant professor of the history of science at King's College, Halifax, Nova Scotia, and one of the world's leading experts on the nonscientific writings of Isaac Newton,

Snobelen had come to London to act as a consultant for an hour-long BBC-2 TV documentary titled *Newton: The Dark Heretic*. As the documentary ends, he is shown standing beside the stacks in the Jewish National and University Library in Jerusalem, holding one of three Newton manuscripts known to carry the end-of-the-world date of 2060.

Several days before the show was aired, Snobelen, interviewed by the religion correspondent for the *Telegraph,* had casually let slip Newton's date for the end of the world. No one could have been more surprised than the Canadian historian of science when, next morning, he saw the interview splashed across the front page of the *Daily Telegraph* with Newton's apocalypse date screaming out of the title.[1]

Several months later, in an article in which he tried to explain why a simple prediction by Isaac Newton would cause such a furor worldwide, Snobelen, who is a cofounder of the online Newton Project and director of its Canadian branch, suggested that "Newton's prediction became entangled in real history unfolding in early 2003." Snobelen cited the war in Iraq; nuclear-capable missile rattling in North Korea; India, and Pakistan; and the global epidemic of SARS. "In the context of these times of jittery nerves," he concluded, "it is perhaps not surprising that the 2060 story resonated so well with the public."[2]

Anyone learning of Newton's prediction in 2016 would have found much more cause for alarm. Conflict still rages in Iraq and Afghanistan and a bloody no-holds-barred civil war is tearing Syria apart. A ruthless terrorist army fond of public beheadings, ISIS, lays waste to large tracts of the Middle East as it fights to build a radical Islamic caliphate spanning the Arab world. These wars are driving millions of desperate refugees into Europe, where their presence is straining the once-solid superstructures of these long-established states. North Korea is rattling its nuclear-capable missiles more loudly than ever. A new virus, zika,* is

---

*The zika virus causes microcephaly, a condition in which babies are born with small heads and incomplete brain development.

spreading through South and Central America and threatening North America.

An even more dire problem looms over us. Newton believed the physical cause of the end of the world might be a *diluvium ignis,* a flood of fire. Today, global warming, aka climate change, is beginning to profoundly affect our planet. The Earth's north and south polar ice caps, along with Greenland's ice sheath, are melting so fast that our planet's ocean levels are higher than they've ever been before. Experts warn that if we don't drastically scale back man-made carbon emissions, the Earth will be 4° to 7°C (7.2°–12.6°F) warmer in 2100, a level of heat that mankind will find difficult to tolerate.

In April 2016, the National Oceanic and Atmospheric Administration (NOAA) announced:

> For 2016 year to date (January–March), the average temperature for the globe was 2.07 degrees F above the 20th-century average. . . . This was the highest temperature for this period in the 1880–2016 record, surpassing the previous record set in 2015 by 0.50 degrees F. The globally averaged sea surface temperature for the year to date was also highest on record, surpassing the same period in 1998 by 0.42 degrees F, the last time a similar strength El Niño occurred.[3]

Mankind has caused the problem by allowing carbon to escape into the atmosphere, and only mankind can fix the problem. Except that, in the United States, deregulated capitalism, which realizes trillions of dollars a year from the production of fossil fuels, is unalterably opposed to halting global warming and even to admitting there's a problem. Moreover, a billionare apostle of deregulated capitalism (and denier of global warming), Donald Trump, has just become the forty-fifth president of the United States. "Our economic system and our planetary system are at war,"[4] sums up Naomi Klein in her magisterial *This Changes Everything: Capitalism vs. the Climate* (2014). NOAA warns that if global warming goes on at the present rate, dire consequences await mankind soon after the middle of the twenty-first century.*

Which takes us to about the year 2060.

Isaac Newton was arguably the leading expert of his day on the prophetic texts of the Bible. His 323-page *Observations upon the Prophecies of Daniel, and the Apocalypse of St. John,* published posthumously in 1733, would go into eleven editions. The 1936 Sotheby's auction brought to light hundreds of pages of additional *Observations*-related treatises and drafts of treatises. Some revealed the heretical nature of Newton's religious thinking; others made strikingly original comments about the Book of Revelation. Among the papers were three citing the 2060 end-of-world date.

In the decade before Newton was born, something akin to a Copernican revolution had taken place in the world of seventeenth-century English biblical-prophecy interpretation. The beloved and learned Cambridge don Joseph Mede had devised an innovative, quasi-scientific system for interpreting the Books of Daniel and Revelation. This system was the "Bible code" that, half a century later, Isaac Newton adopted, with modifications and elaborations, for his *Observations.* This "Newton Code" was superior to anything that had come before; in this chapter, we'll take a brief look at it. First, though, we'll look at the mysterious author of Saint John's Apocalypse/the Book of Revelation.

Probably about AD 90, a certain John, minister of seven proto-Christian churches in Asia Minor, was seized by Roman soldiers in Ephesus, put in chains, bundled aboard a boat, and transported seventy miles across the Aegean to the tiny Greek island of Patmos. Here he was flung into prison. His crime (or so he tells us) was vigorously defending "the Word of God and the testimony of Jesus Christ." This John was the Christian

---

*In January 2017, NOAA reported that the Earth was even hotter in 2016 than it was in 2015—in fact, the hottest it's been in 125,000 years.

Jew who would channel the Book of Revelation from his prison cave.

That is practically all we know about the historical John, though there are plenty of myths. Isaac Newton cites one: "John was put by Nero into a vessel of hot oil, and coming out unhurt, was banished by him into Patmos."[5] Another says that while John was on his way to Patmos a storm blew up and a passenger was washed overboard. John prayed for help and the passenger was immediately washed back on board. For the rest of the trip, John preached to his suddenly attentive Roman guards. When he arrived, the governor of Patmos personally removed his chains and escorted him to his cell.[6]

Tourists can visit John's cell today. Its entrance is hewn into a rocky mountainside sloping sharply down to the Aegean. Inside, flickering candles light up places on the wall framed in silver where the prophet rested his head or put his hands when he got up. On the stone floor, a railing traces out the spot where John lay down to sleep. On the ceiling, a painting shows him kneeling in ecstatic wonderment as Jesus dictates the Book of Revelation.[7]

There was nothing romantic about John's imprisonment on Patmos. This was a penal settlement, a mini Devil's Island—a craggy contorted volcanic rock honeycombed with tiny caves that had been converted into prison cells. There was no escape except by sea. Water had to be brought in by boat. One scholar surmises that John may not even have had writing utensils and wrote the Book of Revelation from memory after his release, likely in AD 97, on the death of the emperor Diocletian.

Historian Will Durant wrote in 1944: "It seems incredible that the Apocalypse and the Fourth Gospel should have come from the same hand."[8] In fact, we know today that they did not. The author of Revelation was not the John, son of Zebedee and apostle of Christ, who wrote the Fourth Gospel (though the apostle John apparently lived long enough to do so; Sir Isaac repeats the ancient scuttlebutt that "in his old age he [the apostle] was so infirm as to be carried to church, dying above 90 years"[9]).

Today we know that the apostles were illiterate and that the apostle

John didn't even write the Gospel that bears his name. This was penned by an unknown scribe, probably in Rome in 100 BC. This anonymous author didn't write the Book of Revelation either. Princeton's Harrington Spear Paine and Professor of Religion Elaine Pagels tells us that as early as the third century AD Bishop Dionysius concluded that the author of Revelation couldn't have been the author of the Gospel of Saint John. Pagels writes:

> He [Dionysius] points out differences that literary critics have noted ever since; for example, that John of Patmos often mentions his own name but never claims to be an apostle; that the tone of his writing, the style, and the language, which is "not really Greek" but uses "barbarous idioms," are distinctly different from those of the fourth gospel.[10]

Though we know little about John, we can surmise that he was an extremely angry man. After all, he had seen the Romans trample on everything Jewish and Christian. They had killed Jesus (perhaps John's parents had told the story to the young boy). They had demolished the Temple of Jerusalem (Newton thought John must have been a soldier in the Jewish-Roman war). In the seven cities in which John carried out his ministry he had watched with mounting rage as the Romans raised up huge temples decorated with frescoes celebrating the Roman victory over the Jews.[11]

Rage dominates almost every part of the Book of Revelation. Pagels calls the book "wartime literature" and describes its contents as "not stories and moral teaching but visions, dreams, and nightmares."[12] She tells us that it "speaks to something deep in human nature."[13]

That "something deep" is our terror of the unknown. John of Patmos is (among many other things) the Donald Trump of the first century BC. He terrifies us with his images of bloody wars to come even as he promises that if we are patient Christ will rescue us.

All the images in the Book of Revelation are surrealistic. Fury, hatred, and violence twist them into ever more grotesque samples of surrealism. As the book opens, Satan has lost a battle wth God in heaven and now leads Rome in an attack on Christianity (or that is how the earliest readers of Revelation saw it). Monstrous beasts made up of pieces of other animals stamp, howl, bite, and destroy. Many-headed and multi-horned, they claw their way out of bottomless pits or stalk ashore from the sea like Godzilla. Around their feet, armies as vast as the sea, stained with blood, attack under pitch-black skies shot through with thunder and lightning; they crash together like ocean waves, subside, attack again. Stars fall; the sun goes out; angels scatter poison from the skies.

The Four Horsemen of the Apocalypse stream by furiously; then comes the Whore of Babylon, astride a red dragon, gorgeously and garishly attired. A woman clothed in the sun appears and gives birth to—the Christ child? Is normalcy entering? No; a dragon rears up before her, so gigantic he has to flick one-third of the stars from the sky with his tail to have room to stand; the woman flees in terror. As these hallucinatory images streak by, our Earth staggers in agony, is turned upside down and consumed by fire; fissures open; mountains are drowned, and evil triumphs.

The black fury of Revelation troubled British novelist D. H. Lawrence, who believed the soul of John of Patmos must be "ruled not by love but by that dangerous psychic poison diagnosed unsparingly by Nietzsche and known by its French word *ressentiment*."[14] Classics scholar Michael Grosso agrees, adding that "Tacitus, the Roman writer, may have had this resentful side of primitive Christianity in mind when he said of the new cult that it was based on 'hatred of the human race.' . . . One thing is clear: you cannot fault John for inconsistency; vindictive rage reigns supreme until the last verses of his text."[15]

Carl Jung wondered why the Christ of Revelation was not the Christ of sweetness and light of the Gospels and decided that the beautiful world of Jesus as presented in the New Testament was all the time repressing the part of life that was bound up with evil. The pendulum had to swing back, in Revelation, to a God that (as in the Old Testament) could aggressively combat sin, sinners, and evildoers, making sure that they were all properly punished. The God of Revelation was the dark side of God—his anima, so to speak. (Jung believed that Revelation was not the last word, because the Holy Spirit would eventually manifest as a new incarnation called the Paraclete.)[16]

The Book of Revelation proclaims, writes Grosso, "the coming of a supernatural overturning of the existing order, a cosmic cataclysm [that] would generate a new heaven and new earth."[17] Good emerges from the cataclysm as John beholds the Heavenly City of Jerusalem and witnesses the Second Coming of Christ and the resurrection of the 144,000 who have remained faithful to the Word of God.

These are universal apocalyptic themes, but Revelation often sounds startlingly topical: demon-angels turn the sea "into blood like a dead man's," "every living thing in the sea will die"; and the heat of the Earth will be unleashed upon the wicked as earthquakes rumble and giant hailstones fall. If John is describing global warming, then the Koch brothers are the Antichrist.

Earlier ages also found Revelation topical. Its first readers identified the Roman emperor Nero (AD 37–68) as John's Antichrist. Nero accused the Christians of setting the Great Fire of Rome in AD 64 and had many of them tortured to death. (The emperor was loved, feared, and hated by his subjects. Some things stuck out, like his having murdered his own mother, and a hundred years after his death people still feared he would somehow come back and persecute them.)

When violent schisms shook the early Catholic Church, some decided the Antichrist was Athanasius, the fourth-century Catholic archbishop who upheld the doctrine of the Trinity; others singled out the heretic Arius as God's ultimate enemy. When the Protestant Reformation came, many raged against its leader, Martin Luther, as the Antichrist. The finest thinkers of seventeenth-century England regarded the pope as God's archenemy and excoriated the Catholic Church with a raw vitriol that amazes us today, seeing as it comes from the lips of men of unusual erudition and eloquence. (Newton himself was no shrinking violet when it came to lambasting the Vatican with barracks-room invective.) The centuries rolled on, and the Antichrist became Napoleon, then Adolf Hitler, then Franklin D. Roosevelt (for introducing the Social Security Act), then communism, then the World Bank, then the Federal Reserve—it seems as if the list will go on forever.

Many of the books promoting these ideas were lightweight, with little hard thinking and less scholarship. Most were confined to fundamentalist and evangelical circles; these were big circles, but, by and large, in the twentieth century the Antichrist, the Apocalypse, and John's Revelation seemed to drop below the radar of the average fairly well-educated citizen.

Then, in 1997, a book came along that aroused worldwide interest in the prophetic books of the Bible. This was *The Bible Code* by American journalist Michael Drosnin, which became a runaway global bestseller. It used computers as a search tool, which accounted for some of its popularity. But it was clear that man's age-old anxiety to know the future was still there and that anyone who knew how to provoke it could attract a lot of attention and make a lot of money.

Drosnin's *Bible Code* came under withering fire from the critics. Its promoters hastily ransacked history to find a way of giving the book respectability. They retrieved from the dustbins of the past Isaac Newton's *Observations upon the Prophecies of Daniel, and the Apocalypse of St. John*. Look, they told the critics, Isaac Newton invented a Bible code. If this towering genius did this, isn't that a good enough reason for taking Michael Drosnin's Bible code seriously?

But in fact it wasn't. Critics who went out and bought Newton's volume discovered that the great mathematician's "Bible code" was so much more complex and sophisticated than Drosnin's Revelation-inspired computer-game text that it put the latter to shame. In fact, Newton's book wasn't a Bible code at all. Working with painstaking care over many years, Newton had used layer upon layer of diverse quasi-scientific methodologies to arrive at his complex conclusions. But, while admiring his book, modern-day prophecy mavens decided it was just too much work to get through. Interest in Newton's *Observations* dried up even as sales of Drosnin's *Bible Code* soared.

But the world had rediscovered Newton's learned and beguiling text. Many now saw that interpreting the Book of Revelation was a serious matter, requiring a scientific approach, and that the easy technology-enhanced ways of modern times had gotten in the way of that truth.

Let's briefly examine Drosnin's *The Bible Code* (which was followed by two sequels: *Bible Code II: The Countdown*, in 2003, and *Bible Code III: Saving the World*, in 2010, both of them also remarkably successful). That will help us see all the more clearly, when we broach the subject farther on in this chapter, the unique qualities that Newton's *Observations* possesses.

When the U.S.S.R. invaded Czechoslovakia in 1968, a young mathematician named Eliyahu Rips sat down in a public square in Riga, Latvia, and set himself on fire to protest the Soviet Union's act of aggression.

He was rescued by a crowd and sent to a Soviet psychiatric hospital, where he spent much of his time solving the "dimension subject conjecture," a problem that had bedeviled mathematicians for decades.

Out after two years, Rips immigrated to Israel, arriving with a reputation as a man of unshakable integrity and as a brilliant theoretical mathematician specializing in "group theory." This was why his colleagues took him seriously when, in 1991 and now a professor of mathematics at Hebrew University, he announced he had discovered a code in the Hebrew Bible describing events that took place hundreds of years after the Bible was written.

Rips's "Bible code" was based on the assumption that God had dictated to Moses on Mount Sinai the first five books of the Old Testament—the Torah—in a single string of 304,805 uninterrupted letters.

Rips believed—as had many eminent rabbis of the Middle Ages, including the great sage R. Moses ben Nahman, aka Ramban or Nahmanides (ca. AD 1195–1270)—that not a word of the Torah had been changed since Moses's time.

Words and messages had anciently been encrypted in the Torah using what Rips called equidistant letter sequencing (ELS); that is, if we decide, say, on the number three, and read the letters of the Torah consistently skipping two letters, we will spell out prophetic messages. We can't do this with the Torah per se; we have to do it when the Torah is installed on a computer in such a way as to leave out all spaces, including the spaces for vowel sounds.

This mathematician Rips did; and, searching the Book of Genesis for the names of post-Torah rabbis, he came up with thirty-two. Rips also discovered the birth dates of the rabbis embedded in the text around them.

Rips came up with "future"—that is, post-Mosaic—events predicted for our time. The Hebrew words for *Sadat, president, shot, gunfire, murder,* and *parade,* all lying in one sequence, were, he decided, a prediction of the assassination of Anwar Sadat in 1981. Then he came across the name of General Norman Schwartzkopf, the American who commanded the coalition forces in the 1991 Gulf War.

In 1994, Rips's findings were published in the peer-reviewed journal *Statistical Science.* The journal decided that his conclusions were a "challenging puzzle," not a proven fact. Some claimed Rips et al. had changed the spellings of the rabbis' names ex post facto to effect more matchups with the birth dates, others that he had exaggerated the odds against discovering the rabbis' names and matching them with birth dates.

In 1995, the American journalist Michael Drosnin appropriated Rips's computerized Torah/ELS program and almost immediately came up with words predicting the assassination of Israeli prime minister Yitzhak Rabin in a year; Rabin was in fact assassinated a year later. Drosnin capitalized on this experience with *The Bible Code,* which published in 1997, skyrocketed to number one on the bestseller list. This book inspired a host of imitators and competing Torah/ELS software programs; "doing" the Bible code became all the rage. In 2002, Drosnin published *The Bible Code II.* Though its predictions were completely off target and it was badly reviewed, it too became a bestseller.

Many objections can and have been raised against the predictive methodology hyped by Drosnin. No one believes any more that the Torah was dictated to Moses by God in one fell swoop; for one thing, it tells of events that took place after Moses died and predicts events that you might have expected Moses would try to forestall. "If they knew what was going to happen, why didn't the people shape up?" asks mathematician Randall Ingermanson.[18] Many more objections were leveled, and these days we hear little about Drosnin's Bible code.

Finding out how Isaac Newton cracked the codes of the Book of Revelation is a harsh dousing in reality for biblical-prophecy groupies looking for easy answers. But it is a dousing as instructive as several university courses. Did Newton, using the myriad and intricate methodologies he employed to unseal Revelation, actually lay bare the future history of mankind? The global reaction to the "leaked" apocalypse date of 2060 suggests that people want an answer to that question. And so we will address it. But the seventeenth century was remarkably different from the twenty-first, and it's necessary to begin by filling in some background.

The age of John of Patmos was an age of universal belief in God, if not the God of the Jews or Christians, then the gods of the pagan world. People weren't necessarily better than they are today, but God's presence was felt, feared, and worshipped.

Fifteen hundred years later, men and women in general still believed, if often unreflectively, in the existence of God. The human mind hadn't yet been hardwired to deal with the idea that he didn't exist; for almost everyone, grappling with such an idea was an impossibility, like talking about what would happen if spring turned into winter.

If the people of Newton's time believed in God, that didn't keep them from being constantly affrighted by the treacherous and unpredictable world in which they lived. In 1642, civil war erupted across Britain; a king was beheaded; the war raged on and off for nine years and brought suffering into many homes. The Great Plague of 1665 killed

70,000 people, half of them Londoners. In 1666, the Great Fire burned to the ground two-thirds of the city and left 200,000 people homeless. Disease was rampant; by midcentury the death rate in London exceeded the birth rate, only the stream of migrants from the countryside making up the loss. The practice of medicine was still largely confined to the precepts laid down by the ancient Roman Galen (AD 130–200), with plenty of astrology, magic, witchcraft, herbal remedies, and spreading of excrement on wounds thrown in. Medicine would only begin to change late in the century when Harvey's discovery of the circulation of the blood was made known.

London was a free-for-all of crime. Daring robberies were committed in broad daylight. When families left London for a week they locked up their furniture in upholsterers' warehouses for security. Thieves stole into balls and snipped jewels off the necks of ladies. One day a felon guilty of thirty murders would be hung; the next, a starving waif who stole sixpence. Men were half-hung and then saw their insides cut out before their eyes. Public whippings, pilloryings, bear baitings, and cockfights were standard forms of entertainment. Though crusaders worked for justice, seventeenth-century England tolerated a norm for cruelty that we can scarcely countenance today.

Many hoped, as much as they feared, that the world was coming to an end; Christ would come and save them. The lower classes despaired of God's protection. Mostly illiterate, yearning for certainty, they searched for omens, paid for horoscopes, consulted fortune-tellers, ransacked their dreams for prophecies, worked out lucky and unlucky days, and tied bundles of talismans to their person.

It's not surprising that these troubled times encouraged a belief in millenarianism. This was the notion that the kingdom of Christ would be established on Earth in the very near future, perhaps even tomorrow, and that it would be followed by a thousand years of peace. Christ had told his disciples he would return in their lifetime. Saint Paul wrote a few years later: "The appointed time has grown very short. From now on, those who have wives should live as though they had none . . . and those who buy anything as if they did not own it. . . . For the present shape of the world is passing away" (1 Cor. 7:29–31).

In the second century AD, the prophet Montanus preached the doctrine of millenarianism so forcefully that, a century after his death, "some believed the New Jerusalem seemed already hovering over the earth in readiness for its descent, and Tertullian records how the soldiers of Severus' army had seen its walls on the horizon, shining in the light of dawn, for forty days, as they marched through Palestine."[19]

Near the end of the fourth century, when the church had become almost as powerful as the empire, the pope officially repudiated millenarianism. He didn't want men and women waiting around for Christ to return and tell them what to do; he wanted them to live in accordance with the official strictures of the Roman Catholic Church. Millenarianism went underground and reemerged as a potent force only after Martin Luther and the Protestant Reformation effectively challenged the all-encompassing power of the Catholic Church. "Luther and some of the other reformers identified the Pope with Antichrist and attempted to find a prediction of the Reformation in the Apocalypse," writes Professor Peter Clouse. "The expositors of the seventeenth century further modified their views of the Apocalypse until many of them became millenarian."[20]

Millenarianism became especially popular in England, where John's Book of Revelation was a source of intense interest for a group of thinkers who wanted to know exactly when the Kingdom of Christ would come.

The upper classes of England were often as frightened by life and its uncertainties as were the poor and the illiterate they looked down upon with disdain. The prophetic books of the Bible became the fortune-tellers and crystal balls of aristocrats, university dons, physicians, and barristers. The greatest minds of the day—Isaac Newton, John Locke, Henry More, Robert Boyle, and many more—studied Daniel and Revelation with the same seriousness that physicist Stephen Hawking breathes into the study of black holes today, or linguist Noam Chomsky brings to his scathing comments on American democracy. Beneath the dreaming spires of Oxford and Cambridge, in the great manor houses of the countryside, in the living quarters of cathedrals, natural philosophers and Anglican divines wrestled with the Whore of Babylon or galloped beside the Four Horsemen of the Apocalypse. They counted the horns on the Beast from the Sea and sounded the depths of the bottomless pit, watching anxiously as the dragon's tail swiped one-third of the stars from the sky. All strove mightily to undo the Gordian knot that was the Book of Revelation.

Most of these men were unmarried. The muse of prophecy, waltzing her way through the seven scrolls in Revelation as seductively as Salome ever danced the dance of the seven veils, was the mistress of them all. She beckoned, and they followed; they held back, and she let slip another secret. These men of note circled each other warily, each jealous of what she might have whispered to the other.

Scholars Buchwald and Feingold report that

> Newton is said to have refused to see Richard Bentley [Master of Trinity College and preeminent classical scholar] for an entire year because the latter dared to inquire whether Newton "could *demonstrate*" that a prophetic day [such as were found in Daniel and Revelation] denoted "a *year* in their completion." According to William Whiston, Newton was offended by Bentley's challenge because he interpreted the request "as invidiously alluding to his being a *mathematician;* which science was not concerned in this matter."[21]

The same Bentley argued so vociferously with his fiancée over a passage in Revelation that she broke down in tears. Was this because he'd spoken too harshly to her? Or because she disagreed with him? Or was it because the future Mrs. Bentley had suddenly realized that a marriage to Mr. Bentley would be a ménage à trois—an eternal triangle of herself, her husband, and John of Patmos? Whatever the case, she married him anyway.[22]

In August 1680, in a letter to mathematician John Sharp, philosopher Henry More described how much he had enjoyed an evening spent discussing Revelation with Isaac Newton.

> For after his reading of the Exposition of the Apocalypse which I gave him, he came to my chamber, where he seem'd to me not only to approve my Exposition as coherent and perspicacious throughout from the beginning to the end, but (by the manner of his countenance which is ordinarily melancholy and thoughtful, but then mighty lightsome and cheerful, and by the free profession of what satisfaction he took therein) to be in a manner transported.[23]

This letter gives us a glimpse of a more tender side to Isaac Newton, though later More would be pained to discover that Newton hadn't changed his views one bit after reading More's exposition.

Newton was close to the Swiss mathematical genius Nicholas Fatio de Duillier, who understood the *Principia* in depth and shared Newton's passion for biblical prophecy. Fatio had a fantastic linguistic gift and was said to speak fifty-two languages; you might have thought this would help him interpret prophecy, but it only helped him to go overboard. He saw the entire Bible as a Book of Revelation, believing the serpent in the Garden was the Roman Empire, Eve was the pristine Christian Church, and Adam was the clergy caught in between. Fatio glimpsed Apocalypse, martyrs, and the papacy behind almost every paragraph of Job, Psalms, Proverbs, and much more. His damn-the-torpedoes approach to the prophetic texts was one of the few strains in his relationship with Newton, who chided him gently in a letter: "I am glad you have taken the prophecies into consideration & I believe there is much in what you say about them, but I fear you indulge too much in fancy in some things."[24]

Not all eminent Europeans, especially busy generals and anticlerical philosophers, took the Book of Revelation seriously. In 1712, the mathematician William Whiston contacted Prince Eugene of Savoy (comrade in arms of the Duke of Marlborough) to tell him the prince's victory over the Turks at Corfu, along with the Peace of Karlowitz, were foretold in Revelation 9:15. In response, "the prince sent Whiston fifteen guineas and a note thanking him for bringing it to the prince's attention that he 'had the honor to be known to St. John.'"[25] In France, the hater of Catholicism and iconoclastic writer Voltaire wittily remarked: "Sir Isaac Newton wrote his comment upon the Revelation to console mankind for the great superiority he had over them in other respects."[26]

"Zeus himself intends a prophet's revelations to be incomplete, so that humanity may miss some part of Heaven's design."

These words are spoken by Phineus, the blind seer in Apollonius of Rhodes's epic poem *The Voyage of Argo*. The seer had once disclosed Zeus's intentions, and the god had punished him by blinding him.*[27]

Did Zeus know, as Doc Emmett Brown puts it in the hit 1980s film trilogy *Back to the Future*, that knowledge of the future can cause a rupture in the space-time continuum and change the present and the past? We don't know whether Newton believed this, but we do know that he believed God intended that the prophecies in Daniel and Revelation should be understood only after the prophesied events had gone by; God had designed his hieroglyphs in such a way that we can only know the future once it has become the past.

That being said, there was a good reason for interpreting biblical prophecy even after the prophesied events had taken place. It was to

---

*Zeus seems to have really hated prophets who blabbed about the future. Apollonius writes that in addition to blinding Phineus the god gave him a lingering old age and "even robbed him of such pleasure as he might have got from the many dainties which neighbours kept bringing to his house when they came there to consult the oracle. On every such occasion the Harpies swooped down through the clouds and snatched the food from his mouth and hands with their beaks, sometimes leaving him not a morsel, sometimes a few scraps, so that he might live and be tormented. They gave a loathsome stench to everything. What bits were left emitted such a smell that no one could have borne to put them in his mouth or even to come near." (Apollonius, *The Voyage of Argo*, trans. E. V. Rieu [New York: Penguin, 2006].)

demonstrate the existence of God. Scholar Matt Goff writes: "Since the Bible was thought to be divinely inspired, one can 'prove' that human history is all carried out according to a divine plan by showing how historical events correlate to biblical prophecy.... This work [Revelation] attempts to demonstrate that God orchestrates human history."[28]

Prophecy wasn't about the prophet, Newton insisted. It was about God. God had dictated the future history of mankind to John so that, once that future history had become the past, you could match it up with the actual facts of history and demonstrate that there was a God. John should not get a swelled head over this.

In Newton's words:

> The folly of Interpreters has been, to foretell times and things by this Prophecy [John's], as if God designed to make them prophets. By this rashness they have not only exposed themselves, but brought the prophecy also into contempt. The design of God was much otherwise. He gave this and the Prophecies of the Old Testament, not to gratify men's curiosities by enabling them to foreknow things, but that after they were fulfilled they might be interpreted by the event, and his own Providence, not the Interpreters, be then manifested thereby to the world. For the event of things predicted many ages before, will then be a convincing argument that the world is governed by providence.[29]

But Newton and his fellow prophecy enthusiasts were only human, and they couldn't resist trying to find out what the future might hold for the seventeenth century, in particular, when Christ would return and the End Times begin. But the real thrust of Newton's monumental examination of Revelation was to unveil the future events God had dictated to John so that Newton could demonstrate that they jibed with the actual facts of history.

Newton prepared himself for the task of interpreting Daniel and Revelation with a diligence that puts to shame today's biblical prophecy mavens looking for a trip to the future as easy as hitting the keys of a computer. Richard Westfall writes:

> First of all, with his customary thoroughness Newton established a proper text for Daniel and Revelation. He compared twenty different editions of Revelation and two manuscript versions, scouring "individual passages such as he could find them in ancient commentators" such as Cyprian, Irenaeus, and Tertullian. Rather early, he composed a *Variantes Lectiones Apocalypticae* ("Variant Apocalyptic Readings") which proceeded through Revelation verse by verse, indicating variant readings from his many sources.[30]

The texts of Daniel and Revelation are peppered with surrealist hieroglyph-like images. Newton called these symbolic images prophetic figures or prophetic hieroglyphs. He believed that in the beginning of the world mankind shared a universal language written in hieroglyphs. All prophets worldwide shared a second-generation hieroglyphic language. The hieroglyphs of ancient Egypt were a first cousin to this prophetic language, which was, as scholar Michael Murrin explains, a "common symbolic discourse, resembling a code which could be broken or a forgotten language which could be recovered."[31]

Newton wrote:

> As critics for understanding the Hebrew consult also other Oriental Languages of the same root, so I have not feared sometimes to call in to my assistance the eastern expositors of their mystical writers. . . . For the language of the Prophets, being hieroglyphical, had affinity with that of the Egyptian Priests & eastern wise men & therefore was anciently much better understood in the East then it is now in the west.[32]

When we watch Isaac Newton scouring the ancient literatures of the world for help in interpreting Daniel and Revelation, he comes across to us as an extremely buttoned-down seventeenth-century version of Indiana Jones racing in hot pursuit of the lost Ark of the Covenant or the ever-vanishing Holy Grail. There's a touch of the romantic in the source books Newton uses. One is the *Oneirocriticon,* or *Interpretation of Dreams,* written by one Achmet, who preferred to be called the son of Sereim, dream interpreter to Mámún, Caliph of Babylonia. We know today that this extraordinary text, from which Freud borrowed the title for his own *Interpretation of Dreams,* was actually written in the tenth century AD by an anonymous Byzantine Christian.

The *Oneirocriticon* was based on the *Oneirocritica* (also translated as *Interpretation of Dreams*) written in the middle of the second century AD by the Hellenistic Greek Artemidorus of Daldis. Legend has it that Artemidorus copied many of his dream interpretations from the ancient monuments of Egypt, Persia, and India. Some of these "dream graffiti" are said to have come from the "Dream Book" of Ashurbanipal, king of Assyria from 669 to 626 BC.

Ashurbanipal's great loves were war and learning. Pillaging the homes of the more literary of his conquered adversaries helped him build a library of thirty thousand cuneiform tablets. The king's "Dream Book" was discovered in the ruins of this library early in the nineteenth century—or not: no trace of the book remains today. Ashurbanipal's journal of nocturnal adventure was said to be the final link in a chain of dream books stretching back to 5000 BC. (The king supposedly wrote on a tablet found in his library the tantalizing words, "I have examined stone inscriptions from before the Flood."[33])

Another sourcebook Newton used derived from the shifting cultures of the Middle East two centuries before Christ. By the fourth century BC, the use of Hebrew as the vernacular language in Palestine was in steady decline. By the second century, Aramaic was the language of choice, and Palestinians no longer learned Hebrew. Aramaic translations of the Hebrew Bible began to appear, usually taking the form of semi-paraphrases in Aramaic with plenty of commentary.

These new texts were called *targumin* ("translations," sing. *targum*).

Newton used the *Targum Jonathan ben Uzziel,* aka *Targum Onkelos,* which was probably written in the first century AD though the final version, a Latin translation used by Newton, wasn't completed until the fourth century AD.

Rabbi Jonathan ben Uzziel was a man after Newton's heart; it's said this prodigious rabbi studied Torah so intensely that "the birds flying over him were burnt to death."[34] The mathematician used the rabbi's *targum* under the title of *Chalde Paraphrastas (Chaldean Semiparaphrases).*

Here is Newton using the *Oneirocriticon* and the *Chalde Paraphrastas* as he works his way toward the meaning of the prophetic figure "locust":

> Locusts are generally referred to a multitude of enemies—If any king or Potentate see Locusts come upon a place, let him expect a powerful multitude of enemies there: & look what hurt the Locusts do the enemy will do mischief proportionally. Ind. Pers. Ægypt. in Achmet. c. 300. . . .
>
> 13. Wild beasts also by reason of their feeding upon vegetables, & preying upon one another signify Kingdoms of the Earth with their armies. A particular Beast, as in Daniels prophesies, signifies a particular kingdom, & Beasts in general kingdoms in general. Come ye, assemble all the Beasts of the field, come to devour. Chal. Paraphr.[35]

Newton believed the fact that God himself had created the prophecies of Daniel and Revelation meant that they were eminently decipherable. "If it cannot be understood," he queried, "then why did God give it? Does he trifle?"[36] They were simply expressed with simple meanings. "Truth is recognized as such by its simplicity and harmony," Newton wrote, admonishing us to

> choose those constructions which, without straining, reduce things to the greatest simplicity. . . . Truth is ever to be found in simplicity,

and not in the multiplicity and confusing of things. As the world, which to the naked eye exhibits the greatest variety of objects, appears very simple in its internal constitution when surveyed by a philosophic understanding, and so much the simpler, the better it is understood, so it is in these visions. It is the perfection of all God's works that they are done with the greatest simplicity.[37]

The language of biblical prophecy was consistent from the first page of the Bible to the last: "John did not write in one language, Daniel in another, Isaiah in a third, and the rest in another," Newton declared; all the prophets "wrote in one and the same mystical language."[38]

There were no garish cosmic extravaganzas in the Book of Revelation. What Newton found instead were commonsensical down-to-earth people and places. The prophets consistently employed an "analogy between the world natural & [the] world politique";[39] everything in Revelation could be translated into a political or social entity. Manuel explains that Newton worked out a dictionary of historical, political, and ecclesiastical equivalents for the images and symbols in prophetic literature. Once an appropriate political translation of any given "prophetic hieroglyph" had been determined, the same meaning had to apply whenever it appeared in a book of prophecy. The tests of truth were constancy and consistency.*[40]

If you were looking for the wild and fantastic, you would be disappointed. Everything worldly (for example, men, beasts, insects, greenery, and so on), and everything cosmic (for example the sun, the stars, the planets, and so forth), always referred to a political entity. "Heaven" always meant a mundane throne or court or honors bestowed by a court. "Eye" did not have the mystical aspect that is imputed to the Masonic eye on the American dollar; it always meant worldly knowledge. "Beast," though it may make us think of monsters, always referred to "a body

---

*For a list of twenty-two prophetic hieroglyphs/figures from Newton's dictionary, the reader is invited to go to appendix A.

politique & sometimes a single person, . . . or an army whereby kingdoms are usually founded and upheld." ("If any man interpret a Beast to signify some great vice," adds Newton, "this is to be rejected as his private imagination.)[41] "Woman" always meant a church—and when we know this we know that we will never find anything risqué in Newton's interpretation of the Book of Revelation.

Our interpretations must never be personal. Newton tells us we should "rely rather upon the traditions of the ancient Sages than upon the suggestions of private fancy."[42] Propriety, reason, simplicity, harmony: these qualities define the character of a prophetic text, and they also define the character of the prophet and of the interpreter of prophecy. Newton believed, writes Manuel, that divinely inspired prophets were never

> enthusiasts, ranters, men who spoke with tongues. . . . [The prophet was, rather,] immensely learned, of impeccable moral virtue, a man who had devoted himself to years of study, and who when properly prepared was the perfect vehicle for God's word. . . . [He was] a supremely rational man, a man worthy of receiving a message from the Divine Reason through the agency of the prophetic spirit.[43]

The most famous of all the seventeenth-century interpreters of divine prophecy, Joseph Mede, came close to fitting this description. Mede (1586–1639) entered Christ's College, Cambridge, when he was fifteen, and in the course of his stay there (he became a don) mastered biblical scripture, philology, world history, mathematics, physics, botany, anatomy, astrology, and the Semitic languages. He must have had a gift for connecting with people as well as for connecting with Christ through the Book of Revelation, for as a Cambridge don he was constantly surrounded by admiring pupils (the poet John Milton was one of them), and he carried on an immense correspondence with most of the learned men of Europe. He died at age fifty-three, probably from overwork.

Mede wrote what is today arguably the most unread masterpiece in the history of English literature. This was the *Clavis Apocalyptica* (*Key to the Apocalypse*), which appeared in Latin in 1627. Lord Protector Oliver Cromwell was greatly taken by this book and ordered that it be translated into English and a copy placed near the pulpit of every Puritan Church in England and Scotland. The translation, titled *Key of the Revelation Searched and Demonstrated,* appeared in 1643, and did end up in many churches.

So Mede's *Clavis Apocalyptica* was certainly read in its time—and by no one more avidly than Sir Isaac Newton—but when the Age of Reason dawned and the idea that God had dictated the prophetic books of the Bible became untenable, Mede's book began to lose favor and was quickly relegated to the attics of the scholars who had once pored over it with delight.

Mede invented two laws for interpreting biblical prophecy. The first was the law of synchronistical necessity. To understand what this law is all about, we have to leave Joseph Mede for a moment and look at the Book of Daniel, which Mede, and then Newton, regarded as Revelation's twin and an indispensable guide to deciphering the Book of Revelation.

> And suddenly in the midst of the darkness a vast hall appears, illuminated by golden candelabra. Candles so lofty that they are half lost in the darkness, stretch away beyond the lines of banquet tables, which seem to extend to the horizon. . . . On the pavement below crawl the captive kings whose hands and feet have been cut off; from time to time he flings them bones to gnaw. Further off sit his brothers with bandages across their eyes, being all blind.[44]

Thus does the French novelist Gustave Flaubert describe in bloody, surreal, and exaggerated detail the splendor and misery of the banquet hall of King Nebuchadnezzar of Babylonia, who ruled from 605 to 562 BC.

One of the "brothers with bandages across their eyes" might well have been Zedekiah, vassal king of Jerusalem from 597 to 587 BC. In 587 BC, Nebuchadnezzar conquered Jerusalem, destroyed the Temple of Solomon, took Zedekiah captive, forced him to watch as his children were slaughtered, and had his eyes put out. Blind, stumbling, and shattered, the defeated vassal was led off to Babylon in chains.

Nebuchadnezzar had conquered Jerusalem once before. On March 16, 597 BC, he had subdued the capital of Judah, pillaged the city and the Temple of Solomon, and forced the Jewish king Jeconiah (or Jehoiakim) to endure the same long march to Babylon that Zedekiah would endure ten years later. With the king came "all Jerusalem, and all the princes, and all the mighty men of valor, ten thousand captives, and all the craftsmen and all the smiths . . . except [for] the poorest people of the land" (2 Kings 24:14). Among the captives was a handsome young prince named Daniel. He and other Jewish royals were accepted into Babylonian society and treated almost as equals. Daniel acquired a reputation as a soothsayer and dream interpreter.

The Book of Daniel tells us that, in about 575 BC, King Nebuchadnezzar had a dream so disturbing that he summoned all the soothsayers in Babylonia to interpret it. Nebuchadnezzar administered to them a test; if they could not themselves redream the king's dream, they were not capable of interpreting. Daniel redreamed the dream, thus passing the test, and told the king the following: He had seen a gigantic metal statue rise up, its head made of gold, its chest and arms of silver, its belly and thighs of iron, and its legs of brass. Its feet were a mixture of bronze and clay. A boulder rolled down a hill and pulverized the statue. The boulder then became a huge mountain.

Daniel told Nebuchadnezzar that the dream foretold the rise and fall of four world empires: Nebuchadnezzar's, the Median, the Persian, and the Macedonian. The boulder symbolized a messianic kingdom that would put an end to all worldly powers. (Later readers of Daniel found that it made more sense to interpret the sequence of empires as Babylonian, Persian, Greek, and Roman.[45])

Much later, during the reign of the king's son Belshazzar,

Daniel had several visions of his own that reinforced his reading of Nebuchadnezzar's dream. Newton describes it in the *Observations;* in the first vision "the prophecy of the four empires is repeated, with several new additions." Daniel sees in rapid succession a lion with eagle's wings, a bear with three ribs in its mouth, a leopard with four heads and four wings, and a fourth "beast" with "huge iron teeth." Each beast makes clearer which kingdom is intended: the lion with wings suggests the winged bulls of ancient Babylonian statuary; the three ribs in the bear's mouth are the three kingdoms conquered by the Medes and Persians (Babylon, Lydia, and Egypt); the leopard with four wings conveys the tremendous speed with which Alexander the Great conquered almost all the known world, thereby laying the foundations for the Hellenistic Greek empire; and the beast with the iron teeth is Rome.[46]

Daniel's second vision built on his first. He saw a ram with two horns and a he-goat with a horn between its eyes square off against each other. The he-goat subdued the ram with its horn. That horn was broken but became four horns, from one of which a smaller horn appeared and then grew very big. Daniel 8:18 goes on: "As for the ram which you saw with the two horns, these are the kings of Media and Persia. And the he-goat is the king of Greece; and the great horn between his eyes is the first king." Newton believed this dream foretold the coming of the Roman Empire, the world-historical event whose history will be foretold in the Book of Revelation. (Daniel had yet another vision, which later scholars believed foretold the birth of Christ and the coming of Christianity. We will discuss this vision in chapter 9, "The Conversion of the Jews.")

Because the four empires were actual historical entities that took the path described for them by Daniel, in ages to come most scholars, one of them Isaac Newton, believed the Book of Daniel sucessfully foretold the future and proved the existence of God. Today, scholars no longer believe the Book of Daniel was written in the sixth century BC, or even that there was a Daniel. Instead, they believe Daniel was written between 167 and 163 BC by an anonymous scribe to "predict an evil end for Antiochus IV Epiphanes, a Hellenistic king of Syria who set up an altar—and possibly a statue—to Zeus in the Temple in Jerusalem. To the Jews, this was an unthinkable abomination," writes Richard Smoley.[47] And Northrop Frye writes that "the Book of Daniel turns into Aramaic halfway through, and could no more be written by a contemporary of Nebuchadnezzar than a book that turned from Latin into Italian could be Julius Caesar's."[48]

The author of Daniel knew exactly how to draw in his audience. He placed his drama in a setting that was charged with meaning and emotion for the Jews: the court of Nebuchadnezzar at the time of the Babylonian Captivity. And he wove Jewish myth and legend into a heady narrative that could not fail to delight a Judaean audience: God himself predicting the collapse of mighty empires that had oppressed the Jews.

Finally, explains Goff, "in declaring that the Book of Daniel was written during the Captivity, he made it appear as if the prophecies had *come true,* and that the Book of Daniel was therefore truly the word of God and they, the Jews, should rest assured that He existed and that He was compassionate."[49]

To this day, many fundamentalist and evangelical churches, such as the Church of Seventh Day Adventists, still believe the Book of Daniel was written in the sixth century BC. They regard Daniel as a fount of universal wisdom and believe his book contains solutions to any problems that might beset mankind. In 2013, when the Tea Party threatened to shut down the federal government over Obamacare, Seventh Day Adventist churches across the United States offered sermons on "Daniel: God's Health Care Plan Revealed." Thus did Daniel join hands with the Republican Party to help maintain the profitability of Big Pharma and U.S. medical insurance companies.

Now we can return to Mede's law of synchronistical necessity. Mede, and then Newton, believed the Books of Daniel and Revelation were a single prophecy. Daniel tells the future history of the world down to the early Roman Empire and the birth of Christ, and Revelation picks up from there (both books surge ahead to Apocalyptic times, though Daniel only briefly). The author of Daniel backs up this single-prophecy hypothesis by quoting God's words: "But thou, O Daniel, shut up the words, and seal the book, *even* to the time of the end: many shall run to and fro, and knowledge shall be increased" (12:4). Mede asserted that Daniel's sealed book was the scroll Christ unseals in Revelation. Newton sums up: Daniel's words were "shut up & seal[ed] till the time of the end."[50]

Therefore references in Revelation can be interpreted in terms of references in Daniel; this is the law of synchronistical necessity. Daniel tells us that "horns" refer to kingdoms, and Mede and Newton apply this information to the various birthings of horns in Revelation. Important prophetic time spans in Daniel are also repeated in Revelation. Most important is "time, [two] times, time-and-a-half." Mede maintained that this equaled 1,260 years, because a "time" meant a day and a day meant a year (1 + 2 + ½ = 3½ days = 3½ years = 1,260 days = 1,260 years); he applied this same interpretation in Revelation.

Mede also invented the law of homogeneal necessity, which states that when prophesied events take the same amount of time to transpire, they are the same event; before Mede, such events were thought to succeed one another. Mede argued, for example, that the four events of the Woman driven into the Wilderness (and who lived there for 1,260 days); the Seven-Headed Beast Restored (restored and empowered for 42 months); the Outer Court Trodden Down by the Gentiles (for 42 months); and the Witnesses on Earth prophesying in Sackcloth (for 1,260 days) were the same event, since 42 months = 1,260 days. Each event has the same temporal length, said Mede, and therefore they are the same event, or aspects of the same event.

The contents of a biblical prophecy couldn't be known until the events it prophesied had gone past. So Newton (and Mede) studied the actual history of the Middle East and Europe from the time of John of Patmos to the seventeenth century. It was a question of matching these events up with prophecies so that they could show that these events were the very same events that John had predicted in Revelation. Then they had proved the existence of God; they had accomplished what they had set out to accomplish.

Richard Westfall explains that

> to vindicate the dominion of God, he [Newton] must demonstrate as well that the facts of history have corresponded to the words of prophecy. . . . He ransacked the ancient historians and chroniclers, pagan and Christian alike, plus orations, letters, the Theodosian Code, anything that could help him establish the order of events. . . . In one short passage of ten pages, in which Newton was concerned to establish that peace broke out [between Romans and barbarians] for a short period beginning in 380, he cited Zosimus, Theodoret, Cedrenus, Baronio, Marcellinus, Ammanianus Marcellinus, Socrates (the historian), Sozomen, Prudentius.[51]

Westfall cites eighteen additional sources that Newton drew on to write these ten pages. He evaluates Newton as a historian.

> No one, I think, would call Newton a great historian. He approached history with an *a priori* pattern of interpretation, and he produced indigestible catenae of quotations instead of readable narratives. His goal was rigor rather than *belles lettres,* however, and I suspect that no one would sneer at him on that score. He brought the standards of scientific demonstration to historical research. He pursued evidence relentlessly. I seriously doubt that any historian has ever attained a firmer grasp of the facts relating to the barbarian invasions of the fifth and sixth centuries.[52]

CHAPTER THREE

NEWTON'S GOD

Tis the temper of the hot and superstitious part of mankind in matters of religion ever to be fond of mysteries, and for that reason to like best what they understand least.
ISAAC NEWTON, "AN HISTORICAL ACCOUNT OF
TWO NOTABLE CORRUPTIONS OF SCRIPTURE,
IN A LETTER TO A FRIEND," NOVEMBER 1690

On November 14, 1690, Isaac Newton sent John Locke (1632–1704), England's leading philosopher, a 25,000-word letter whose contents were so heretical that if word had gotten out about them, Newton would surely have lost his professorship at Cambridge. It wouldn't have mattered that the *Principia Mathematica* had been out for three years and Newton was being hailed as the greatest natural philosopher of his time and perhaps of all time.

There were some in England who, if they'd gotten wind of the contents of this letter, would have done their utmost to have Newton thrown in jail. A thousand miles away at the Vatican, there were cardinals who, if they'd had the slightest hint of what it said, would have whispered in the pope's ear that it was time for the blasphemous genius from England to be burned at the stake.

When the judicious and dispassionate philosopher John Locke, who lived on the luxuriant estate of Lord and Lady Masham near Oxford, read the letter in its entirety, he must have paced the famous gardens of that illustrious manor house with a look of more than usual thoughtfulness on his face. He'd be in great trouble himself if people found out he'd received this letter.

Locke had little knowledge of mathematics, but men of genius had explained the *Principia Mathematica* to him while he was in exile in Holland. The philosopher recognized the greatness of Newton's achievement, and when he returned to England he and Newton became friends. Not close friends, for Newton was almost incapable of that; but the two worlds these men had created with exemplary genius, modern psychology on the part of Locke, and physics and mathematics on the part of Newton, did not really intersect, so there could be no ideological clashes when the two men met, but only mutual, appreciative admiration.*

They shared the same religious views (Locke wrote extensively about "simplifying" Christianity); Newton trusted the preternaturally fair-minded Locke; and so it is not surprising that Newton sent Locke this 25,000-word letter—actually a treatise—in which Newton sought not so much the great philosopher's approbation so much as confirmatory, supplemental knowledge.

Locke was in agreement with everything Newton wrote in the letter, including the inflammatory parts. But as he read what was so

---

*Locke was done a fine service in his career by Catharine Trotter Cockburn (1679–1749), one of the female geniuses whom modern researchers are beginning to extricate from the male-dominated history making and history writing of the past. Cockburn knew French, Latin, and Greek; had learned much philosophy; and at age twenty-one had three blockbuster plays running simultaneously on Drury Lane in London. Locke's *An Essay Concerning Human Understanding* had been published in 1689 to mostly negative reviews because it seemed to threaten the authority of the church. Some years later Cockburn published anonymously *A Defence of Mr. Locke's Essay of Human Understanding*. When her identity became known six months later, all her plays closed in an apparent blacklisting. Locke sought out Cockburn and gave her books and a great deal of money in gratitude. Her *Defence of Mr. Locke's Essay* is today considered a first-rate work of philosophical analysis. Locke's essay has gone on to be acclaimed as a highly influential world masterpiece. Isaac Newton had an encounter with another one of the unsung female geniuses of his time, Anne Finch, the Viscountess Conway (1631–1679) (see chapter 15). On Cockburn, see Moore and Bruder, *Philosophy: The Power of Ideas*, 341.

clearly heretical he must have felt on his cheeks the heat of the fires that had so many times in the past burned up the lives of good men who had accepted these beliefs. But the immensely intelligent Locke was brave and loyal. Before the week was out, he posted a lengthy reply to Newton. At some point on the journey to Cambridge the reply crossed paths with a second letter Newton had sent Locke at the end of November. This second letter was only 7,000 words long, but its contents were as blasphemous as those of the first letter. Locke was again in agreement. He again posted a lengthy reply.

What were the contents of the two letters Newton sent Locke?

Newton titled the first, which was actually two letters, "An Historical Account of Two Notable Corruptions of Scripture, in a Letter to a Friend." The second letter, which Newton left untitled, gave accounts of twenty-five additional "notable corruptions of Scripture."

It took extraordinary courage for Newton to write these letters, and even a little uncustomary recklessness. The twenty-seven passages in the New Testament that he demonstrates in these letters to be "corruptions," intentional or not, of the original text were passages the Roman Catholic Church considered to be proof of the doctrine of the Trinity—that the Father, the Son, and the Holy Ghost are one and the same.

Tens of thousands of people had been executed by the Roman Catholic Church over the centuries for refusing to swear that God and Jesus Christ were one. A mere seventy years or so divided Newton from the last, cruel, fires of the Spanish Inquisition—the same number of years that divides us from the Second World War. Even in the lifetime of Newton's father, people had been burned alive at the stake for making statements about Jesus Christ that the Roman Catholic Church considered to be heretical.

Today you can say anything you want about Jesus Christ.

In *The Last Temptation of Christ* (1953), Nobel Prize–winning Greek novelist Nikos Kazantzakis portrays the son of God as a lusty womanizer.

In *The Power Tactics of Jesus Christ and Other Essays* (1967),

American psychotherapist Jay Haley describes him as, variously, a man building a mass movement to topple an entrenched power structure, a schizophrenic, and a sociopath.

A few years ago, a flurry of books appeared asserting that there had never been a Jesus Christ. This supposed son of God was pure myth, a god cobbled together, perhaps mostly unconsciously, from bits and pieces of earlier pagan gods like the Titan Prometheus, who was chained to a mountain crag (or as a philosopher of myth might say, crucified) for stealing fire from the Olympic gods and giving it to mankind.

The latest take on Jesus is that he existed but was somebody else, a minor Middle Eastern king named Monobazus who ruled Edessa/Palmyra from AD 57 to 71 and was a military leader in the Jewish revolt against the Romans in AD 66 to 70.[1]

Things could hardly have been more different in Isaac Newton's time. Thirty years before Newton was born, in 1612, two English clergymen, Bartholomew Legate and Edward Wightman, were burned alive at the stake for refusing to swear that Jesus was the equal of God. In 1646, when Newton was four, an Englishman named Paul Best was sentenced to death by hanging for refusing to affirm the doctrine of the Trinity. He spent time in jail but was granted a reprieve at the last moment.

There were fates almost worse than death for some said to have blasphemed the divinity of Christ. In October 1656, a Quaker minister named James Nayler rode into Bristol on the back of a donkey. He was reenacting Christ's entry into Jerusalem on an ass, an event that is commemorated on Palm Sunday every year. Nayler was arrested, locked in the stocks, and given three hundred lashes for committing this "horrid blasphemy." His tongue was pierced with a red-hot iron and his forehead was branded with a *B*. Then he was thrown into prison, where he spent three years in solitary confinement.

A law was passed during Oliver Cromwell's Commonwealth, and reframed and renamed at the end of the century as the Blasphemy Act of 1697, that barred from office, sentencing to "three years' imprisonment on a second conviction all those, educated as Christians, who denied the doctrine of the Trinity, the truth of Christianity, or the divine authority of the Bible."[2] This law did not exclude executing a person for denying the Trinity: In 1697, seven years after Newton sent his dangerously heretical letters to Locke, Thomas Aikenhead, a twenty-year-old Edinburgh University student, was hanged for publicly ridiculing the scriptures and the Trinity.[3] He was summarily tried the day after he made these statements and executed the day after that.

Newton's lifelong "anti-Trinitarianism" was molded at an early age. He rejected the doctrine of the Trinity—that the Father, Son, and Holy Ghost are one—while a nineteen-year-old student at Trinity College in Cambridge. He told no one; if he had done so, this extraordinarily promising student would surely have been expelled. As Newton scholar Stephen Snobelen writes, "Newton lived in an age when heresy was not only a religious crime, but also a civil offense and a social outrage. When he converted to anti-Trinitarianism at Cambridge, he opposed a triad of legal structures: civil, ecclesiastical and academic."[4]

Newton decided that believing God and Christ were one was a form of idolatry—the "betrayal of God" we spoke of in chapter 1. God had told Moses, "Thou shalt have no other gods before me," and it seemed to Newton that to elevate Christ to a position of equality with God was, if not to put another God before him, at least to place another God beside him, and that this—since it deflected attention from the single-minded concentrated adoration of God—was equally idolatrous, even blasphemous.*

Newton said this to very few people during his lifetime, but there is ample evidence of this belief, found in the two letters to Locke and in

---

*The great Russian novelist Leo Tolstoy (1828–1910) shared Newton's beliefs to the letter, though it's unlikely he knew about them. Replying to the edict of excommunication promulgated against him by the Russian Orthodox Church in 1901, the author of *War and Peace* and *Anna Karenina* wrote, "I believe that the will of God is most clearly and intelligibly expressed in the teachings of the man Jesus, whom to consider a God and pray to, I esteem the greatest blasphemy." (Steiner, *Tolstoy or Dostoevsky*, 263.)

other parts of his nonscientific writings unknown before 1936.

Over the years Newton acquired a comprehensive knowledge of the Bible equal to that of anyone in England. A leading Anglican bishop told philosopher Richard Bentley in 1699 that Newton "knows more than all the rest of us about Scripture."[5] The bishop would not have wanted to know some of the things Newton had come to believe. The more the great scientist studied Holy Scripture, the more he became convinced that very early on the church fathers had rewritten certain passages of the New Testament to falsely portray Christ as the equal of God. Richard Westfall writes that by the late 1680s Newton had become convinced that "a massive fraud, which began in the fourth and fifth centuries, had perverted the legacy of the early church. Central to the fraud were the Scriptures, which Newton began to believe had been corrupted to support trinitarianism."[6]

Newton set himself the task of weeding out these corruptions. He brought all of his immense intelligence and learning to bear on this project. Newton scholar Justin Champion describes the working draft of the letters to Locke, which has come down to us.

> The pages of Newton's piece are littered with marginal references to biblical manuscripts and printed editions of all varieties: Slavonic, Ethiopick, Syriac, Arabic, Armenian, Turkish, French, Latin, and Greek. Specific and important codices were cited by name, library, and shelf mark; encouragement to ocular examination accompanied many of these citations. Codices such as Vaticanus, Claromontanus, Bezae, and Alexandrinus were the staple of orthodox biblical critics: Newton supplemented these with the codices Lobiensis, Tomacensis, Buslidianus, and Rhodiensis.[7]

What passages in the New Testament was Newton concerned with? In his first letter to Locke—which was actually two letters—Newton deals with the corruptions in just two biblical passages, 1 John 5:6–8 and 1 Timothy 3:16.

In the New Testament of Newton's day, 1 John 5:6–8 read:

> This is he who came through water and blood—Jesus Christ; not by water only, but by the water and the blood. And the spirit is the one who testifies, because the spirit is truth. For there are three who give testimony in heaven: the Father, the Word and the Holy Spirit; and the three are one. And there are three who give testimony on earth: the Spirit and the water and the blood; and these three are one.

Newton decided that the phrase "in heaven: the Father, the Word and the Holy Spirit; and the three are one. And there are three who give testimony on earth" had been fraudulently inserted into 1 John 5:6–8, possibly at the time of the Council of Nicaea (AD 325), and probably by the church father Athanasius, then very young and an assistant to the archbishop of Alexandria. Why was this change made? Here is how Newton explains it to Locke:

> Now this mystical application of the spirit, water and blood to signify the Trinity, seems to me to have given occasion to somebody either fraudulently to insert the testimony of the three in heaven in express words into the text for proving the Trinity, or else to note it in the margin in his book by way of interpretation, whence it might afterwards creep into the text in transcribing.[8]

Newton further tells Locke that the phrase in question—which biblical scholars have come to call the *comma Johanneum,* or, "John's phrase"—did not appear in earlier texts: "[It] is not read thus in the Syrian Bible. Not by Ignatius, Justin, Irenaeus, Tertullian, Origen, Athanasius, Nazianzen Didym Chrysostom, Hilarious . . ."[9] Newton had scrutinized more than thirty versions of the New Testament; he cites close to one hundred commentaries, by the church fathers and others, that seemed to refer to a 1 John 5:6–8 that is significantly different from the text that appeared in the Bibles of Newton's day. In conclusion, writes

Snobelen, "Newton characterizes the *comma* as a sham foisted on the text by the orthodox party and that the sense of the text was vastly superior without it."[10]

But Isaac Newton was not the first to question the authenticity of the *comma Johanneum*. A small, frail, brilliant, and extremely courageous man from Holland had noticed this discrepancy a hundred years before and had had to pay a price for having noticed. This was the Dutch Renaissance humanist, Catholic priest, theologian, popular writer, philosopher, and linguist Desiderius Erasmus, and it is important to understand him and exactly what he did if we are to understand the courage Newton showed in sending John Locke his letters.

Erasmus was born in Rotterdam, Holland, in 1466, dying in 1536. He was the illegitimate son of a priest by a washerwoman. The sickly child was raised in poverty, attending a one-room school where 275 boys were taught by a single master. At age eight he saw two hundred prisoners of war broken on the rack outside the gates of Utrecht by order of the local bishop. This made Erasmus a pacifist for life; and, though he became a devout Catholic, for the rest of his life he never accepted church doctrine uncritically.

His parents died when he was eighteen. Erasmus was left with nothing but a questioning mind and a thirst for learning. He became a priest—the only trade that enabled thousands of intelligent men of his time to be able to eat every day. His intellectual brilliance and practical competence were soon recognized, and he became secretary to a bishop who soon sent him to the university in Paris. Erasmus became a sort of itinerant, "Have genius, know Latin and Greek, will travel," scholar and writer, working for a number of different church officials as he made his way across Europe. Eventually he arrived in England, and ultimately at Oxford University, where he spent extended periods of time at the home of Sir Thomas More, the celebrated author of *Utopia*.

Erasmus was a born writer in the popular vein; he knew how to connect with the man and woman on the street. He had to earn a living, his creative powers needed expression, and he was becoming increasingly indignant at the injustices of the world; all of this drove Erasmus to turn out book after book.

At first his books weren't much more than collections of dialogues and adages and how-to techniques. Basically they told people, in an engaging, down-to-earth, humanistic fashion, ways they could nourish their souls. The prolific author-priest became an internationally best-selling author; in some years, one-tenth to one-fifth of all the books sold in the capital cities of Europe were written by Erasmus.

The Dutch scholar-priest's most enduring work is *The Praise of Folly* (*Moriae Encomium*) (1514). In Erasmus's lifetime it went into more than forty editions and was translated into twelve languages. The book is narrated by the Goddess of Folly, who has no trouble proving how widespread her influence is. She provides the reader with a nearly endless catalog of human folly, and it is through this device that Erasmus is able to gently satirize practically everything in the world.

This even included the doctrine of the Trinity. In *The Praise of Folly*, Erasmus/Folly describes a priest delivering a sermon on a new way of interpreting the Trinity.

> [T]o wit, from the Letters, Syllables, and the word Itself; then from the Coherence of the Nominative Case and the Verb, and the adjective and Substantive: and while most of the Auditory [audience] wondered, and some of them wondered that of Horace, "what does all this Trumpery drive at?" at last he brought the matter to this Head, that he would demonstrate that the Mystery of the Trinity was so clearly expressed in the very Rudiments of Grammar, that the best Mathematician could not chalk it out more clearly.[11]

Perhaps it was because he was a priest and known to be an ardent Catholic, or perhaps because he was so universally popular—or maybe because, no matter what he said, he never actually advocated changes in the orthodox church—but Erasmus and his *Praise of Folly* somehow managed not to bring down the wrath of the Vatican. In fact (as Erasmus himself tells us), Pope Julius II "read the *Moriae* in person but he laughed. His only comment was, 'I am glad that our Erasmus himself was in the book.'" A contemporary letter by a French abbot tells us that Julius's successor, Pope Leo X, read the whole book with "evident delight" as did all his bishops, archbishops, and cardinals.[12]

This would soon change. What happened to Erasmus next would prove to be a grim warning to those who were starting to read the New Testament with the fresh, clear, rational eyes that Isaac Newton was to apply to the text with laserlike intensity a century later.

Erasmus published his most important book, a Greek-Latin parallel New Testament titled the *Instrumentum Novi Testamenti,* in 1516. Sales of this work were greatly helped by the popularity of Erasmus's other books; at one point, 300,000 copies of his parallel New Testament were in circulation.

The New Testament was originally written in Greek. Late in the fourth century AD, Saint Jerome translated it into Latin. Jerome's Latin text came to be called the "Vulgate" Bible, because Latin was the everyday, "vulgar" language of the Roman people.

Erasmus had originally intended his Greek-Latin parallel New Testament to contain the original Greek version from which Jerome had worked. In making a new translation of this version into Latin, Erasmus used a number of other ancient, sometimes different, manuscripts of the Greek New Testament. The Dutch scholar-writer examined each word anew, sometimes coming up with a different translation from Jerome's. Erasmus's Latin New Testament ended up being an updated—in fact, a corrected—version of Saint Jerome's Vulgate.

In making this new translation, Erasmus discovered that the phrase known today as the *comma Johanneum* was not present in 1 John 5:6–8 in any of the earlier versions of the Greek New Testament he consulted. So when he published his 1516 parallel New Testament, he left the phrase out of both the Greek and Latin translations.

Perhaps the Vatican had resented Erasmus's satirical runs at the church in *The Praise of Folly* more than the world realized—more, even, than the Vatican itself knew. At any rate, Erasmus's innocent deletion of the *comma Johanneum* from both texts triggered a torrent of abuse. The Vatican was incensed at Erasmus for sundering what it viewed as a fundamental defense, directly from the Word of God, against the heresy of anti-Trinitarianism. The English theologian Edward Lee, later archbishop of York, hinted darkly that the Dutch humanist was a heretic in disguise and accused him of indolence in not looking harder for an early Greek version of the New Testament that actually contained the *comma Johanneum*.

In May 1520, Erasmus replied: "What sort of indolence is that, if I did not consult the manuscripts which I could not manage to have? At least, I collected as many as I could. Let Lee produce a Greek manuscript in which is written the words lacking in my edition, and let him prove that I had access to this manuscript, and then let him accuse me of indolence."[13]

This didn't make the church any happier, and near the end of 1520 a Franciscan friar at Oxford somehow came up with a previously unknown Greek copy of the New Testament containing the *comma Johanneum*. Erasmus had to bow to the Vatican and include the phrase in the third, 1522, edition of his parallel New Testament. (He had excluded it from the second, 1519, edition.)

This did not mollify the church. In his annotations to his translation, Erasmus had observed that Christ was rarely explicitly called "God" in the New Testament. In the third edition he intimated that he thought the *comma Johanneum* in the text found at Oxford was a forgery. Erasmus insisted that his comments were philological, not doctrinal. Nevertheless, at a conference in Valladolid, Spain, in 1527, the Vatican hierarchy drew up a list of charges against him. His most determined critic at the conference was the Spanish priest-theologian Diego López Zúñiga (Stunica), who lobbied then, and lobbied until his own death in 1631, that Erasmus be burned at the stake.

The Dutch humanist took this list of charges so seriously that he responded by listing eighty places in his writings where he explicitly stated his belief in the doctrine of the Trinity. But the humanist's *Instrumentum Novi Testamenti* was repudiated by the church at the Council of Trent in 1559, and Jerome's Vulgate edition of the New Testament was adopted as the "official" Vatican text. Thus 1 John 5:7, with its *comma Johanneum*, remained up front and center as an important biblical affirmation of the dogma that Jesus, God, and the Holy Spirit are one.

A few brave men followed in Erasmus's footsteps, often with tragic results. The young Spanish nobleman Miguel Servet (1509/1511–1553) was passionately devoted to religion; hearing Erasmus's name often, he became intrigued, read the Dutch humanist's works, and immersed himself in the study of scripture. In 1531, Servet published, under the name Michael Servetus, his first important work, *On the Errors of the Trinity*.

The title was a dangerous one. But Servetus seemed to make it clear that he did not want to repudiate the doctrine of the Trinity. We would not be wrong in saying that he regarded this doctrine as the keel of a ship—a keel that had once kept the ship of the church beautifully upright but that had, on account of a great many stormy voyages, become so heavily encrusted with mud, slime, and barnacles that the ship listed this way or that from time to time. Servetus wanted to clean all the muck off the keel so the ship could sail briskly forward, upright, and in its original pristine glory. (Servetus didn't think anyone remembered exactly what the original keel had looked like. He wanted to find out.)

The daring, if not reckless, Spanish theological writer went on to become a medical doctor; included in his later work is a treatise on anatomy in which he anticipates Harvey's discovery of the circulation of the blood. But the Vatican had never quite gotten over Servetus's first book, *On the Errors of the Trinity*. In 1553, the still very young writer completed his masterpiece, the seven-hundred-page *Restitution of Christianity;* he did this even as the Inquisition was beginning to pursue him from city to city. When he arrived in Geneva he didn't bother to hide himself and was quickly rooted out by the severe inventor of Calvinism, John Calvin. This provoked dire consequences, because Servetus had been embroiled in a vicious controversy with Calvin for over a month. The virulently anticlerical Voltaire, whose rallying cry was *"Ecrasez l'infame!"*—"Crush the infamy [of religion]!"—completes the story: "[Calvin] was cowardly enough to have him [Servetus] arrested, and barbarous enough to have him condemned to be roasted by a slow fire—the same punishment which Calvin himself had narrowly escaped in France. Nearly all the theologians of that time were by turns persecuting and persecuted, executioners and victims."[14]

Along with the *comma Johanneum,* Newton grappled in depth with another suspect passage in his first "Notable Corruptions of Scripture" letter to John Locke. This was a passage in 1 Timothy 3:16 that read, "And without controversy great is the mystery of godliness: God was manifest in the flesh."

Richard Westfall tells us that "Newton found that early versions did not contain the word ['God'] but read only, 'great is the mystery of godliness which was manifested in the flesh.'"[15] Reading through the religious debates of the fourth and fifth centuries, he found many instances where this passage, if it had included the word "God" as in his own day, could have been used effectively by the orthodox Trinitarian Church as doctrinal ammunition against the anti-Trinitarian heretics; but it never was. Newton could only conclude that this doctrinal ammunition hadn't been used because the word "God" hadn't yet been added to the passage.

In his second letter to Locke, written in late November 1690, Sir Isaac anatomized and deconstructed twenty-seven more "corruptions" in the New Testament. He prefaces his demolition job in this way:

> Having given you an historical account of the corruption of two texts of scripture, I shall now mention some others more briefly. For the attempts to corrupt ye scriptures have been very many & amongst many attempts 'tis no wonder if some have succeeded. I shall mention

those that have not succeeded as well as those that have, because the first will be more easily allowed to be corruptions, & by being convinced of those you will cease to be averse from believing the last.[16]

The twenty-seven corruptions were in the following passages.

John 3:6	Acts 13:41
Philippians 3:3	2 Thessalonians 1:9
1 John 5:20	Acts 20:28
Luke 19:41	1 John 3:16
Luke 22:43–44	Jude 1:4
Matthew 19:16–17	Philippians 4:13
Matthew 24:36	Romans 15:32
Ephesians 3:14–15	Colossians 3:15
Ephesians 3:9	Apocalypse 1:11
Apocalypse 1:8	2 Peter 3:18
Corinthians 10:9	Romans 9:5
Jude 1:5	Hebrews 2:9
1 John 4:3	Philippians 2:6
John 19:40	

The reader will find the text of these passages, with comments on Newton's deconstruction of each, in appendix B, "Further Corruptions Found in the New Testament."

Was Newton correct in labeling these passages corrupt and in fact deliberately corrupted? Regarding the *comma Johanneum*, the eminent dissident Catholic theologian Hans Küng writes:

> Historical-critical research has unmasked this sentence as a forgery which came into being in North Africa or Spain in the third or fourth century, although the Roman Inquisition [still part of the ongoing Catholic Church, according to Küng] was still vainly attempting to defend its authenticity at the beginning of our

century.... Although there are so many triadic formulas in the New Testament, there is not a word anywhere in the New Testament about the "unity" of these three highly different entities [the Father, the Word, and the Spirit].[17]

With respect to the opening lines of John, "In the beginning was the Word, and the Word was with God . . . and the Word was made flesh," Küng remarks that the author/authors "by no means say that Jesus was eternal like God, but only that he was created by God at a certain time as the incarnation of the Word."[18]

Over the past decades, Bart D. Ehrman has been diligently discussing biblical corruptions in a clear and straightforward manner that has brought him many readers. Ehrman recently left the church; we'll probably never know the extent to which this was prompted by his growing awareness, based on his research, of the fraudulent nature of certain passages in the New Testament. In *The Orthodox Corruption of Scripture: The Effect of Early Christological Controversies on the Text of the New Testament* (1993), Ehrman tells us that "proto-orthodox scribes of the second and third centuries occasionally modified their texts of Scripture in order to make them coincide more closely with the christological views embraced by the party that would seal its victory at Nicaea and Chalcedon."[19]

Stephen Snobelen wrote in 2002 that the recent work of Bart D. Ehrman "reveals a growing awareness on the part of contemporary textual critics that proto-Trinitarian forces were responsible for a large number of doctrinally-inspired textual corruptions of the New Testament."[20]

If for Newton, God and Jesus were not one and the same, then what were they? What was the nature of their relationship?

First, regarding God, Voltaire summed up Newton's beliefs as follows: "Sir Isaac Newton was firmly persuaded of the Existence of God; by which he understood not only an infinite, omnipotent, and creating being, but moreover a Master who has a Relation between himself and his Creatures."[21] The key word is "master." Newton never speaks of a compassionate God or a loving God; he speaks of a God of dominion, a God of providence. This is the God who made his presence known by acting in the world (see chapter 18). In the "General Scholium" that ends the *Principia Mathematica,* Newton writes:

> The supreme God is ... all eye, all ear, all brain, all arm, all power to perceive, to understand, and to act; but in a manner not at all human, in a manner not at all corporeal, in a manner utterly unknown to us. As a blind man has no idea of colours, so have we no idea of the manner by which the all-wise God perceives and understands all things. He is utterly void of all body and bodily figure, and can therefore neither be seen, nor heard, nor touched; nor ought to be worshiped under the representation of any corporeal thing.[22]

To have in mind, while worshipping, a description of God, or to have in view, while worshipping, an image of God, is seriously misleading, since the nature of God cannot be grasped by the senses. Fixating on an image of him, whether in the imagination or in the real world as a painting or a statue or whatever, is, Newton believes, idolatrous. It is demeaning to God. Newton continues:

> We have ideas of his attributes, but what the real substance of any thing is, we know not. In bodies we see only their figures and colors, we hear only the sounds, we touch only their outward surfaces, we smell only the smells, and taste the favours; but their inward substances are not to be known, either by our senses, or by any reflex act of our minds; much less then have we any idea of the substance of God. We know him only by his most wise and excellent contrivances of things, and final causes.[23]

If we wish to grasp what is God, the only way we can do so is by becoming aware of his activities in the world ("his most wise and excellent contrivances of things"). These are ubiquitous; we should show "reverence and adore him on account of his dominion."[24]

Second, regarding Jesus, Newton was not a Trinitarian. He did not believe God and Christ were the same.

But Newton believed that, if Christ was not God, he was very nearly God. Newton belonged to a heretical sect of Christianity called Arianism, of which John Locke, William Whiston, the founder of modern chemistry Robert Boyle, and many other outstanding Englishmen were adherents. Arius (AD 256–336) was a Libyan presbyter and theologian, descended from Berbers, who lived in Alexandria, Egypt. Arius spent decades warring with Athanasius, the fiery orthodox bishop of Alexandria, over the nature of the relationship between Jesus Christ and God. In the next two chapters, we will describe in detail the phases of that war. Suffice it to say for the moment that Athanasius finally triumphed; because of him, the idea of the Trinity—that God, Christ, and the Holy Spirit are one—became the cornerstone doctrine of the orthodox Roman Catholic Church.

Arius was a subtle, probing, and responsible theologian who has had a powerful impact on thinkers through the ages (see chapter 4). He believed God was "uncreated and unbegotten, eternal, without a beginning and unchangeable." Christ was all of these except one: he had been created by God. Arius qualifies this: Christ was "not simply created in time" but "before all time." The Libyan bishop said, famously, "There was once [a time], when he [the Son] was not."

Newton Project director Rob Iliffe writes:

> The fourth-century theologian Arius opposed the doctrine of the Trinity, maintaining that Jesus was a figure distinct from and inferior to God. According to most accepted church histories of Newton's time, Arius's views were rightly condemned as heretical by the contemporary champion of trinitarianism, Athanasius (later

Saint Athanasius). Newton, however, turned this version on its head, maintaining that it was Athanasius who was the heretic while poor honest Arius had been repeatedly misrepresented and slandered by the wicked trinitarians from his own lifetime right up to Newton's.[25]

So Newton did not believe in the doctrine of the Trinity; he did not believe that the Father, the Son, and Holy Spirit were one.

But he believed that a Christ who was so very nearly identical with God was a powerful being whom God must have deployed in a plenitude of powerful ways. Arius believed God had created Christ so the Son could make manifest the physical universe, which was something God could not do directly. Christ created the material reality in which we live in his capacity as the link between the physical nature of the universe and the spiritual nature of God. Spirit and matter are different and opposed; they cannot "touch"; they require an intermediary who partakes in the nature of each. This intermediary is Christ, who is both God (though not quite *the* God) and man; it is through this Chrst that spirit and matter interact.

For Newton, Christ is God's agent, "His Man in the Physical Universe." He is the Deity's district manager, the executive officer of physical creation. In *The Janus Faces of Genius,* Newton scholar B. T. J. Dobbs writes:

> Newton's Christ is a very unorthodox Christ indeed but one whose many duties keep him engaged with the world throughout time. A part of his function is to insure God's continued relationship with His creation. Even though Newton's God is exceedingly transcendent, He never loses touch with His creation, for He always has the Christ transmitting His will into action in the world.[26]

There existed in the Newton papers at one time an unusual collection of essays, and perhaps even an entire book, called *On the Church,* whose contents suggest that Newton at the end of his life was still elaborating on a highly unusual (and certainly heretical) concept of Christ. The Newton scholar (and also playwright and drama scholar) David Castillejo was able to see these papers in the late 1970s. He drew from them some astonishing conclusions. Castillejo wrote:

> the fragments that have survived on Christ are . . . astounding. Newton believed that Christ dominates both the Old and the New Testament. All appearances of Jehovah in the Bible are in fact appearances of Christ: it was Christ who walked in the Garden of Eden, who gave Moses the Ten Commandments, who appeared to Abraham as an Angel, who fought with Jacob, who gave the prophecies to the prophets. He is the Prince Michael mentioned by Daniel, and he will come to judge the quick and the dead. The Jews are ruled by Christ in an absolute monarchy, and he is their lord.

Castillejo continues:

> What Newton has done is push God right out of the Bible and back into nature, where he is everywhere present, immovable and invisible. Conversely he has brought Christ forward as a moving, living and commanding being, who crosses back and forth over the boundary of immortality, and dominates the whole of the Jewish and Christian religious experience. It is an amazing transformation of traditional opinion. Newton's whole book is in some way a commentary on the true and false relationship between Christ and God.[27]

The identity of Jesus Christ, the extent to which he participated in the essence of God—these questions were not absolutely essential for Newton when it came to having a correct understanding of the practice of Christianity. Newton saw the Christian faith as having a double aspect: it could be "meat for men," and it could be "milk for babes."

"Meat for men" involved acquiring a profound and comprehensive knowledge of Christianity, historically, philosophically, and indeed in all of its facets. "Milk for babes" was the more important aspect. It meant that, to be a decent Christian, you had to do only two things: adore God, and love your neighbor. Newton wrote: "In matters of religion the first & great Commandment hath always been: Thou shalt love the Lord thy God with all thy heart & with all thy soul & with all thy mind. And the second is like unto it: Thou shalt love thy neighbor as thy self. On these two hang all the Law & the Prophets (Matthew 22:27)."[28]

Newton's devastating critique of the "orthodox corruption of scripture" was only a part of his massive assault on the Roman Catholic Church's insistence, even to committing fraud, on the absolute primacy of the doctrine of the Trinity. A key element of that assault was the great mathematician's bitter, brilliant, and devastating examination of the life and character of the celebrated champion of the doctrine of the Trinity, Athanasius, archbishop of Alexandria.

In the two chapters that follow, we won't only have a look at Athanasius's conduct at the paradigm-destroying Council of Nicaea of AD 325, but we'll also take a tour through many other aspects of Athanasius's life and discover that he in his way was as corrupt as many of the corruptions that Newton found in the New Testament. We'll do all this looking through the prism of one of the least known and most shocking documents in the entire history of religion: Isaac Newton's own "Paradoxical Questions Concerning the Morals and Actions of Athanasius and His Followers."

What happened to the two letters that Newton sent to Locke? Did they have any effect on the history of theology, particularly regarding biblical text?

Locke made a copy of the first letter and, thinking he was complying with Newton's half-expressed wishes, sent it, without revealing the author's name, to a friend of his, an M. Le Clerc, in Amsterdam. He asked Le Clerc to have the letter translated into French and published. In his reply dated April 11, 1691, Le Clerc agreed that it should be published and offered to translate it into Latin or French.

Locke informed Newton of what he had done—and Newton

panicked. He insisted that the letter be suppressed. Locke wrote Le Clerc a second time, and Le Clerc complied, merely having the letter deposited in a private museum in Holland, that of the Remonstrants (a group of religious dissenters).

In 1754—twenty-six years after Newton's death—the letter was published in London. A historian published the text to a wider audience in 1785, in the middle of a heated controversy over the text of John 1:7–8. The contents of the second letter weren't published until the mid-twentieth century, when the text appeared in the third volume of the University of Cambridge Press's *Correspondence of Isaac Newton*.

It's not easy to assess the general effect of Newton's letters to Locke. Most of the corruptions he pointed out have by now been corrected (though not in the official Roman Catholic Church edition of the Bible). Already during the last third of Newton's life a number of scholars had begun to make quietly devastating statements about the number of "variant" readings between manuscripts of the Bible. Diarmaid MacCulloch writes that "by 1707 one distinguished mainstream English biblical scholar, John Mill [the father of philosopher John Stuart Mill], reckoned these to be around thirty thousand in number. Some of these variable readings could plausibly be considered as later interpolations in the interests of Trinitarian belief."[29]

What about Trinitarianism in relation to the other two great monotheistic religions of the world, Judaism and Islam? In Judaism there is only one God, Jehovah, and he does not have a son. Islam too eschews any thought that God could have a son. MacCulloch writes: "In a much-discussed and not conclusively understood verse [4.171] of the Qur'an, God is represented as telling the Christians 'believe in God and his messengers and do not speak of a "Trinity."' . . . God is only one God, He is far above having a Son."[30]

Now let us a look at the ferocious struggle in the fourth century AD that made the doctrine of the Trinity part of Christianity—a struggle that Newton documented endlessly even as he endlessly reviled it. The archbishop Athanasius will take center stage as antihero and archvillain.

CHAPTER FOUR

BLOODBATH IN A BOGHOUSE
Murder in the Fourth Century AD, Part 1

If it hadn't been for the determination of Bishop Athanasius of Alexandria, the doctrine of the Trinity probably would not have survived long enough to be declared, near the end of the fourth century AD, the cornerstone doctrine of the Roman Catholic Church. That's why the Vatican reveres Athanasius, now Saint Athanasius, calling him the Father of Orthodoxy.

Isaac Newton believed that the doctrine of the Trinity was erroneous and that Athanasius had not only led Christianity down the wrong path but also had blood, real blood, on his hands. He believed that the Father of Orthodoxy was a murderer, a rapist, a slanderer of his peers, a falsifier of documents, a rewriter of history to serve his own interests, and truly one of the worst men in the world.

In the 1690s, Newton labored over draft after draft of an incendiary 43,000-word, twenty-four-point legal brief setting out in minute detail the evidence to prove that Athanasius had committed all these crimes. The document bears the unexciting title of "Paradoxical Questions Concerning the Morals and Actions of Athanasius and His Followers." Newton didn't publish it, and in his lifetime few knew that it existed. Some time after Newton's death, the manuscript found its way into the Clark Library in Cambridge, where it was kept under lock and key; Newton's heirs didn't want the world to know that their illustrious ancestor had been a nonconforming anti-Trinitarian. Edward Gibbon, at work in Lausanne, Switzerland, on *The Decline and Fall of the Roman Empire,* wrote to Cambridge for permission to come and read "Paradoxical Questions" in the library, but permission was denied; his assessment of Athanasius in his masterpiece might have been significantly different if he'd read Newton's withering words.[1] Only when Newton's nonscientific writings were auctioned off in 1936 did "Paradoxical Questions" become available to the general public and begin to figure in discussions about Sir Isaac Newton.

"Paradoxical Questions Concerning the Morals and Actions of Athanasius and His Followers" is written in a dry-as-dust style that contains not a single surplus adjective or adverb. It's a bare-bones string of sentences as spare as a page of equations. It's easy to believe that Newton didn't want anyone to read it, for he makes no concessions whatsoever to the reader. Perhaps his unforgiving style was meant in part to deter people from discovering his heretical beliefs.

Even with a cursory reading, Newton's hatred of Athanasius shows through. Richard Westfall writes that in "Paradoxical Questions" he "virtually stood Athanasius in the dock and prosecuted him for a litany of sins, . . . [seeking to show] not only that Athanasius was the author of 'the whole fornication'—that is, of trinitarianism, 'the cult of three equal gods'—but also that Athanasius was a depraved man ready even to use murder to promote his ends."[2]

That being said, "Paradoxical Questions" is a story—a scoop!—that would whet the appetite of any journalist, provided that person knew enough about its implications to want to work his or her way through its rebarbative phraseology. For, if you look hard beneath the surface, you see that Sir Isaac has written a sordid drama of murder, fraud, character assassination, and interfaith conflict in the best tradition of religious holy wars. It's what you might get if Agatha Christie had woven a detective thriller out of the more sordid episodes of the New Testament. The plot includes:

- The ignominious death of a controversial prelate in a public latrine in Constantinople. This prelate might have been struck down by God, or he might have been murdered by his enemies, or the whole thing might not have happened at all.
- The forgotten other Council of Nicaea—namely its dark twin, the Council of Tyre—convened in AD 335. This council, attended by as many bishops as had attended Nicaea, dealt with human sin and folly rather than the divine Word and came perilously close to reversing the most important decisions made at the Council of Nicaea.
- Rape; a severed hand; a second murdered bishop; falsified letters; the wholesale rewriting of ecclesiastical history; "Words, Whips, Clubs, and all methods of Cruelty and Severity, not sparing even the devoted Virgins, whom they suffered the very Gentiles to strip naked"; and numerous other elements smacking of the surrealist, the macabre, and the horror entertainment of the Grand Guignol of Paris.
- A notorious ecclesiastic trial of which history has given us two different versions, in some places three, and which transports us to the same subjective, relativistic, and shifting world as the one created by Japanese filmmaker Akira Kurosawa's masterpiece *Rashomon*.
- A celibate monk so horribly tormented by imaginary visions and demons, against all of which he resisted nobly, that for almost 2,000 years he has been a poster monk promoting the miseries and splendors of the monastic life.
- The most famous collectors in the ancient world of the bones of saints and all other holy relics.
- A bloody battle between Roman legionnaires and ecstatic Christians inside a church in Alexandria, during which "the drawn swords shone by candle light and Virgins were slain and trodden under foot."
- And much more.

The story, for those who can penetrate beneath its ultra-austere surface, is bizarre and macabre after the manner of the Spanish surrealist filmmaker Luis Buñuel, who hated the Catholic Church but must have known everything about it, because a severed hand appears out of nowhere in his film *The Andalusian Dog* (1929), and a saint on top of a sixty-foot pole is tempted by an alluring devil in his film *Simon of the Desert* (1965).

Newton's "Paradoxical Questions Concerning the Morals and Actions of Athanasius and his Followers" is a detective story, with Isaac Newton in the role of a sleuth relentlessly pursuing what must be the coldest cold case in history, one that took place thirteen hundred years before the investigator was born.

The action of "Paradoxical Questions" unfolds on a wide canvas, including three big cities: (1) Alexandria, the Egyptian port that boasted three miles of colonnades, nine miles of wharves, the body of Alexander the Great (preserved in honey in a glass box), and a hundred bitterly warring religious and philosophical factions; (2) Constantinople, across the Bosporus from the ruins of Troy, turned into the eastern capital of the empire by Constantine, and today, in a richly expanded version, Istanbul, Turkey, with a population of 14 million; and (3) Phoenician Tyre, the ancient seaport from which, according to legend, Noah launched the ark and Saint Paul sailed for Rome. And there is the town, that of Nicaea, in Asian Turkey, today a sleepy resort town, but in AD 325 the bustling, agitated site of the paradigm-busting Council of Nicaea.

From time to time the action skids to a stop in the deserts of Egypt, where Athanasius is exiled several times and where he writes incendiary letters and a brief incendiary biography of a saint.

The dramatis personae consists of:

- Saint Anthony, a desert monk;
- Constantine, an emperor of Rome;
- Constantius, Constantine's son, also an emperor of Rome;

- Arius, a heretic, the father of Arianism;
- Athanasius, a Trinitarian, the Father of Orthodoxy;
- Major and minor prelates whose names take on resonance as Newton's narrative proceeds: Bishop Alexander of Alexandria, Bishop Alexander of Constantinople, Hosnius, Melitius, Arsenius, Macarias, and others;
- A number of women, only two of them having names: Helena, the dowager empress, Constantine's mother; and Irene, Constantine's sister. The rest are unnamed virgins and unnamed prostitutes;
- The Alexandrian mob in at least two incarnations.

Richard Westfall writes that, for Newton:

> the corruptions of Scripture came relatively late. The earlier corruption of doctrine, which called for the corruption of Scripture to support it, occurred in the fourth century, when the triumph of Athanasius over Arius imposed the false doctrine of the trinity on Christianity.
>
> He became fascinated with the man Athanasius and with the history of the church in the fourth century, when a passionate and bloody conflict raged between Athanasius and his followers, on the one hand, and Arius, on the other. . . . Once started, Newton set himself the task of mastering the whole corpus of patristic literature [literature of the church fathers].[3]

Westfall lists the myriad church fathers Newton studied to prepare his case against Athanasius. Irenaeus, Tertullian, Cyprian, Eusebius—the list goes on and on; probably Newton knew a hundred sources, and we're left in no doubt that, whatever his bias, he knew everything there was to know or could be inferred about the Father of Orthodoxy.

Saint Anthony, the archetypal tortured monk who virtually invented the system of monasteries, appears near the beginning of Newton's story, because Athanasius as a teenager may have met him in the desert and been influenced by his teachings. But Newton reserves the bulk of his discussion of Anthony for the end of "Paradoxical Questions," and it has made more sense to deal with that part of Athanasius's story in a separate chapter (see chapter 6, "The Temptation of Saint Anthony"). Following, however, are some introductory words about Saint Anthony from Newton, which he bases on the historian Sozomen.

> [Anthony] received letters from Constantine the great, lost his parents in his youth, distributed his father's lands amongst his townsmen, gave the rest of his goods to the poor, conversed with all wise men and imitated what was best in each, ate only bread & salt & drank only water, dined at sunset, often fasted two days or more, often watched all night, slept on a mat & frequently on the bare ground, never anointed nor bathed himself nor saw himself naked, was meek, prudent, pleasant, foreknew things, but dissuaded the monks from affecting it, spent his time in working, came often to the cities to defend the injured, interceded for them with the Presidents & great men who delighted to see & hear him, but immediately returned to the wilderness saying that as fishes cannot live on dry land so monks in cities lose their virtue.[4]

Athanasius was born in Alexandria around 295 and died in 373. A tenth-century Arabic chronicle of Coptic patriarchs says his parents were pagan, and he converted his widowed mother to Christianity when he was a teenager.[5] At about that time, his family was driven into the desert by Diocletian's persecutions of the Christians. It may have been then that the zealous, rigid, fiery youth met the pious, charismatic Anthony and fell under the enchantment of his teachings on the Trinity.

Athanasius returned to Alexandria in 313, immersed himself in the study of the scriptures, and caught the attention of Bishop Alexander of Alexandria, who made him his personal secretary when Athanasius was only twenty. In 325, Alexander brought Athanasius to the Council of Nicaea to argue the case for the doctrine of the Trinity. Athanasius succeeded admirably, if abrasively. He was on his way to becoming a forger.

The adult Athanasius "allied ruthlessness to an acute theological mind,"[6] writes Diarmaid MacCulloch. This self-appointed guardian of the Trinity could "frame a memorable phrase," asserting, for example, that the equality of Son and Father was "like the sight of two eyes."[7] Edward Gibbon tells us Athanasius's mind was "clear, forcible, and persuasive" but "tainted by the contagion of fanaticism."[8] Paul Johnson goes a step further, revealing that the archbishop of Alexandria (whom Athanasius became) was "a violent man, who regularly flogged his junior clergy and imprisoned his expelled bishops."[9] Newton makes the dark side of Athanasius his entire focus of his inquiry in "Paradoxical Questions." He is, as we will see, not unpersuasive in making the case that the dark side of Athanasius *is* Athanasius. All the brilliance with which Newton wrote the *Principia* he aims with laser sharpness at the archbishop of Alexandria and his deeds. We will find cherry-picking and deductive logic in a wicked and furious partnership.

The Council of Nicaea was convened because of the fiery rhetoric of the Alexandrian prelate Arius (AD 250/256–336), who declared that Christ was divine but not as divine as God. A swarthy, volatile Libyan, not ordained until he was more than fifty, Arius "provoked and infuriated opposition in Alexandria, including that of his bishop, Alexander."[10] Arius preached anywhere he could, subtly promoting his beliefs at the grassroots level by composing short, simple, sometimes racy songs that the workers in the mills and taverns, on the docks, could learn by heart. This was an early form of subliminal advertising! Arius was an admirer of Plato and spent much of his time trying to make Christianity presentable to pagan philosophers. But, despite his obvious rationality, he was hounded by accusations (from his enemies); seventeenth-century scholar William Cave picks up one: "Arius was a man 'of a subtle and Versatil Wit, of a turbulent and unquiet Head, but which he vail'd with a specious Mask of Sanctity.'"[11]

In AD 322, Emperor Constantine the Great (272–337) succeeded in welding the western and eastern parts of the Roman Empire into a single unified whole. He had made Christianity the official religion of Rome in 313, believing it could serve as a crucial binding agent for the empire.

Constantine was a ruthless politician and military strategist who wasn't without feeling or genius. He was tall and athletic, with a bull neck, a square face, blue eyes, and a peevish mouth. Indifferently educated, Constantine campaigned much of his life and spoke Latin, Greek, Pict, Gaulish, Frankish, and at least one Asiatic dialect. He told the court chronicler, Eusebius Pamphilus, that at the battle of Milvian Bridge, near Rome—the last battle he had to win to become emperor—he saw "a cross of light in the heavens, above the sun, and bearing the inscription, CONQUER BY THIS."[12] Constantine decided he owed his victory to Christ—that the Christian deity was a god of battles who could be relied upon to protect him as long as he strove to be a decent Christian.

The emperor lapsed badly when he murdered his first wife and one of his sons (perhaps for sound reasons of state) and tried to make up for it by showering Christians with churches, high office, and wealth and buying up entire towns and cities to make sure the inhabitants accepted the new religion. Increasingly he regarded himself as honorary bishop in chief and, says MacCulloch, "regularly delivered sermons to his no doubt slightly embarrassed courtiers."[13]

The pagan and Christian in Constantine combined to make him an extravagantly ambitious collector of religious relics. Constantine had pieces of the one true Cross packed into a hollow porphyry column on top of which bestrode a statue of himself. He made a nail from the Cross into a bit for his horse and slipped another nail in his tiara.[14] The emperor oversaw the construction of the Church of the Twelve Apostles in Constantinople, placing in it twelve coffins ready to house the greatest relics of all: the remains of the twelve apostles. Peter and Paul had been buried in Rome; he transferred their bones to the church. The only coffin that got filled, however, was the thirteenth, his own, which he'd placed in the center of all the others.[15]

▲

In AD 320, his second year as ruler of a united Roman Empire, Constantine decided he had to deal with Arius's heretical teachings, which were threatening to split the Christian Church. Constantine couldn't afford this; Christianity really was helping bind the empire together. For months he tried to be conciliatory, even sending the Spanish bishop Hosnius to Alexandria to try to effect a reconciliation. But nothing availed, and Constantine, despairing, convened a conference at his summer palace at Nicaea, in Asia Minor, for the summer of 325.

The emperor agreed to pay all transportation and lodging costs along the speedy Roman roads. He would pay all expenses during the conference. He would personally attend the sessions. And, at times, he would—begging the bishops' indulgence, of course—say a word or two himself. The all-powerful Constantine the Great clearly meant business. The bishops acquiesced.

In the summer of 325 the sun beat down on the shadowless scorched streets of Nicaea more pitilessly than ever, as if trying to burn through to the truth of every man and woman there. People of every color, trade, shape, size, in tatters, dressed richly, full of hope, fear, despair, surged through. Often they were brushed aside by clattering chariots that drove away wild dogs feeding on animal guts tossed into the center of the street from the butcher shops. The polyglot uproar was shot through with screams, laughter, and the bellowing of animals. The stench of animal dung, vomit, urine, and garbage mingled with the sharp aroma of fermented sauces and rotting fish. The smell of animal fat rose from burning altars.

Inside the walls of the imperial residence, the heat and stench of the streets yielded to cool breezes from swaying fans and the musky odor of perfumes daubed on by the bishops. There were 318[*] of them in all,

---

[*]Edward Gibbon gives the figure of 380. Eusebius speaks of 250 bishops. Later Arabic manuscripts put the figure at 2,000. Athanasius (in *Ad Afros*) gives a figure of 318.

sitting around the white marbled walls like wary, startled birds of prey dressed in purple robes. Many of them bore the scars of Diocletian's persecutions: an eye gouged out, a thumb sliced off, a leg dragging behind because its hamstring had been cut. At the opening ceremony Constantine, moving piously among these martyred priests, bent forward impulsively to kiss an empty eye socket or the flat stump of a thumb. Behind him there trod cautiously Bishop Hosnius, the council director, very tall, almost ninety years old, and bearing a thick, red, scythe-like scar that ran from the tip of his ear under his eye to his nose; this he had received during the Cordova persecutions.

Constantine kept his word and attended regularly. Gibbon writes, "Leaving his guards at the door, he seated himself (with the permission of the council) on a low stool in the midst of the hall."[16] Eusebius, court chronicler and fawning flatterer, describes the emperor as "clothed in raiment which glittered as it were with rays of light, reflecting the glowing radiance of a purple robe."[17] Years later his nephew, the emperor Julian, remarked scornfully that Constantine "made himself ridiculous by his appearance—weird, stiff eastern garments, jewels on his arms, a tiara on his head, perched crazily on top of a tinted wig."[18] However he was garbed, that summer at the council he "listened with patience and spoke with modesty," writes Gibbon.[19]

The debates began. The bishops didn't discuss the canon of the New Testament, as is often said. They debated regulations governing the clergy. The first they agreed to conveys the flavor of the whole: "If anyone in sickness has undergone surgery at the hands of physicians or has been castrated by barbarians, let him remain among the clergy. But if anyone in good health has castrated himself, if he is enrolled among the clergy he should be suspended and in future no such man should be promoted."[20] Other regulations included ordering the clergy not to live with any women except a mother, a sister, or an aunt. They tried to establish a proper date for Easter.

The council now entered into a debate that would have direct bearing on the future of Christianity. This debate glittered with dialectics, smoldered with antipathies murmured in Christian piety, and was often confused, abstract, oversubtle, and generally mystifying. Arius, Athanasius, and their lieutenants argued fiercely over whether God and Christ were one or just a tiny bit different. The days were consumed in dialectic. Constantine grew impatient. The bishops trembled at their politely smiling emperor who had the power of life and death over every one of them. They hurried the process. Presumably at the urging of Hosius, Constantine suggested the Greek word *homoousios* (ὅμοούσιος), meaning "consubstantial," or "of one substance," be added to the final agreement to convey the appropriate relationship between God and Jesus Christ. The vote was called. It occurred to those Arians who were resisting that the word *homoousios* was just barely ambiguous enough to admit of a razor's edge of difference between God and his Son. The final agreement, incorporating ὅμοούσιος, passed with just two dissenters. Arius was one. When the council ended he was excommunicated, his writings burned, and he was sent into exile in the remote Roman province of Illyricum.

The battle to make the Trinity the central doctrine of the church had only just begun. It would end at the Council of Constantinople in 381, when the word ὅμοούσιος was officially accepted into the Nicene Creed, and it became a criminal offense to be an Arian.

Hundreds of thousands of books, pamphlets, sermons, and tracts have been written about the use of the word ὅμοούσιος as it was first conceived at the Council of Nicaea. Newton wrote one of those tracts. It is probably the least known and the most brilliant ever written. Newton called his twenty-three-question-long treatise "Queries Regarding the Word ὅμοούσιος." Following are the first two questions. They are the most important when it comes to understanding how Newton felt about the Christian Church. (All twenty-three queries can be found in appendix C.)

> Query 1. Whether Christ sent his apostles to preach metaphysics to the unlearned common people, and to their wives and children.[21]

Newton enlarged on this query in this statement in the "Irenicum":

> If you would know the meaning of the several names given to Christ in preaching the Gospel, you are to have recourse not to Metaphysicks & Philosophy but to the scriptures of the old Testament. For Christ sent not his disciples to preach Metaphysics to the common people & to their wives & children, but expounded to them out of Moses & the Prophets & Psalms the things concerning himself & opened their understanding that they might understand the scriptures & then sent them to teach all nations what he had taught them. And the Apostle bids us beware of vain philosophy.[22]

And the second query was as follows:

> Query 2. Whether the word ὁμοούσιος ever was in any creed before the Nicene; or any creed was produced by any one bishop at the Council of Nice for authorizing the use of that word.[23]

Not even the most clairvoyant of the bishops assembled at Nicaea could have known that, in the Book of Revelation, John of Patmos and/or Jesus Christ had foretold the moment when the Roman Empire and the Christian world would start on the path to partnership. But Isaac Newton was certain that he had found that moment hidden among the prophetic hieroglyphs, and he tells us in his "Two Incomplete Treatises on Prophecy" that "the chief character of the sixth Dynasty was the dethroning of heathenism & enthroning of Christianity. And this is represented at the opening of the sixth seal by a description of the end of the heathen world, in the repetition of the Prophesy by the casting of the Dragon that old Serpent out heaven & exalting the Man-child up to the throne."[24]

Frank Manuel tells us that Newton's "Paradoxical Questions"

> is a sharply argued historical brief, not untouched by irony, in which

St. Athanasius too [like Jerome in Newton's *An Historical Account of Two Notable Corruptions of Scripture*] is demonstrated to be a manipulator of sacred records and a bearer of false witness against Arius. Newton's attack on Athanasius took the form of what lawyers call a "consciousness of guilt" argument, designed to show that a series of actions performed by Athanasius were those of a man who knew he had committed a wrong and thereby tacitly admitted it.[25]

Newton was trying to prove that the doctrine of the Trinity was wrong by proving that Athanasius was awful. But over and above this purpose, never far from his thoughts, was Newton's overwhelming need to perform a detailed and complete autopsy on the body of early fourth-century Christianity, the better to know what toxins already inhabited it, what diseases were pending, and how it was that the religion of Noah, of the Jews, then of Christ had already, at that early date, become so vile and corrupt that it would turn into the Roman Catholic Church. If he knew all the stages of the disease, all the twists and turns of that process of corruption, then perhaps one day he would know how, if not to turn back the clock, then perhaps, just maybe, to ameliorate some of the problems of the body religious of his own time.

His being an Arian, however, made that extremely difficult.

Newton's first paradoxical question, which seems very far away from the Council of Nicaea but actually stems directly from it, is "Whether the ignominious death of Arius in a boghouse was not a story feigned & put about by Athanasius above twenty years after his death."

The foul odors of Nicaea were as nothing compared to the stench of Constantinople, that city, formerly Byzantium, that Constantine had made the new capital of the empire in AD 329. That stench was never more apparent than in the flagstone square fronting the Forum of Constantine. Here horses were paraded all day long, leaving reeking piles of dung that armies of slaves scurried to sweep up but never quite disposed. There rose up in the center of the square a 120-foot-high reddish porphyry marble column. This column was surmounted by a bronze statue, seven rays emanating from its head, that had originally resembled Apollo but had been reworked to resemble Constantine. The column was hollow and packed with pieces of "the one true Cross" that the dowager empress Helena, Constantine's mother (and once a tavern girl in Bythnia), had sent back from Jerusalem. These most precious relics of the Passion of Christ must (if they really were inside the porphyry column) have lent the proverbial odor of sanctity to the forum square, even if they did nothing to relieve the stench.

On one side of the public square there stood a public latrine that bordered on being a public menace. With regard to ancient latrines, a modern-day scholar of ancient Roman plumbing tells us that foul-smelling mephitic gases, expanding through the sewers, could explode at any moment and send sheets of flames sweeping up through the toilet seats. Sharp-toothed scampering rats poked their heads up everywhere.* If you had any open wounds you ran the risk of contracting diarrhea, gonorrhea, tuberculosis, or worse from a previous occupant of the seat.[26]

History informs us that in the summer of 339 a terrible accident took place in this public latrine that stood on one side of Constantinople's forum square. Many historians report that early that summer Constantine the Great summoned Arius to a private audience at his royal palace in Constantinople. Standing uneasily before the emperor, trying not to notice the pagan pomp and splendor of the palace, Arius thought he knew what this was all about, and he had come prepared: concealed beneath his robe was a copy of his creation, the Arian manifesto.

Constantine told the notorious heretic that if he swore an oath that he had renounced his heresy, then he, the emperor, would guarantee

---

*Apparently private latrines weren't much better: "The Roman author Aelian writes of a wealthy Iberian merchant who, puzzled by the gradual disappearance of the pickled fish stored in his well-stocked pantry, discovered it was being eaten bit by bit every night by an octopus that came up through the toilet." (Koloski-Ostrow, "Raising a Really Big Stink," 43.)

his reception back into the orthodox church and even allow him to celebrate Mass at the church of Constantinople the next morning. Arius swore this oath; and a rumor has sullied his memory down through history that at the same moment that he swore he clutched beneath his robe his Arian manifesto denying the oneness of God and Christ.

Next, writes Newton, *"the emperor dismissed him with these words: 'If thy faith be right thou hast well sworn, but if impious and yet thou hast sworn, God will condemn thee for thy oath'"* (emphasis in original).[27]

An emissary raced to the church of Constantinople to tell Bishop Alexander (not to be confused with Bishop Alexander of Alexandria) about the unexpected visitor who would be arriving to take communion the next morning. Shortly afterward, anyone entering this cathedral would have been startled to see a bundle of bright crimson robes lying, as if flung there, at the foot of the altar. Quick jerks would have revealed the outline of agitated legs: the bundle of robes would have Alexander, in tears, prostrate on the marble floor, arms outstretched in supplication, beseeching God to "take Arius away" lest this man who was polluted with heresy might, if he were received into the church of Constantinople, pollute the church as well.

The next morning, crossing the forum square with his assistants, on the way to communion at the church, Arius must have been in a celebratory mood. He would have skirted the snorting horses and the piles of manure with more than his usual alacrity. Perhaps he pointed out to his retinue the statue of Constantine on the porphyry column and thanked God for the emperor's beneficence.

Passing by the multi-seat public latrine, Arius was seized by the need to relieve himself and disappeared into the boghouse (outhouse).

Did the pieces of the one true Cross in the porphyry column rising up from the center of the square transmit evil rays to Arius that morning? After all, he was a heretic, and he was about to invade a bastion of orthodoxy! Certainly the latrine didn't lack for means for a vengeful God to punish a blasphemous servant: gas explosions; a poisonous rat bite; a contracted, lingering, fatal disease—any one of these would do the trick. Arius's assistants waited for him outside the latrine. Time passed. Arius did not reappear. They rushed in. The great heretic lay dead on the floor in a pool of his own blood and excrement; his bowels had burst open. In the words of Isaac Newton, Arius, entering the latrine, "falling headlong burst in sunder & died upon the ground, being deprived both of communion & life."[28]

Edward Gibbon writes, "On the same day which had been fixed for the triumph of Arius, he expired; and the strange and horrid circumstances of his death might excite a suspicion that the orthodox saints had contributed more efficaciously than by their prayers to deliver the church from the more formidable of their enemies." Gibbon continues:

> We derive the original story from Athanasius, who expresses some reluctance to stigmatize the memory of the dead. He might exaggerate; but the perpetual commerce of Alexandria and Constantinople would have rendered it dangerous to invent. Those who press the literal narrative of the death of Arius (his bowels suddenly burst open in a privy) must make their choice between *poison* and a *miracle*.[29]

We realize to what extent we are, with this event, in the tragicomic world of the surreal, the absurd and the macabre, when MacCulloch tells us that "the orthodox tradition of public worship contains hymns of hate directed toward named individuals who are defined as heretical" and that such a hymn of hate was written about Arius—a hymn in which "in celebration of the First Council of Nicaea, the liturgy describes with relish (and a malevolent theological pun) the wretched end of of Nicaea's arch-villain in fatal diarrhea on the privy."

> Arius fell into the precipice of sin,
> Having shut his eyes so as not to see the light,
> And he was ripped asunder
> by a divine hook so that along with his entrails
> he forcibly emptied out

> all his essence [ousia!] and his soul,
> and was named another Judas
> both for his ideas and the manner of his death.[30]

Did the archbishop Alexander or the emperor Constantine take matters into their own hands and have Arius murdered? Was Athanasius the culprit?

Sir Isaac Newton thought none of the above. He didn't believe Arius had been murdered at all, not on his way to communion nor at any other time. Newton believed Arius had lived for ten more years at least.

Newton is indicting Athanasius for bearing false witness against Arius—in modern parlance, he is blaming him for character assassination. We'll recall that the first of the paradoxical questions asks, "Whether the ignominious death of Arius in a boghouse was not a story feigned & put about by Athanasius above twenty years after his death." Newton was convinced the answer was yes. He is accusing Athanasius of fabricating the whole thing many years later, when anyone who might have been connected with such an event would be dead or couldn't be expected to remember what had happened or not happened. Athanasius had invented the story to prove that God preferred orthodoxy to Arianism. He had done this in about 361, while he was in exile in the Egyptian desert.

Newton asks us to reflect on Athanasius's situation during this Egyptian exile. To do this, we need to briefly fill in his life from 328 onward.

From his ascension to the episcopate of Alexandria in 328, up to his exile in Egypt, Athanasius's behavior was characterized by total defiance of the emperor, violent denunciation of Athanasius's foes, and fiery preaching of the doctrine of the Trinity. The bishop of Alexandria was either despised or adored, and outside of Egypt he was mostly despised. Imperious, intransigent, self-righteous, and abrasive to the point of inflicting physical pain, he pushed his doctrinal enemies to the limit, virtually forcing them to accuse him of vile crimes (whether or not they had taken place, and Newton certainly thought they had), including murder (of someone other than Arius). Constantine had died in 327. His three sons took power. The youngest, Constantius, emerged as sole ruler of the empire. Constantius was an avowed Arian, and Athanasius found himself in more hot water than ever.

In council after council, the champion of orthodoxy was never quite able to shake off the charges his enemies had made against him. He was exiled to France for two years and Rome for three years, and he had to cool his heels in the Egyptian desert for three extended periods. Through all of this he never stopped writing, never putting down his pen even during (as we'll see in chapter 5) a bloody battle with Roman soldiers in his own church. As a result of that battle he fled Alexandria, disappearing into the desert and not reemerging for six years.

This final exile cast Athanasius into despair. But his situation wasn't without its perks. Five thousand mostly Egyptian monks, all of whom revered him, vied with each other to serve the dethroned archbishop as bodyguards or secretaries or messengers or cooks—as hosts providing him with every creature comfort. His sojourn wasn't without a touch of what passed for romance in the land of the monks: he was regularly hidden in the house of a twenty-year-old virgin, celebrated for her beauty, who concealed him in "her most secret chamber." Edward Gibbon writes, "As long as the danger continued, she regularly supplied him with books and provisions, washed his feet, managed his correspondence, and dexterously concealed from the eye of suspicion this familiar and solitary intercourse between a saint whose character required the most unblemished chastity and a female whose charms might excite the most dangerous emotions."[31]

Still, all this was hardly enough to console the exiled archbishop, who wondered with increasing desperation how he could stem the tide of Arianism he saw rising all around him. Newton asks us to imagine Anthanasius (and the following words probably convey how Newton really felt) turning this way and that like a cornered rat. What could he do against the heretics? he asked himself despairingly. He had always fought back ruthlessly; there had to be a way to fight back now, from here, from this desert where he was. After all, the survival of the doctrine of the Trinity was in his hands. And, where that was the issue, there couldn't be any such thing as an immoral act.

That is how, says Newton, Athanasius decided to rewrite history (though it's unlikely the exiled archbishop explained it to himself in these words). He engaged in an orgy of letter writing. His gorgeous virgin assistant must never have known how many lies she helped him draft. The letters went out to orthodox monks and bishops near and far. In a separate epistle, Athanasius shared his bogus information with his friend the historian Serapion. In these letters, Athanasius told the world, for the first time, the "true" story of the death of Arius. And that is the story that is told above.

Athanasius explained to one and all that he had heard the story from his assistant Macarius, who was inside the cathedral in Constantinople when Archbishop Alexander threw himself down on the floor and begged God to "take Arius away" if he was indeed a heretic. Athanasius explained to his friends that it has taken him this long to tell the story because he hadn't wanted to make religious-political capital out of the lamentable death of that poor man, Arius—for, after all, we are all appointed to die, are we not?

But now that he had told the story Athanasius felt it was his duty to point out that Arius's death was obviously the answer to Alexander's prayers (and Constantine's warning). Could anyone, knowing the manner of Arius's death, ever doubt that he was struck down by the Lord because his wrong understanding of Christianity was not to be tolerated any longer?

The dethroned archbishop was sorry to say that the recipients of this letter couldn't check out the details with Macarius, who had told Athanasius the story in the first place, because Macarius was no longer among the living. Moreover, seeing as the contents of this letter were of a delicate nature, Athanasius begged the reader not to show it to anyone else—and, once having read it, to destroy it or return it to Athanasius.

Newton tells us that the contents of this letter, which would later make up a part of Athanasius's autobiography, were taken up some sixty to ninety years later by several Greek historians who incorporated them into the accepted record of the times.

This letter, as important as were its particular contents, was only one of many scurrilous rewrites of history that Athanasius sent out during this trying period. (He had sent a good many out before then.) In "Paradoxical Questions," Newton examines all these documents at great length. But, before we can discuss the many other suspicious events for which, according to Newton, Athanasius would be forced to supply alternative accounts, we have to put flesh on the bones of our account of Athanasius's life from 328 until his exile in Egypt. And to do that we'll have to reach much further back than 328—even further back than Athanasius's student days in Alexandria—to the emperor Diocletian's persecution of the Christians.

CHAPTER FIVE

THE SEVERED HAND

Murder in the Fourth Century AD, Part 2

We now begin our account of the second half of Isaac Newton's furious assault on the integrity of Archbishop Athanasius of Alexandria. And we find again, as we found in the previous chapter, proud prelates striving to be Christlike—and to assert their wills—in an only-recently pagan world where Christianity, still in a molten state, struggles within itself to find a new expression. There stems from this disjointed universe, as we saw in chapter 4, seemingly surrealistic scenes and alternate clashing versions of reality. Our dramatis personae expands to include (along with a prostitute and several mobs) two new protagonists: Eusebius, the commanding and dressy archbishop of Nicodemia, and the emperor Constantius, Constantine's son and heir and a bumbling but honest Arian whom Newton admiringly promotes to the head of the class. As for Isaac Newton, he is still the ace detective, still pursuing to the bitter end and exposing the deceptions of Athanasius as he seeks to impose idolatry and apostasy upon the world.

And there is more murder and mayhem, principally in the form of a severed hand.

In the previous chapter we left Athanasius busily rewriting history with the protection of legions of desert monks and the special help of a beautiful young assistant. We begin this new chapter by making a brief leap backward in time, to a searing sequence of events that Athanasius had witnessed as a boy that had made a tremendous impact on him. They also profoundly affected the church, for they were the cause of the first real schism in Christianity.

In February 303, the emperor Diocletian unleashed a final bloody persecution on the Christians. For nine years, rack, scourge, iron hooks, and red-hot beds were the frequent companion of the followers of Jesus who refused to renounce their faith and sacrifice to the pagan gods of Rome.

Newton remarks with uncharacteristic inexactitude that Diocletian's soldiers slaughtered 144,000 Christians in Egypt alone, exclaiming indignantly, "What think you then was done throughout the whole Roman world?"[1] Edward Gibbon put the number of homicides at closer to 2,000, adding that thousands more were permanently maimed, and cautioning that "the biases of the ancient historians, their fierce allegiances, their isolation, difficulties in communications and in finding reliable witnesses, makes it almost impossible to know exactly what happened."[2]

Nine years after Diocletian launched his pogrom, in the summer of 311, Archbishop Peter of Alexandria—then a prisoner of the Romans in a communal jail cell in Alexandria—decisively drew a curtain* across the center of the cell and addressed a disheveled and starving group of clerics sitting around him: "Let those who are of my opinion come to me! Let those who are of Meletius join Meletius."[3]

The opinion in question was whether Christians who renounced their faith rather than be tortured or killed by the Romans should be allowed to rejoin the church once the persecution was over. Melitius, who sat at the other end of the cell, was Peter's second-in-command, a bishop who did what the archbishop didn't have time to do in the archbishopric of Alexandria (and, for that matter, all of Egypt, of which

---

*The curtain was a himation; that is, a contemporary cloak. It must have been a very big cloak or a very narrow jail cell.

Peter had the charge) and kept him aware of all the rest. Between the two prelates huddled a crowd of Egyptian deacons and bishops, rounded up by the Romans during the summer and awaiting martyrdom by execution or extradition to the salt mines of Palestine.

Meletius believed the lapsed Christians should be allowed back into the church only after a period of soul-searching and repentance, if at all. Peter ("a kindly man, and like a father to all," writes the historian Epiphanius) believed these reluctant traitors to the faith should be welcomed joyfully back into the church. "Let that which is lamed not be turned out of the way; but let it rather be healed," he declared. "Peter spoke for mercy and kindness," writes Epiphanius, "and Melitius and his supporters for truth and zeal."[4]

But most of the imprisoned churchmen weren't as merciful as Peter and quickly joined Melitius on the other side of the curtain. A small minority gathered around Peter. Newton sums up:

> When he [Melitius] & Peter & other martyrs & Confessors were in prison together there arose a dispute about the reception of lapsed persons, Peter out of mercy being for a speedy reception & Melitius & Peleus & many other martyrs & confessors out of zeal for piety being for a competent time of penitence before they were received so that the sincerity of their penitence might first appear.[5]

That night—or perhaps it was a few nights later; we don't know for sure—Peter was taken outside the walls of Alexandria and executed. Legend has it that he bent his head, offered his neck to the soldiers, and said quietly, "Do what you have been commanded to do." But none of the soldiers could bring himself to do what he had been commanded to do. Finally the legionnaires put a fund together, chose a soldier by lot, and paid him to decapitate Peter.[6]

Melitius, along with many of the other prisoners, was sent to the salt mines at Phaeno in Palestine. And this was where the split, or schism, in the Church of Egypt that had been developing out of the heated debates between Peter and Meletius finally took concrete form. Gibbon writes: "The confessors who were condemned to work in the mines were permitted, by the humanity or the negligence of their keepers, to build chapels and freely to profess their religion in the midst of those dreary habitations."[7] For the next two years, Melitius preached to his fellow prisoners—and, it is possible, inadvertently to the pillars of salt surrounding him, as if they were among the sinners who, like Lot's wife, stole a backward glance to watch God annihilate the sin-laden cities of Sodom and Gomorrah. Melitius preached, planned, won recruits, and, when the persecution of the Christians ended and he was a free man, founded the Church of the Martyrs, which was not exclusively devoted to martyrs but insisted that Christians who had lapsed during the persecution undergo a lengthy period of repentance, following which their readmittance was almost, though not entirely, assured.*

Alexander, who was with Athanasius at the Council of Nicaea, had succeeded Peter as archbishop of the Trinitarian Church of Egypt. Alexander would be succeeded by Athanasius. The schismatic Church of the Martyrs, whose members called themselves Melitians, would become an increasing thorn in the side of Alexander and then of Athanasius. This was in part because the brainchild of Meletius had begun to incline itself toward the beliefs of Arius. Added to this, a third church, that of the Eusebeans, led by Eusebius in Bythnia—already semi-Arian and not enthusiastic about the doctrine of the Trinity—had begun to align itself tentatively with the Church of the Martyrs.

Then, in 313, a miracle happened. It was engineered by Emperor Constantine. Hans Küng writes: "To the great delight of Christians, in 313 this cool master of realpolitik, with his co-regent, Licinius, granted unlimited freedom of religion to the whole empire. In 315 the punishment of crucifixion was abolished, and in 321 Sunday was introduced as a legal festival and the church was allowed to accept legacies."[8]

---

*Athanasius has a different account, writing that "Peter deposed Melitius for cause at a council, and that Melitius retaliated by starting the schism." (Epiphanius, *Panarion*, 317, n. 6.)

Christians rejoiced in their sudden freedom. But it was the freedom to quarrel as well as to grow. Over the next dozen years, the Christianity of the East increasingly assumed the form of what the French call a *panier de crabes*—a "basket of crabs"—that is, an organization whose people metaphorically claw and crawl over one another to get to the top. The followers of Melitius, the bishop who had laid the foundations for a new church while toiling in the salt mines of Phaeno; of Eusebius, the most prolific writer of his day and an ecclesiastical fashion plate who wore a scarflike omophorion that swept down from his shoulder to his knees; and of Arius, the sly heretic who embedded subversive doctrines in the ditties he sang to dockworkers—the followers of these three not only fought singly or together against Alexander's Orthodox Church of Egypt and the idea that God and Christ were one, but they often quarreled hotly with each other over points of doctrine that were of burning concern to them but mean nothing to us today.

This forced a brilliant and now disgruntled Emperor Constantine to convene the Council of Nicaea in 325. That council had ended in a shotgun wedding. The emperor held the shotgun and performed the ceremony. The blushing brides were Alexander and Athanasius; the scowling grooms were Arius, Melitius, and Eusebius. The wedding vow, sung out joyfully by the brides and muttered grimly by the grooms, was the first draft of the Nicene Creed.

Once Constantine, whom the bishops feared greatly because he held the power of instant life or death over all of them, had dissolved the council and headed home, the marriage fell apart, though nobody dared admit it let alone ask for a divorce. The participants began to quarrel all over again, and this time their disputes were envenomed by the conclusions the council had reached about the relationship of God and Jesus. As Newton writes, "when the Council of Nicea was ended there burned an implacable fury of contention among the Egyptians . . . this contention was about the Nicene decree of the word homousios."[9]

Athanasius, who had been such a trustworthy secretary to Alexander at Nicaea, was named a deacon of the orthodox church at Alexandria in 326. He immediately began to co-opt the functions of the aging archbishop. This made him more and more the target of doctrinal and personal attacks, and he returned these attacks with a vengeance. Newton writes: "Athanasius, in the controversy between the Clergy of Alexander [about the Son of God], inflamed differences, thereby to throw out part of the Clergy & make room for himself & his friends: & when he had thus gotten to be Deacon, the reputation & interest he had got with his friends by that controversy served him to invade the Bishoprick."[10]

Alexander died in 328. Athanasius immediately began to maneuver ruthlessly to become the new archbishop. He succeeded, but his election was contested and it still is today. And here the steamy, contentious, volatile city of Alexandria itself with all its mobs enters our drama as a character, for it played a role in that election.

Straddling East and West, the city of Alexandria had a population of 800,000 in 328, about the same as San Francisco today. The ancient seaport, founded by Alexander the Great in 334 BC, boiled over with an explosive mix of Greeks, Egyptians, Jews (one-quarter of the population), Italians, Arabs, Phoenicians, Persians, Ethiopians, Syrians, Libyans, Cilicians, Scythians, Indians, Nubians, and others—a wild superabundance of ethnic diversity that has not been seen in the world again even to our day.

The emperor Hadrian (AD 76–138) wrote of the Alexandrians that they were "very seditious, very vain, and very quarrelsome. The city is commercial, opulent, and populous. No one is idle. Some make glass; others manufacture paper; they seem to be, and indeed are, of all trades; not even the gout in their feet and hands can reduce them to entire inactivity; even the blind work. Money is a god which the Christians, Jews and all men adore alike."[11]

The city was a "continuous . . . revel of dancers, whistlers, and murderers,"[12] declared Dio Chrysostom, while Voltaire insisted that the Alexandrians were of a "contentious and quarrelsome spirit, joined to cowardice, superstition and debauchery."[13] This quarrelsome people had a near-magical capacity for forming and reforming into ferocious howling mobs, accomplishing some piece of work (usually a destructive one) in one part of the city, and then melting away into the landscape as if they had never existed. It took the merest peccadillo to incite one of these collective rampages: the shortage of a fruit in the marketplace, a perceived snub, a plebeian elbowing a nobleman in the baths.

Usually, though, the Alexandrian mob was bought, and the principal buyers were the politicians and the ecclesiastics. But—caveat emptor!—the mob could easily turn on its new owner, and it often did. When it was paid by the bishops it was often joined by a horde of desert monks who, skinny as poles and dressed in black tatters, shouting hymns and brandishing cudgels, swarmed up over the sand dunes surrounding the city.

These monks, who spent most of their days praying to God, were notorious for the violent anger they often displayed when they got off their knees. In 415, a gang of the desert monks would murder the beautiful Neoplatonist philosopher and mathematician Hypatia (370–415). Alexandria was celebrated for the brilliance, variety, and sheer number of its schools of rhetoric and philosophy. It was rare to see a woman among the thinkers who frequented these schools, but Hypatia was first among equals. Bishop Cyril of Alexandria complained that this upstart woman, a pagan, was fomenting trouble between his church and the civic authorities. Cyril exhorted a mob of desert monks then roaming the streets to solve this problem. They seized her in the street, dragged her into a temple, and then stripped her, killed her, mutilated her body with broken glass, and set her corpse on fire. They were never punished.

In the last week of February 329, a vociferous Alexandrian mob, which could have included the grandfathers of the desert monks who had murdered Hypatia a century later, surged down a side street that radiated out from two huge avenues that met and crossed at the center of Alexandria. Barking dogs fled before the mob; the homeless, stretched out on the side street, rolled out of the way. The masses of men wove giddily between statues of the dog-headed god Anubis and pointy-eared Hermes and orange Egyptian pylons stretching up to the sky; spears glinted and swords rippled with the light of swaying torches; the mob sang hymns and shouted imprecations and chanted. Some parts of the mob shouted, "The black dwarf! The black dwarf!" This was the people's nickname for the short, swarthy, scowling Egyptian Copt who answered to the name of Athanasius. The mob was in the pay of those who supported him, and possibly in the pay simply of Athanasius himself.

They arrived at their first destination: the Church of Saint Theognis. Just inside gleamed the coffin in which Alexander lay in ecclesiastic state. There was almost nobody there, and the mob didn't stop for long. A few dropped to one knee, bowed their head, and quickly rose.

To the left, some distance down another side street, there was a muted clamor; screams, shouts, and the clashing of swords filled the turgid air. The mob turned away from Saint Theognis and, racing down the narrow side street, came out into a large square in which a multitide of force, packed tightly together, contended for mastery. Thick knots of men swayed, grappled, struck out at each other with broadswords; some fled, others fell, new fighters entered the fray. The pandemonium was atrocious; but above it could be heard, sometimes indistinctly, other times quite clearly, the refrain, "Oh, wickedness! Is he a bishop or is he a boy?"

These were the anti-Athanasian factions, who in some cases were very carefully chosen Melitians, Arians, or Eusebeans, in others, pure howling mob. It would be some time before the facts of this night were sorted out—and there must have been other nights like this—but those who hated Athanasius had brought forward the objection that he was not yet thirty-five years old, and a candidate for archbishop had to be thirty-five.

This was problematical; for centuries it was believed that Athanasius really *was* thirty-five when he became archbishop. But others, including Newton, believe that Athanasius falsified the records some years later and that he was born sometime between 296 and 298. Whatever the case may be, that night the rival camps were battling in the square to try to gain entrance to the Church of Saint Dionysius, which rose up majestically on one side of the square, or to keep others from gaining entrance. There were wild rumors as to what was going on in the chuch, and some didn't know exactly what they were fighting for. But it would become clear, if not that night then the next, or the next, that the Church of Saint Dionysius had become a jail; that its jailers were a mixture of paid mob and supporters of Athanasius; and that the jailed were a dozen bishops whose vote was needed to make Athanasius archbishop.

This, if we are to believe the somewhat contradictory records, had been going on for days, and the imprisoned bishops were starving, exhausted, and enraged. But there came a night—perhaps it was that very night—when the bishops agreed to make Athanasius archbishop, and they were liberated from the church that had become their prison. The mob dissipated, having played out its role, and Alexandria became merely a backdrop, and not an actual participant, in the story of the slow and tormented climb to power of now Archbishop Athanasius.

In the twelfth of his paradoxical questions, Newton introduces a document that, he says, was released by the Council of Alexandria fifteen years after the election of Athanasius and that claimed to be a letter put together by the people of Alexandria during those tumultuous nights in late February 329. The letter asserts that Athanasius was elected directly after Alexander died. It reads in part, "*we* [the city fathers] *& the whole City & Province are witnesses that all the multitude & all the people of the Catholick church* (that is [adds Newton] all whom they would acknowledge to be catholick) *being assembled as with one soul & body, cried out with great acclamations desiring that Athanasius might be Bishop of the Church*" (emphasis in original).[14]

Newton believed this letter was a forgery, likely written by Athanasius at the time of the Alexandrian synod. Or, he states elsewhere, it was written by Athanasius (without any townspeople) directly after a group of bishops had ordained him in a secret place. But Newton seems ultimately to have believed that a small number of bishops had been forced by the mob and its puppet masters to make Athanasius the new archbishop, just as has been described above.*

The Black Dwarf, when he became archbishop, was far more abrasive and polarizing than he had been as deacon. He was quarrelsome and vindictive. He kept all the factions agitated. He kept his flock of Christians regularly informed, and he chastised them regularly. He was, though Newton scarcely mentions it, one of the great polemical theologians of his time. Here, for example, he explains that the three aspects of the Trinity are not inferior to one another but that God is a Trinity, "not only in name and linguistic expression, but Trinity in reality and truth. Just as the Father is the 'One who is' (Ex 3:14), so likewise is his Word the 'One who is, God over all' (Rom 9:5). Nor is the Holy Spirit non-existent but truly exists and subsists."[15]

Athanasius treated God, Christ, and the Holy Ghost with exquisite tact and tenderness, and human beings at best autocratically and at worst violently. His abuses mounted steadily, if we are to believe Sir Isaac Newton. They reached a high point in probably 333. He stalked into a Melitian Church of the Martyrs in Mareotis, a suburb of Alexandria, shattered the communion cup, overturned the communion table, and (surely with the aid of others) laid waste to the entire church.

Did this really happen? It was the Melitians who leveled the accusations. When Athanasius denied them, they dug up the past, charging him (as they had before) with the use of "violence and bribery to achieve episcopal election while below the canonical age."[16] Melitians, Eusebeans, and Arians alike joined in the assault on Archbishop Athanasius, piling up a heap of charges: the Egyptian prelate was guilty of immoral conduct, of illegally taxing the people, of plotting treason

---

*There are other versions of the story of the succession of Athanasius. One says it took much more badgering of the bishops, perhaps three months' worth, before he was named archbishop; another, that a far greater number of bishops were held prisoner; and a third that for a long time—during which Athanasius basically disappeared—there was only an interim archbishop of Alexandria.

against the the emperor. Athanasius railed furiously at his accusers. They railed back and threw in more chips, accusing him of raping a virgin when she hid him in her home to save him from the mob and of murdering a Melitian bishop and making supernatural use of a part of his body. Newton sums up:

> When Athanasius succeeded in the Bishopric of Alexandria he was accused of tyrannical behaviour towards the Meletians so as in the time of the Sacrament to break the communion cup of one Ischyras a Meletian Presbyter in Mareote & subvert the communion table & cause the Church to be speedily demolished, & some time after to kill Arsenius a Bishop the successor of Melitius in Hypsalita.[17]

With charges and countercharges flying all around, it was time to ask the emperor Constantine for mediation.

But, first: How is it that these Christian leaders, dedicated to the imitation of the life of Christ, could act so vilely? Basil of Caesarea (329/330–379), a more or less contemporary bishop-monk, wrote that the progress of Christianity at the time was like "a naval battle fought at night in a storm, with crews and soldiers fighting among themselves, often in purely selfish power struggles, heedless of orders from above and fighting for mastery even while their ship foundered."[18]

MacCulloch observes: "It may seem baffling now that such apparently rarefied disputes could have aroused the sort of passion now largely confined to the aftermath of a football match."[19] And the journalist and historian Paul Johnson perhaps goes to the root of the problem: "The venom employed in these endemic controversies reflects the fundamental instability of Christian belief during the early centuries, before a canon of New Testament writings had been established, credal formulations evolved to epitomize them, and a regular ecclesiastical structure built up to protect and propagate such agreed beliefs."[20] To all the accusations that the Melitians aimed at Athanasius, Athanasius responded by blaming the Melitians; they had lied or they themselves had committed these same crimes and others. Isaac Newton's second paradoxical question asks, "Whether the Meletians deserved that ill Character which Athanasius gave them." Newton thought absolutely not; we will see why shortly.

The Melitians pleaded with Constantine to put Athanasius on trial. The emperor stalled for a year. He had an excuse: he was busily trying to make some Roman laws a little less unforgiving. Constantine understood that a changed religion demands a changed social system,[21] and he changed as much as he could, though this could not be a great deal, since the emperor was locked in the traditional armor of Roman sternness. Constantine promulgated a law forbidding the separation by sale of slaves who were man and wife; he prohibited divorce except on statutory grounds; he created other laws making relations between Roman men and women a little easier.[22]

But he could fob off the enemies of Athanasius for only so long. Constantine was bold, brilliant, fearless, and frightened. For all his breathtaking palaces, his bedrooms lined with statues of the gods (and now of Jesus Christ), for all the soft rustling silk of his bedsheets, all the beauty of his compliant concubines—despite all that, the emperor had terrible dreams. He had slaughtered tens of thousands in his wars; he had tortured multitudes; worst of all, he had murdered his first wife and his eldest son and his favorite sister's husband. He wondered if the God who had given him victory at the Milvian Bridge wasn't having second thoughts about favoring a man whose hands were red with the blood of his own family.

He trembled at the thought of offending Jesus Christ and listened to his bishops. He convened a council at Constantinople. There, he would try Athanasius for his alleged crimes. The council got under way. Athanasius, insulted, sulky, frightened, didn't turn up. Furious, the emperor canceled the council.

The next year found the emperor campaigning on the farthest frontiers of the empire. He subdued, bullied, slaughtered defiant tribes, signed treaties and abrogated others. One day an ecclesiastical letter arrived from Alexandria. In the murky brightness of a German sun, on an autumnal day, he read it with growing anger. Eusebeans, Arians, Melitians—all listed new crimes by Athanasius and pleaded with the emperor to return and punish the evil archbishop. The impertinence of their words enraged Constantine. Then he remembered, once again, that he couldn't afford to offend Jesus Christ, whom he had made the supreme God of the Roman Empire. The emperor wrote back to Alexandria. On his return to Constantinople he convened, for the summer of 335, a Council at Tyre in Phoenicia. He ordered Athanasius to attend and told him if he, Athanasius, saw any difficulty in this, then he, Constantine, would send a legion to help him out.

Athanasius agreed to come.

In Phoenician Tyre, in the summer of 335, the stench of the famous imperial dyeing factories dominated all other odors just as it had as far back as anyone could remember. That July, bright purple robes (likely colored with that same Tyrean dye) suddenly sprouted in the streets, competing for attention with the black goatskin tunics and yellow cotton robes of the milling Tyrean crowds. These were the same bright purple robes that had outshone the workaday garb of the townspeople of Nicaea a decade before. The Christian bishops of the eastern Roman Empire, more than three hundred strong, were gathering for a council once again.

The Council of Tyre was no mere kangaroo court, hastily put together to try, convict, and hurry Athanasius off to oblivion. That is how many historians have depicted it; Newton thought they did so to belittle the importance of the council's findings. Newton's paradoxical question number 3 is, "Whether the Council of Tyre & Jerusalem [a side trip to Jerusalem would be a part of this council] was not an orthodox authentic Council bigger than that of Nice." Newton answered that

> it was an ancient Canon of the Church, as well as a necessary one, that no man should be received [taken back into the church] by a

less number of Bishops than those by which he had been ejected. [Constantine was planning to reinstate Arius at this Tyrean synod.] And therefore the Emperor sent his letters into all the Eastern Empire requiring the attendance of the Bishops that the Council might be full.[23]

Once again, Constantine paid all the expenses. The Great Hall of the municipal palace in Tyre swarmed with Libyans, Egyptians, Syrians, Phoenicians, Ethiopians, every variety of Middle Eastern prelate. These included Melitians, Arians, Eusebeans, and supporters of the Orthodox Catholic Church of Alexandria. Athanasius had swelled the numbers significantly, "bringing," writes Newton, "a great multitude [forty-seven bishops and presbyters] out of Egypt to create disturbance, & behaving himself very turbulently in his trial, as the Council of Tyre in their circulatory letters [post-conference reports] complained."[24]

Against a backdrop of lurid Eastern tapestries sweeping down to the floor (the largest depicted two purple lions devouring each other on an orange desert beneath a yellow moon), the bishops milled about, fussing over each other's health, pouncing on each other's doctrinal errors, sharing self-martyrdom stories (such as "Pachomius went 53 days without closing an eye," or, "Eustachius carried thirty-eight pounds of bronze on his loins"), flexing their rhetorical muscles, and complaining bitterly that there were too many heretics about in the hall. The number of one-eyed, one-thumbed, hobbling bishops—grim witnesses to Diocletian's persecutions—had greatly diminished over the years, but they were still there, and the soldiers posted at the doors glanced away from them uneasily.

The Council of Tyre picked up where the Council of Nicaea left off. It followed the Nicaean Synod as the night follows the day, and the Council of Tyre was certainly the night. It was the black twin of the luminous Council of Nicaea. The earlier synod had touched the gleaming gates of heaven with its lofty debates on the nature of the Trinity; the Tyrean council would scrape the sulfuric walls of hell with its parade of human sin and folly including murder, rape, sedition, the destruction of church property, and much more. But the political and ecclesiastical stakes at this second council were just as high; they might decide with what doctrine, and by whom, the Christian Church would lead the world to salvation.

At the front of the hall, Constantine's representative, Dionysius, disinterested, swathed in a black panther cloak, sat enthroned on the dais. Behind him, a Syrian tapestry swept down from the ceiling depicting three satyrs playing the pipes at the edge of a tide as they chased three wide-eyed, nubile sea nymphs. At Dionysius's side stood the universally learned Eusebius of Caesarea, executive director of the council, clutching a papyrus on which he was scribbling notes and wearing a snow-white omophorion that, all aglitter with four gold crosses and an eight-pointed star, swept down to below his knees.

History has given us two versions of the Council of Tyre—sometimes three!—and Newton chronicles them all in his "Paradoxical Questions." To most modern readers, this council will come across as macabre, even absurdist. As noted above, it was a gathering that could have come straight from the imagination of the surrealist filmmaker Luis Buñuel (1900–1983). And with its multiple versions it might have been a creation of the Japanese filmmaker Akira Kurosawa, that supreme master at dramatizing mankind's tendency to see the world based on personal impressions and feelings and opinions rather than on external facts. (In his 1950 masterpiece *Rashomon,* four witnesses to a murder tell the story of the murder in four completely different ways; one of the witnesses is the victim, whose testimony is channeled through a medium.)

With a certain disdain Dionysius called the proceedings to order. The bishops took their seats. Eusebius delivered a benediction that insulted the Trinitarians even as it praised them. The amens had hardly died away before a shapely, haggard woman dressed in a red pleated tunic was pushed forward by two of the guards.

Her name has not come down to us. The historian Theodoret calls her "a woman of lewd life." We might call her a prostitute. She "deposed in a loud impudent manner" that all her life she had rejoiced in being a virgin, but that, when she had taken him into her home after he was threatened by a mob, Archbishop Athanasius had robbed her of that virginity.[25]

It mustn't have been easy for the assembled bishops, all of them sworn to virginity, most of them hot-blooded sons of Middle Eastern climes, to sit and watch this deposition. It would have provoked stirrings and arousals in them that they'd fought to suppress for years, some of them even resorting (before the Council of Nicaea forbade it)[26] to castrating in imitation of the great church father Origen.* These feelings would have added a certain poignancy, even some anger, and perhaps a soupçon of ruefulness, to the acrimonious shouts of disapproval that must have arisen in the hall, if only briefly, on that first morning of the council.

The shapely, haggard woman dressed in a red pleated tunic had completed her testimony. Dionysius asked Athanasius to step forward and defend himself. But the cleric who now strode purposefully up to the dais wasn't Athanasius; history has left no description of this new witness, but we know he wasn't the Black Dwarf. Standing before Dionysius, the newcomer turned sharply to the woman and shouted: "Have I, O woman, ever conversed with you, or have I entered your house?"

Theodoret tells us that the woman replied, "with still greater effrontery, screaming aloud in her dispute . . . and, pointing at him with her finger, exclaimed, 'It was you who robbed me of my virginity; it was you who stripped me of my chastity (adding other indelicate expressions which are used by shameless women).'"

Everyone in the hall knew this was not Athanasius, although almost no one knew it was an orthodox priest named Timotheus. But the ruse had worked. Theodoret tells us that "the devisers of this calumny were

---

*Origen Adamantius (184/185–253/254), a brilliant scholar and early Christian theologian said to have written a thousand books, some heretical, most now lost, literally castrated himself, basing this action on Matthew 19:12.

put to shame, and all the bishops who were privy to it, blushed."[27]

You might have thought the matter was settled. But the restless shade of Akira Kurosawa, who knew so well how to film human subjectivity, must have been on the prowl in the Great Hall that morning. For history has given us a second version of the story of Athanasius and the woman of lewd life. In this second version, recorded by the historian Philostorgius (368–ca. 439), the woman who is guided to the dais is eight months pregnant, and she tells the court that the father of her child is Eusebius of Caesarea, that very same executive director of the council and prolific author who is standing before her at that very moment. Isaac Newton summarizes:

> When Athanasius, being impelled by the Emperor's threatening [him], came to Tyre, he would not submit to stand in judgment, but sent in a big-bellied woman which he had hired to accuse Eusebius of adultery: hoping that by the tumult which would probably be raised, he might escape being tried. But when Eusebius [not letting on who he was] asked her if she knew the man & whether he was amongst the Bishops then present, she answered that she was not so senseless as to accuse such men of base lust—and by those words [they] discovered the fraud.[28]

These two tawdry mini-soap operas, hardly the kind of drama we expect to find at an ecclesiastical conference, were only the warm-up act for the macabre and surrealist spectacle that was to follow. The transcendent genius of Akira Kurosawa is at work again, for there are two versions of this second melodrama, and, by the time we have come to the end of the second—we find there are three.

Dionysius now called to testify Bishop John, the current leader of the Melitians (the founder, the former preacher in the salt mines of Phaeno, being long dead). History gives us no description of John. We know only that he came up to the dais clutching a leather bag somewhat bigger than a boxing glove. He stopped before Dionysius, reached into the bag, and—we can imagine there was a sudden expectant lull in the room—slowly withdrew an object that was long, white, and withered.

It had five fingers.

"This is a human hand!" exclaimed Dionysius.

And so it was. John held the withered appendage up for all to see, declaring that it was the severed hand of the Melitian bishop Arsenius, who had disappeared the month before. The appendage had been cut off, John continued, by Archbishop Athanasius of Alexandria, "to be made use of in Arts of diabolic Conjurations!"

There must have been an outcry. But it wasn't necessarily as great as the one that had accompanied the appearance of the woman of lewd life. A severed hand was far easier for a fourth-century bishop to deal with than a prostitute. He had seen plenty of severed hands, and severed arms, and severed legs, and even severed heads, both in the aftermath of a battle, where he ministered to the wounded and the dying, or in connection with some local quarrel or awful accident. Many of these clerics had been soldiers themselves in their pre-cleric youth, buckling on armor, flashing swords, and severing the odd limb themselves. In the fourth century, by the time a priest had became a bishop, there was no sort of bodily injury he hadn't seen.

Still, there must have been an outcry, if only one of indignation and confusion (for nothing had been proved yet). And it is at this point that we can imagine a second filmmaker of genius slipping down from heaven to watch the spectacle, namely, Luis Buñuel, in whose celebrated *The Andalusian Dog* (1929) a severed hand appears and flops around for at least a minute. (It's possible that Buñuel, who abominated the Roman Catholic Church and researched its history thoroughly, had read the proceedings of the Council of Tyre. Another of his films is *Simon of the Desert* [1965], about a fourth-century desert monk who lives on top of a column.)

Buñuel's roving, surrealist camera would certainly have lingered lovingly on this severed hand held up so everyone could see it. He would likely have panned in on the startled eyes of the sea nymphs frolicking in the tapestry on the wall behind Dionysius, and let those eyes, expanded to fill the screen, exclaim to the viewer: "We too see the hand. And it is awful! It is absurd! We can make nothing of it! Such is the world!"

But Buñuel's camera could only have lingered on the sea nymphs for a moment, for almost immediately something intervened to draw his attention away.

There was a sudden hubbub at the back of the hall. All eyes turned. Athanasius had entered and was pausing inside the doorway. Beside him stood a tall mysterious figure whose features were hidden by a scarf wrapped around its head. The figure was draped in a voluminous cloak that went down to the ankles. The sleeves were so long that they covered both the hands.

Athanasius took the mysterious figure's arm. He strode forward purposefully, bringing the other along with him. The archbishop of Alexandria wasn't smiling. Athanasius rarely smiled; only the thought of the Trinity ever brought a smile to his lips, and that thought must have been far from his mind now, for he was scowling ferociously. It is impossible to imagine what sort of a hubbub must have been taking place in the hall at that moment. No doubt a few shouted, "Is the Black Dwarf a murderer?" And there may have been a (not entirely relevant), "Oh, wickedness! Is he a bishop or is he a boy?"

They reached the dais. Athanasius turned and unwound the scarf from his companion's head. The features were laid bare. "This," announced Athanasius, "is Arsenius!"

Theodoret writes that the archbishop then "turn[ed] back the man's Cloak, and shew[ed] them one of his Hands; and after a little pause, to give them time to suspect it might be the other hand [that was missing], he put back the other side of the Cloak, and shew[ed] the other."[29]

Those who knew Arsenius knew this was Arsenius. Those who had eyes in their head saw that he had two hands.

▲

Theodoret's version ends here, and you might have thought that Athanasius was totally exonerated from the charge of murdering Arsenius and cutting off his hand. But the immortal soul of Akira Kurosawa is on the prowl again, proud, compassionate, and imperious in his knowledge of the ridiculous lack of objectivity of humankind, for there is a second version to this story and, as that version nears its end, it branches out into a third version.

In this second version death approached the dais again, but this time in a much bigger package. John the Melitian leader made his way to the dais again, but this time he was with a fellow Melitian, and the two dragged between them a huge oblong box.

They arrived at the dais and, kneeling, removed the lid from the box. They reached in. John pulled out of the box a white, withered hand. His companion pulled out a second white, withered hand. These hands were not severed; the two bishops struggled to their feet and hauled up to a semi-standing position an entire corpse, dressed in the robes of a bishop, its eyes closed, its face a dreadful chalk white.

The large oblong box was a coffin. John declared loudly, as the dead-white head lolled between them, "This is the body of Bishop Arsenius, murdered by Archbishop Athanasius of Alexandria!"

Let us not even try to imagine the pandemonium in the hall. A delighted Luis Buñuel would have turned his camera here and there and everywhere, hardly knowing what to settle on in this amazing treasure trove of Hieronymus Bosch delights. Perhaps he would have turned to the other tapestry and panned in on the two purple lions devouring each other on an orange desert beneath a yellow moon, expanding the image until it filled the entire screen and conveyed, in its surrealist way, the message: "Can't you see that Christian bishops are no better than savage lions?"

But Buñuel's attention would have been distracted by another new development, a voice coming from the back of the hall and saying, "I have something to show you."

It was Athanasius.

The archbishop of Alexandria had just come through the door. He strode forward slowly and thoughtfully, wearing the simple white robe of a mere priest. Perhaps there were cries of execration from the bishops; many, still trying to process the sudden appearance of a corpse in their midst, for that matter one of their own brethren, must have stared in silent, astonished, consternation. The bishops parted before Athanasius. He arrived at the dais, stopped, and slowly slipped a letter out of his pocket. He may have taken care to avoid looking at the corpse.

Here the spirit of Kurosawa runs riot: history has given us three versions of this letter.

In the first version, Athanasius explained that the letter he was holding wasn't addressed to himself but to a Melitian presbyter named Pinnes. Then he read the letter aloud. In it, Arsenius assured Pinnes that he was in good health. He repudiated the Arian heresy. He promised never again to bar from his church Christians who had renounced their faith during Diocletian's persecution. He pleaded with Pinnes to persuade Athanasius to allow him back into the Orthodox Catholic Church.

Having read this, Athanasius slipped the letter back into his pocket and asked Dionysius and Eusebius why he, Athanasius, would want to murder a man who was doing everything that they wanted him to do?

In the second version of this story, Athanasius read aloud a letter from Arsenius that was this time addressed to Athanasius. It merely repeated everything Arsenius had said in the first letter to Pinnes.

There is a third version: a letter written to Athanasius by a Melitian presbyter named Ischyras. This was the same presbyter whose communion cup Athanasius had allegedly smashed when he was engaged in the wholesale destruction of the Melitian church at Mareotis.

In this letter, Ischyras absolved Athanasius of all the crimes with which he has been charged, including the murder of Arsenius. The letter declared that it was he, Ischyras, the undersigned, who had originally brought these accusations against Athanasius, and he had done this because other Melitians had beaten him harshly until he agreed to do so. The fact of the matter was, the letter concluded, that Ischyras was totally on Athanasius's side.

This third letter was signed by fourteen witnesses.

Only a genius like Isaac Newton could make sense of all of this, and we will see farther on in this chapter that Newton decided all three of these letters were forgeries.

But the Council of Tyre wasn't over. Athanasius still had to face charges of using violence and bribery to attain his position as archbishop of Alexandria; of sedition (plotting treason against Constantine); and of smashing a communion cup, wrecking a communion table, and laying waste to an entire Melitian church.

But then something entirely unexpected happened.

God must have taken pity on the beleaguered bishops in the Great Hall of the municipal palace at Tyre that day, for now an event intervened that would transport most of them, in a single day, from the suburbs of hell that the Council of Tyre had become to what would seem to them like the suburbs of heaven.

One of Constantine's trusted emissaries, a shorthand secretary named Marianus, arrived early in the morning to read them a letter from the emperor. The imperial will as expressed in this letter sent the bishops scurrying back to their lodgings to pack a few belongings. Then they hurried to Tyre's city square, not far from the palace, where a long line of blue-maned imperial post horses, each hitched to a chariot, pawed the paving stones impatiently in eager anticipation of galloping out of the courtyard.

The bishops had been told to hurry, but the charioteers, lounging beside the chariots, seemed not to have heard. They got to their feet, ambled over to the bishops, taunted them, refused to answer their questions, and then clutched at their purple robes and threatened to steal them. The bishops protested loudly; suddenly Marianus appeared, wielding a whip and screaming at the charioteers to obey the emperor's orders. He threw himself into their midst, lashed two around the shoulders, then drew his sword and ran straight at a third.

The charioteers immediately obeyed. They bundled the bishops into the chariots. Whips cracked; the drivers shouted; with a clattering of wheels the procession of chariots rolled out of the city square. Barking dogs, shrieking children, and a left-behind bishop raced frantically beside it; beggars and peddlers tumbled out of the way. Marianus, on horseback, still brandishing the whip, galloped ahead, behind, beside. They still had not been told where they were going. The procession of chariots thundered through the fortified south gate of the city. At that moment, a lone seagull alighted on top of the round stone tower rising above the center of the gate; as it did so, a falcon, who had been watching it from on high, plummeted down quick as a lightning bolt, seized the gull in its talons, and bore it silently away into the shining sky.

The procession sped south down the coast of Phoenicia; on the right, the blue-green Mediterranean owned the view up to the wide horizon; near the shore, Roman triremes, Phoenician fishing boats, Libyan trading vessels—every sort of seacraft—jostled each other for possession of the beaches, the wharves, the mooring posts. The procession plunged farther down a coastline that grew straighter and more somber as the sun rose high up in the sky and passed the midpoint. Some afternoon mists drifted in; and then, as if the mood had become too oppressive, the line of chariots veered sharply from the main highway and, turning east, started up a wide, solid, slowly rising road.

Now Marianus told them where they were going. The bishops rejoiced, and in their joy some of them seemed to see unfolding around them the momentous events that had taken place along this highway: the slaves of King Hiram of Phoenicia dragging wooden carts packed with cedars of Lebanon to be used in the building of the Temple of Solomon; and then, a thousand years later, Roman legionnaires escorting back from Jerusalem, in wooden cages, shackled, naked, and bleeding Jewish princes who had tried and failed to save the Temple of Jerusalem from destruction.

Their destination was Jerusalem.

The procession arrived at the city as darkness was falling, and mounted by a winding route the wide, low hill called Mount Golgotha ("Skull"),* where Jesus Christ had been crucified.

Some of the bishops had been here before. But they recognized nothing. Everything had changed. Gone was the elegant temple of Diana whose pagan mass had dominated the top of the hill; in its place there rose up, from a base three times broader—stretching, it seemed in the semidarkness, almost to the sky—a massive, ornate, glittering basilica.

The bishops stared in astonishment. The basilica was all sparkling colonnade and dome and rotunda and foursquare pillars, with portico-lined courtyards threading walls and spaces together. This was the Church of the Great Martyrium, completed only a week before and constructed by Constantine on a schedule that would enable the emperor to consecrate it on the thirtieth anniversary of his reign, which was tomorrow.

Marianus directed the bishops inside, where the smoke of swaying censers caressed their nostrils with perfumes so exquisite that all memory of the stench of the dyeing factories of Tyre was expunged from their minds. Surrounded by "numerous ornaments and gifts" whose "costliness and magnificence is such that they cannot be looked upon without exciting wonder" (in the words of the historian Sozomenus),[30] the bishops gazed in awe at the tomb of Jesus Christ.

Isaac Newton would have despised the Church of the Great Martyrium. Already on the day of its consecration it was on the way to becoming an emporium for idolaters. Along with Christ's tomb (and Adam's!), it could lay claim to a piece of the stone that was rolled away from the door of Christ's tomb, a piece of the pillar Christ was tied to when he was whipped by the Romans, and the copper plate, inscribed

---

*The "skull" refers to the skull of Adam, thought to be buried there, and not the skull of Christ.

with the words "This is the King of the Jews that Pontius Pilate had had nailed to Christ's cross"—and more!*

But these relics, still few in number, were pushed well into the background when, the next morning, the Church of the Great Martyrium was consecrated with ceremonies held around its gleaming altar. The towering atrium teemed with hundreds of bishops, some from places so far away (with names like Merv and Gundeshapur and Kashgar)† that local bishops had thought these cities were mythical places in fairy tales (of which they disapproved) that they knew mothers read to their children to help them fall asleep. Athanasius had chosen not to attend this gala event—and his instincts were sound. It was at this ceremony that Constantine, represented by his grave and decorous sister Constantia (some sources claim it was his sister Irene), received Arius back into the orthodox church and allowed him to take communion—a ceremony that might have provoked in Athanasius the same paroxysms of anger that had landed him in the docket at the Council of Tyre.

The historian Sozomenus writes that Constantine brought the bishops from the Council of Tyre to Jerusalem without prior notice because he "deemed it necessary that the disputes which prevailed among the bishops who had been convened at Tyre should be first adjusted, and that they should be purged of all discord and grief before going to the consecration of the temple."[31]

But the bishops had still been brimming over with discord and grief

---

*Today this church, known as the Church of the Holy Sepulcher and around which thickly populated parts of Jerusalem have grown up, is as crammed with the relics of dead saints as Madame Tussuad's Wax Museum is crammed with wax reproductions of late, great human beings. Visiting the church in 1868, Mark Twain makes note, in his *Innocents Abroad,* of a column marking the center of the Earth, of the spot, beneath this column, where God scooped out the dust to made Adam, of the site of Jesus's cross and the crosses of the two thieves crucified beside him, of the sword of Geoffrey of Bouillon, the first Crusader king of Jerusalem—and much more. (In his book, Twain also delivers a hilarious eulogy on his late great ancestor Adam.) (See *Innocents Abroad,* 404–15.)

†The cities of Merv, in Turkmenistan; Gundeshapur, in Iran; and Kashgar, in China; had archbishops long before Canterbury in England acquired its own archbishop in 597.

when they left Tyre, so much so that even if the consecration ceremony in Jerusalem had been able to purge them of this discord and grief, it's likely that, once they got back to Tyre and found themselves in the same quagmire of sin and folly they'd left behind, they would have immediately forgotten, as if they'd never experienced them, the exquisite scents of the Church of the Great Martyrium, the brilliance of that same light of Golgotha that had shone upon their Savior, and the matchless pleasure of being cosseted, spoiled, and loved by the sister of Constantine.

So the bishops returned, and the Council of Tyre reconvened to consider the charges of murder, mayhem, and sedition that had been laid against Athanasius. And Eusebius of Caesarea—as casually as if he were simply adjusting the exquisite white omophorion around his neck and shoulders—sent six bishops, all Melitians, back to the suburbs of Alexandria to interview all the witnesses a second time about the charges that the Melitians themselves had brought against Athanasius.

Not surprisingly, the members of the task force discovered they'd been right all along. They carried out their duties very thoroughly. Newton scholar Rob Iliffe quotes the seventeenth-century historian William Cave, who, relying on the (pro-Athanasius) historian Theodoret, informs his readers that these six bishops "carried themselves like men resolved to go through their Work, endeavoring to exert Confessions by drawn Swords, Whips, Clubs, and all methods of Cruelty and Severity, not sparing even the devoted Virgins, whom they suffered the very Gentiles to strip naked."[32]

The task force returned. The council ruled against Athanasius. But at some point Athanasius had disappeared. He fled to Constantinople, surprised Constantine as he was galloping back from the hunt, and pleaded with the emperor to come to Tyre and try the case himself.

Constantine, first enraged, then amused, agreed to come. The anti-Athanasius factions at the council, hearing of this and thoroughly frightened, reversed their ruling and, out of the blue, convicted Athanasius of having prevented a fleet of corn ships from sailing from Alexandria to Constantinople.

This accusation and ruling weren't as bizarre as they might seem. Bishops were known to threaten to bottle up fleets of corn ships in the port of Alexandria. Will Durant writes: "Egypt's imperial function was to be the granary of Rome. Large tracts of land were taken from the priests and turned over to Roman and Alexandrian capitalists. . . . Every step in the agricultural process was planned and controlled by the state [for the ultimate benefit of the empire]."[33] And, adds Paul Johnson, "to overturn an imperial decision which impinged on Church affairs . . . the bishops of Alexandria, who controlled the seamen's union of the port, threatened from time to time to starve the imperial capital, Constantinople, of its Egyptian grain supplies."[34]

No evidence was produced to support this ruling against Athanasius. But none was needed, since the bishops knew Constantine was looking for an opportunity to rid himself of his troublesome archbishop Athanasius. The emperor exiled the archbishop to the provincial capital of Triers, in Gaul (close to modern-day Luxembourg), where Athanasius would cool his heels for two years in the company of some of the empire's most brilliant, if banished, intellectuals.

Let's leave Athanasius in Triers and return to three questions Isaac Newton has posed in "Paradoxical Questions" and that it is now time to answer (question 4 has two parts).

- Question 4a: "Whether it was a dead man's hand in a bag or the dead body of Arsenius which was laid before the Council of Tyre to prove that Arsenius was dead";
- Question 4b: "Whether it was Arsenius alive or only his letter which Athanasius produced in the Council of Tyre to prove that he was not dead"; and
- Question 5: "Whether the story of the dead man's hand & the living Arsenius was not feigned by Athanasius about 25 years after the time of the Council of Tyre."

Newton believed Athanasius had written all three letters himself. He had written the first two prior to the Council of Tyre, and the third before the Council of Sardica, convened in 343 in Saint Sophia Church in Serdica or Sardica (now Sofia, the capital of Bulgaria).

We'll shortly see why Newton believed this.

The great mathematician and hounder of Athanasius never ceased to believe the archbishop was guilty of all the crimes for which he was charged at the Council of Tyre (except for the crime for which he was actually convicted: holding up the corn ships at the port of Alexandria).

Why was Newton so sure Athanasius was guilty? Sir Isaac never found a smoking gun. He never found a confession by Athanasius.

What he did find was a series of omissions on the part of Athanasius.

Let us explain.

The charges leveled against Athanasius at Tyre dogged him all his life. He could never shake them off. Nobody could either prove them or disprove them. At all the ecumenical councils the archbishop of Alexandria attended for the rest of his life, there was scarcely one at which he did not have to defend himself against one or another of these charges.

In general the councils were called to debate doctrinal issues. At many, the Arians fought to assert the truth of Arius's views over those of Athanasius. In accusing Athanasius of these crimes, they were attacking the doctrine of the Trinity.

But there was a powerful personal element: Athanasius was despised by many.

Newton tells us that, however passionately Athanasius defended himself against the accusation that he had murdered Arsenius, never once, at any of the councils, did he defend himself by providing documentation that, at the Council of Tyre in 335, he conducted the living Arsenius into the Great Hall and presented him to Dionysius and Eusebius. Usually, Athanasius simply cited the letter from Pinnes he had read at the council.

This was true even for the Council of Alexandria, which Athanasius himself convened in 340. This was a fairly small council. There were fewer than a hundred bishops present, most all of them handpicked by Athanasius, and almost all of them from Egypt. In the broad, crowded streets of Alexandria, it's likely that the bright purple robes of the bishops went relatively unnoticed this time. It's possible that not a single one-eyed, one-thumbed, hobbling bishop attended.

Athanasius had convened the council largely to persuade the bishops to rubber-stamp a defense of himself he had written relative to the crimes he was still being accused of committing. Nowhere in that defense—copies of which exist today, and which at the time were received by bishops across the land—does Athanasius ever mention having escorted the live Arsenius into the Great Hall of the Council of Nicaea.

Newton writes: "Athanasius and the Bishops of Egypt when collected in a Council at Alexandria five years after the Council of Tyre knew nothing of it [presenting Arsenius at Tyre], as you may perceive by that letter which that Council wrote in defence of Athanasius against Arius & the Council of Tyre."[35]

Newton cites two occasions, in the period between the mythical death of Arius and the fourth exile of Athanasius, when it would have been greatly to the advantage of the beleaguered archbishop to make the story of the appearance of Arsenius at the Council of Tyre known. Rob Iliffe notes that "Pope Julius was ignorant of this account when he defended Athanasius in 341 nor did the Council of Sardica mention it in 343."[36] The only piece of exculpatory evidence he ever presented was the ostensible letter from Pinnes.

Newton tells us Arsenius was never seen or heard of again after the Council of Tyre. And we should remember, he says, that Athanasius was actually convicted at the end of the council of the murder of Arsenius (though the bishops subsequently rescinded this conviction out of fear of Constantine). Newton writes: "as if the accusers produced before the Council not a dead man's hand but a dead body & Athanasius produced against them not Arsenius alive but his letter only & the accusers were so far from being shamed that the Council notwithstanding the Letters proceeded to condemn Athanasius for the murder."[37]

If it never happened—if Athanasius never presented the living Arsenius at the Council of Tyre—where did the story come from? From the pen of Athanasius, says Newton.

Newton asks us to revisit the year, when Athanasius found himself banished to the Egyptian desert for the fourth time, when the desperate ex-archbishop engaged in the frenzy of letter writing that included the fraudulent account of the death of Arius in a boghouse. It was then, says Newton—in 360—in letters he wrote the monks and in the *Letters to Serapion* that he wrote at roughly the same time, that Athanasius first told the story of the arrival of Arsenius at the Council of Tyre.

The story of the severed hand and the live Arsenius was just another fiction from the pen of Athanasius—fiction Athanasius had convinced himself was fact, or fiction he had convinced himself God gave him the right to create because it would protect Athanasius as he went about his task of persuading the world of the truth of the doctrine of the Trinity.

Newton makes a final point: "He [Arsenius, had he been alive] would not have suffered the whole Roman world for many years to continue in war & confusion about his death, but have speedily shewn himself to the Emperor & to the world to the confusion of all the enemies of his dear friend Athanasius."[38]

It is impossible, in this short space, to deal with all of the paradoxical questions Newton raises about the actions of Athanasius. Let's complete the chapter by dealing with those questions that, because of their wider focus, have the best chance of holding the interest of the modern reader.

In 355, the drama of Athanasius single-handedly at war with the Roman Empire over the doctrine of the Trinity reached another climactic point. Constantine's son Constantius II was emperor (337–361). He was an avowed Arian, which made him and Athanasius mortal enemies. Constantine had usually held himself apart from theological discussions; his son Constantius, though a poor student, nonetheless developed a sincere interest in theology and loved debating the subject. In his novel *Julian* (1962), Gore Vidal gives us a sympathetic portrait of Constantius, seeing him through the eyes of his nephew, Julian the Apostate (who would succeed his uncle in 361). Constantius was, writes Julian/Vidal, "a man of overwhelming dignity [whose] . . . suspicious nature was obviously made worse by the fact that he was somewhat less intelligent than those he had to deal with. This added to his unease and made him humanly inaccessible." Constantius's most attractive feature, writes the narrator, was his "curiously mournful eyes. . . . They were the eyes of a poet who had seen all the tragedy in this world and knows what is to come in the next. Yet the good effect of those eyes was entirely undone by a peevish mouth."[39]

Thus Vidal, in his wonderful novel (which was deeply researched) presents Constantius II as a man of some complexity and subtlety. Edward Gibbon—who, we'll recall, never read Newton's "Paradoxical Questions"—wrote of the emperor that "the reign of Constantius was disgraced by the unjust and ineffectual persecution of the great Athanasius."[40] Most historians share Gibbon's view, though we should be aware that Constantius was a man not without nuance.

It goes without saying that Newton would have completely disagreed with Gibbon. The mathematician-historian presents Constantius as a man incapable of injustice. Newton writes:

> the virtues of this Emperor were so illustrious that I do not find a better character given of any Prince for clemency, temperance, chastity, contempt of popular fame, affection to Christianity, justice, prudence, princely carriage & good government, then is given to him even by his very enemies. . . . [He] reigned in the hearts of his people, & swayed the world by their love for him, so that no Prince could be farther from deserving the name of a persecutor.[41]

Constantius had already had a number of unpleasant encounters with Athanasius. In 348, the emperor ordered him to be present at an ecumenical council he was convening in Rome. Constantius's intention was to excommunicate Athanasius at this council.

This summons precipitated a confrontation between emperor and ecclesiastic that resonated just as deeply—perhaps more so—as had the confrontations between Constantine and Athanasius at the councils of Nicaea and Tyre. Even the Alexandrian mob would come swarming in at one point.

Newton writes that "when the first Messenger [Montanus] brought the Emperor's Letters, Athanasius & his friends were extremely troubled, thinking it not safe for him to go, nor without danger for him to stay. But the advice for his staying prevailed & so the Messenger returned without doing his business." This didn't surprise Newton, who adds, "I must believe also that he who refused to obey Constantine the great was as refractory to Constantius, as [the church historian] Sozomen tells us he really was."[42]

Athanasius's intransigence infuriated Constantius. He wrote an open letter to the Alexandrian people about their refractory archbishop. It's a letter Newton might have liked to have written himself, and one that would have fired Luis Buñuel's imagination. It reads, in part (Newton is quoting), "[Athanasius is] a man who was [had] emerged from the lowermost hell: who, as in the dark, seduced the desirers of truth to lies . . . [a man] convicted of most foul crimes for which he can never be sufficiently punished; no, not though he should be ten times killed . . . ; [a man] who hurt the Commonwealth & laid his most impious & wicked hands upon most holy men."[43]

The emperor sent a second message. This time troops accompanied the messenger to Alexandria. Athanasius again failed to respond. The enraged emperor ordered two Roman legions, one in Syria, the other in southern Egypt, to march on Alexandria and take Athanasius prisoner. If we can believe the historical records (which are bewildering in their diversity), at one midnight late in February 356, five thousand battle-ready legionnaires, led by Syrianus, Duke of Egypt, battered down the door of the Church of Saint Theognis and, bursting in, loosed a hail of arrows at the clerics and church members gathered there for prayer. Newton writes (and here he gets close to pulp fiction): "there was made a great clashing of arms the drawn swords shining by candle light & Virgins were slain & trodden under foot."[44]

Some sources suggest that an outraged (and well-bribed) pro-Athanasius Alexandrian mob stormed into the Church of Saint Theognis that night and, armed with pikes, swords, and cudgels, turned the tide of this unlikely battle in Athanasius's favor. Either that, or the virgins had swords. And the clerics and church members too. Because Athanasius and his supporters, or so at least one account goes, fought off all five thousand legionnaires! Athanasius, jostled hither and yon by the soldiers (who had orders to take him alive), urged his flock to use prayer as a weapon, then managed somehow to get out of the church before the battle was over.

The next morning, anyone entering the church would have been greeted by a most un-Christian-like spectacle: Roman helmets, breastplates, arrows, broken spears, and bloodstained dented broadswords hanging off all the church walls. A cleric would have pointed proudly to these weapons and told the visitor they were war trophies; that the champion of the doctrine of the Trinity had been the champion in a war against the Roman Empire the night before.

This bizarre and improbable epic drama might have been beyond even the ability of Luis Buñuel to film.

For four months after the archbishop of Alexandria's disappearance, a smoldering low-grade guerrilla war raged in Alexandria. A furious Constantius wanted Athanasius dead or alive and mobilized all the resources of the Roman Empire to run down this Black Dwarf blacker than hell.

Edward Gibbon writes that the Roman legionnaires from Syria and southern Egypt dealt with the presbyters and bishops of Alexandria "with cruel ignominy; consecrated virgins were stripped naked, scourged, and violated; the houses of wealthy citizens were plundered;

and, under the mask of religious zeal, lust, avarice, and private resentment were gratified with impunity, and even with applause."[45]

Finally, after four months, the battles ended and the Roman legions withdrew from Alexandria. Athanasius could not be found, and there was every reason to believe that he had taken refuge, again, among the monks of the Egyptian deserts. And Athanasius had indeed been on the desert for some time, hidden (as we saw in the last chapter) behind an impenetrable wall of protecting, adoring monks. He would remain there for six years, until the death of Constantius, rewriting the history of Christianity in the fourth century in a manner that best suited his own interests—and also, as we are about to see, writing (also to suit his own interests) the biography of the most celebrated desert monk who ever lived.

We've seen that Athanasius, at a succession of Church councils from the years 337 to 356, was never quite able to lay to rest the charges of murder and mayhem that had been leveled against him.

The new emperor, Julian the Apostate, a brilliant thinker and military strategist who tried to banish Christianity and bring back pagan worship, hated Athanasius even more than did Constantius. Writing the prefect of Egypt over that official's delay in carrying out the imperial order to expel Athanasius from Egypt (Athanasius having returned after the death of Constantius in 361), Julian (who ruled from 361 to 363) scolded, "it is your duty to inform me of your conduct towards Athanasius, the enemy of the gods. . . . There is nothing that I should see, nothing that I should hear with more pleasure than the expulsion of Athanasius from all Egypt. The abominable wretch!"[46]

But Athanasius's fortunes changed. A year later, Julian was killed in battle against the Persians, probably assassinated by one of his own men who was a Christian. The new emperor, Jovian, reinstated Athanasius. Soon Jovian was gone, but over the next decade Athanasius's fortunes steadily improved. He never ceased vigorously promoting the doctrine of the Trinity. His efforts were posthumously rewarded when, at the

Council of Constantinople, convened by the emperor Theodosius in 381, the doctrine of the Trinity officially became the cornerstone doctrine of the Orthodox Catholic Church.

Athanasius's fourth exile in the Egyptian desert had yielded up a fraudulent account of the death of Arius and a fraudulent account of Athanasius and Arsenius together at the Council of Tyre. Newton believed the archbishop of Alexandria perpetrated yet another fraud during that final sojourn among the adoring desert monks. Newton scrutinized this third deception with especial care, for it had a direct bearing upon an aspect of Trinitarianism that Newton abhorred: the twin idolatries of the worship of saints and the worship of Christ as God.

History believes that Saint Anthony invented these two forms of idolatry. Isaac Newton was very anxious to set history right. He believed, as we will see, that the real culprit was his lifetime archenemy Athanasius. He was ready to invest a prodigious amount of energy into proving it.

CHAPTER SIX

THE TEMPTATION OF SAINT ANTHONY

The celebrated desert monk Saint Anthony (251–356) was well over six feet tall and pencil thin. He was all hope, charity, and the doctrine of the Trinity, but, like so many other great saints, he had very little body in which to keep all of these merits. Still, he survived to 105.

We've already read Newton's description of Anthony in chapter 4. Sir Isaac dwells on the desert saint's austerity, telling us he gave his goods to the poor, "conversed with all wise men and imitated what was best in each, ate only bread & salt & drank only water, dined at sunset, often fasted two days or more, often watched all night, slept on a mat & frequently on the bare ground, never anointed nor bathed himself nor saw himself naked, was meek, prudent, pleasant, foreknew things, but dissuaded the monks from affecting it."[1]

Saint Anthony has been a popular figure among artists for two thousand years. The French novelist Gustave Flaubert (1821–1880), author of *Madame Bovary*, made him the subject of an experimental drama, *The Temptation of Saint Anthony* (*La Tentation de Saint Antoine*), which was completed in 1874. In this work the author sends the saint all over the Middle East, often bringing him face-to-face with the great religious and philosophical sages of the ancient world. Flaubert turns Saint Anthony into a kind of philosophical Everyman: he uses the story of his life to convey, in the words of one critic, the "tragedy and pathos of man's long search for some body of belief or philosophy by which he could explain to himself the strange great phenomena of life and death, and the inscrutable cruelties of Nature."[2]

Flaubert devotes much space to describing Anthony's imaginary jousts with the world of demons. In the words of another one of Flaubert's critics, the French novelist depicted Anthony as, along with much else:

> tempted with piles of gold, with food, with women, with "a black boy" who was "the spirit of lust." Bands of wild animals made as if to attack him and tried to terrify him with their ferocious cries. Monstrous creatures appeared to him, having shapes half animal, half human; learned men came with beguiling heresies. Anthony was at times transported outside of himself and was able to watch his own actions.[3]

Newton also lavished a lengthy paragraph on Athanasius's description of Anthony's seemingly supernatural side, explaining that the life of this desert saint was

> full of prodigious stories such as are: the devil's appearing frequently to Antony in several shapes & bignesses & talking with him & afflicting him & struggling with him & sometimes multitudes of devils appearing in various shapes. Christ's appearing in the form of light to Antony & speaking to him. Antony's remaining as fresh & plump after long fasting as if he had not fasted, his curing diseases, casting out devils, escaping Crocodiles by prayer, frighting away devils & curing demoniacs by the sign of the cross, seeing the soul of Ammon ascend up to heaven, being himself lifted up into the air, & having revelations & by a spirit of prophesy foretelling things.[4]

It is not to honor Anthony but to expose him that Newton lists these experiences. For the great scientist, these visions are all illusion, delusion, and hallucination. Newton did not believe in the supernatural. He didn't believe in the paranormal. He didn't believe in reincarnation, or in the existence of an afterlife, subscribing to a belief called "psychopannychism," which holds that, when we die, our souls sleep until Judgment Day.[5] (Newton cited Ecclesiastes 9:5–10 to support his belief: "The dead know not anything. . . . There is no work nor knowledge nor wisdom in the grave.")

It goes without saying that Newton didn't believe in ghosts. Westfall tells this story:

> Strange noises in a house opposite St. John's College [at Cambridge University] had many convinced it was haunted. The excitement extended over several days, while crowds gathered outside and diverse intrepid scholars and fellows ventured into the house. [Westfall quotes:]
> "On Monday night likewise there being a great number of people at the door, there chanced to come by Mr. Newton, fellow of Trinity College: a very learned man, and perceiving our fellows to have gone in, and seeing several scholars about the door, 'Oh! Ye fools,' says he, 'will you never have any wit, know ye not that all such things are mere cheats and impostures? Fye, fye! Go home, for shame,' and so he left them, scorning to go in."[6]

Probably Newton believed, with the ancient Greek philosopher Epicurus, that evil gods, or demons, are simply the unhappy fantasies of our dreams.

Rather surprisingly, Newton takes a moment to examine monasticism and Anthony from a psychological point of view. Westfall writes that, for him, "monasticism was yet another atrocity spawned by the evil genius Athanasius, who had fostered it to promote his interests. Newton devoted considerable study to it, specially to the feigning of miracles by monks and to their stories of sexual temptations. The latter apparently fascinated him. Rather solemnly, he lectured the monks on how to avoid unchaste thoughts."[7] No modern psychologist would quibble with the following description by Newton, written with Anthony in mind (Newton was celibate himself).

> To pamper the body inflames lust & makes it less active & fit for use. And on the other hand to macerate it by fasting & watching beyond measure does the same thing. It does not only render the body feeble & unfit for use but also inflames it & invigorates lustful thoughts. The want of sleep & due refreshment disorders the imagination & at length brings men to a sort of distraction & madness so as to make them have visions of women conversing with 'em & think they really see & touch them & hear them talk. . . . For lust, by being forcibly restrained, & by [our] struggling with it, is always inflamed. The way to be chaste is not to contend and struggle with unchaste thoughts but to decline them and keep the mind employed about other things.[8]

Perhaps Newton wouldn't have been so offended by, and contemptuous of, Anthony's "paranormal" experiences if the desert saint had said nothing about them. And, for all we know, Anthony kept them to himself. The problem lay, Newton thought, with Athanasius. Sir Isaac explains that the exiled archbishop didn't only compose and circulate letters in which he rewrote history during his final, seven-year-long, sojourn in the Egyptian desert. He also wrote a biography of Anthony. By the time Athanasius wrote his *Life of Anthony,* Anthony had died and couldn't be consulted, but Athanasius had no trouble getting all the information he needed because of the hundreds of monks who were at his beck and call and had known Anthony and loved him.

Why did Athanasius write this book that is still, today, almost our only source of information about Anthony? Not really to honor the ascetic desert monk, insists Newton, nor to praise him, but entirely out of self-interest. Here's what happened: "[Athanasius,] baffled &

deserted by all but the Monks, . . . finding himself reduced to the utmost desperation, & seeing no hopes of recovery unless by extraordinary practices, set himself upon all kind of sophistry & began with writing this life [of Anthony]. . . .[9]

Even in exile, Athanasius was desperately searching for ways to combat Arianism. The *Life of Anthony* was one of those ways. He uses Anthony's extraordinary visions and experiences, which he implies could only have happened to a man who believed in the Trinity, as a kind of religious bait: If you become orthodox Christian after the manner of Anthony, he seems to be saying, then these wonderful experiences will come to you.

Athanasius's description of Anthony's worship of relics and saints and the exciting paranormal experiences that he claimed stemmed from this practice was a particularly effective way of enticing pagans into converting to Trinitarian Christianity; the worship of analogous artifacts, and the experience of paranormal phenomena, were already part of the pagan rites they practiced.

Newton believed that Athanasius's *Life of Anthony*, which became something of a bestseller throughout the Roman Empire, was the first time most people beyond the Egyptian deserts had ever heard of such rites and experiences. He seems to have believed the book had an instant and universal effect. He explains that Athanasius "contrived his religion for the easy conversion of the heathens by bringing into it as much of the heathen superstitions as the name of Christianity would then bear. . . . By this life [of Anthony] Athanasius propagated Monkery & made it overflow the Roman world like a torrent."[10]

It's almost as if Newton believed the supernatural experiences of Anthony as described by Athanasius had what we would today call a "copycat" effect: thousands of people wanted to emulate them and thought that they could since Athanasius described them as actually happening. Everybody started having supernatural experiences. It was like a contagious disease. It spread like an out-of-control virus, so much so that the *Life of Anthony* was the true original [origin] of heathen ceremonies & superstitions which continue to this day in the Greek & Latin Churches.... First it did set all the Monks upon an humor of of pretending to miracles: so that the whole world presently rang with stories of this kind. And hence it came to pass that the lives of all the first & most eminent Monks were filled with apparitions: of Devils, miraculous cures of diseases, prophesies & other prodigious relations . . . And this was the original of those ecclesiastical Legends which are still used in the Church of Rome.[11]

Newton takes pains in "Paradoxical Questions" to debunk the rituals and experiences Athanasius describes in the *Life of Anthony*. He seeks to demonstrate that there is nothing divine or otherworldly about their origin. Perhaps at the outset only the result of an accident, the rituals evolved out of practical necessity.

Newton tells us about a monk named Paul. This desert celibate decided to say three hundred prayers a day. To make sure he reached this number, he carried three hundred stones around with him and discarded one every time he said a prayer. Other monks began to imitate him. Over the years, the stones tended to get fewer in number and smaller in size. Eventually, a string could be threaded through them. This, declares Newton, was "the original [origin] of the Popish custom of prayer by beads." That is, it was the origin of the rosary and the telling of beads.

Newton cites a second example. Anthony's disciple Macarius, "in going over the sands of Egypt . . . found a dead man's skull & upon asking it questions the skull answered that he was in hell & that the damned found some ease by the prayers of the Church."[12]

Newton believed that Macarius, probably starving and exhausted, was merely hallucinating. Sir Isaac believed this strange encounter, and no doubt many others like it, had given rise to the doctrine of purgatory.

Newton places the origin of the practice of worshipping saints squarely on Anthony's doorstep. He says that according to Athanasius when the desert saint's disciple and friend Ammon died, Anthony saw his soul rise up to heaven in the company of angels. Anthony considered Ammon a saint; if the angels saluted such people when they died, then it was right to do such things ourselves.

Newton claimed nobody had ever known about such experiences until Athanasius popularized them in the *Life of Anthony*.

Newton insists that when Athanasius was writing the *Life of Anthony*, "the superstitious or to speak more truly the magical use of the sign of the Cross" was in use in the Western churches and that "Athanasius in his exile learnt it there & now turning it into an enchantment makes Antony in a large discourse to the Monks teach them how by this sign they may drive away the Devil & dissolve all kinds of enchantments & witchcraft. And this I like to be the original of the Greek & Latin Church using it for this end."[13]

Nobody ever thought Newton was an anthropologist or an ethnologist, but in "Paradoxical Questions" he gives us an astonishing description of the birth, growth, and spread of the practice of relic worshipping. He's scornful of the role Anthony played in promoting the worship of saints and relics. But he believes we would probably never have heard of these practices if Athanasius had not made so much of them in his biography of Anthony.

Newton didn't confine the discussion of these matters to his "Paradoxical Questions." He writes in *Observations upon the Prophecies of Daniel, and the Apocalypse of St. John* that

> in propagating these superstitions, the ringleaders were the Monks, and Antony was at the head of them: for in the end of The Life of Antony, Athanasius relates that these were his dying words to his disciples who then attended him. "Do you take care," said Antony, "to adhere to Christ in the first place, and then to the Saints, that after death they may receive you as friends and acquaintance into the everlasting Tabernacles."[14]

Newton states elsewhere that Anthony had never said these things on his deathbed at all but that Athanasius had put the words in his mouth to show that these beliefs, which were actually those of Athanasius, had had the support of the highly esteemed and wholly trusted Saint Anthony.

In a whole other matter, Newton missed a tremendous opportunity of catching Athanasius out in a whopping lie and thereby further discrediting him. The exiled bishop of Alexandria wrote in the *Life of Anthony* that the desert saint could neither read nor write. For centuries everybody believed this, including Newton. But we know today that that wasn't true and that Athanasius must have known it wasn't true.

The idea that an illiterate man could know and preach the scriptures as well as Anthony did had a certain allure during the early history of Christianity. The ancient world accepted it as fact. Saint Augustine (354–430) found the notion thrilling; the idea was in his mind at almost the same moment that he himself was suddenly converted to Christianity.

Isaac Newton had no trouble believing Anthony was illiterate. He quotes Augustine: "Antony without any knowledge of letters had the scriptures by heart."[15] Newton also cites at least half a dozen of Anthony's contemporaries and near contemporaries who believed the desert saint was illiterate; they include Baronius, Bellarmin, Socrates, Cassian, Sozomen, and Gregory of Nazianzen.

When Anthony was asked how he could manage without the solace of books, he replied that "the nature of things" [the natural order] was his book. The church fathers believed that the desert saint was so acutely attentive to what was read or said to him that it came naturally to him to remember it all without the benefit of "letters." When his questioners asked him why he had never learned to read or write, he in turn asked them, according to Newton, "which was oldest, the mind or letters & which was the cause of the other? To which, when they replied that the mind was older than letters & the inventor of them, he answered that he therefore who has a sound mind has no need of letters."[16]

But there is a growing number of scholars today who believe Anthony really could read and write. The Princeton religion scholar Elaine Pagels tells us in *Revelations* that

> letters long attributed to Anthony often had been disregarded as pseudonymous [by an unknown author], since their content conflicts with much that is found in Athanasius's classic *Life of Anthony*. Like many others, however, I am persuaded by the analysis offered in Samuel Robertson's book *The Letters of St. Anthony: Monasticism and the Making of a Saint* (Minneapolis: Fortress Press, 1995) that the letters he identifies there are most likely to be genuine.[17]

Although Athanasius states categorically that the desert saint was an "illiterate and simple man," Pagels is of the opinion that Anthony emerges in his letters as a "sophisticated and fiercely independent teacher." The Princeton scholar believes Athanasius put many of his own likes and dislikes into Anthony's mouth to give them added credibility. He makes Anthony into someone who "hates Christian dissidents as much as he [Athanasius] did—and who, like the bishop himself, calls them not only *heretics* but 'forerunners of Antichrist.'"[18] But this is not Anthony talking. It is Athanasius.

In making Anthony illiterate, Athanasius is implying that book learning and intellectual debate have little place in the true, orthodox practice of Christianity. Knowledge other than that of the scriptural text should be left to God and the bishops. It's up to ordinary Christians to accept the authority of the bishop, believe what he tells them to believe, and do what he tells them to do. There's no need for reading, and Anthony, the exemplary Christian, knew this.

Newton would have been delighted to have this further proof of Athanasius's duplicity.

If we lament the fact that certain apocryphal texts, like the Book of Enoch, are missing from the Bible, then we only have Athanasius to blame. It was the archbishop of Alexandria and author of the *Life of Anthony* who, in his *39th Festal Letter* (written in 367), laid out the canon of the Bible as we have it today. Athanasius didn't have the authority to impose his canon, but his influence was such that it was widely accepted at the time and quickly became the canon of choice. Athanasius insisted that the books he excluded from the Bible should never be read. He said they were "empty and polluted," written by people "who do not seek what benefits the church."[19]

It is to Athanasius that we owe the inclusion of the Book of Revelation in the New Testament; all other would-be canon creators excluded it. And, says Pagels, Athanasius had a unique take on the identity of the Great Beast in Revelation, identifying it as the "demonically deceived Christian."[20] He insisted that "Babylon," or the "Great Whore," was heresy personified—the archetype of heresy, we might say today. (And represented every single person in the world who did not believe exactly what Athanasius believed.)

Newton's discussion of how Athanasius used Anthony, if it is accurate (and modern scholars seem to feel it is, though they are primarily concerned with Athanasius's having lied about Anthony's illiteracy) is disturbing in that it greatly broadens the sphere of Athanasius's fraudulent activities. The orthodox Christianity of the fourth century AD, Newton is saying, was built on many lies, and Athanasius was the biggest liar.

CHAPTER SEVEN

THE GREAT APOSTASY

Abram Sachar writes that the Temple of Solomon, built about 1000 BC,

> grew until it overshadowed Jerusalem. It became more than an object of worldly greatness. It was a symbol of peace, of social justice. Ethical meanings were read into its blocks and stones; allegories were found hidden in its measurements! No iron, it was said, went into the construction of the Temple, for iron is a weapon of war. The Temple site, the rabbis taught, was chosen because on it two brothers had shown for each other a divine, self-sacrificing love. These and other legends clustered about Solomon's work of pride until the Temple rivaled Sinai in its religious significance.[1]

The Temple of Solomon, which was destroyed by the Babylonians in 587 BC, in Newton's time had become a cult, a science, a dream, and the seventeenth-century equivalent of a mega-computer. It was not of earthly design. Just as Moses had received the divine plan of the tabernacle from Jehovah on Mount Sinai, so had David received the divine plan of the temple from God and passed it on to his son Solomon, assuring him that "all this . . . the Lord made me understand in writing by his hand upon me, even in all the works of this pattern" (1 Chron. 28:19). The Temple of Solomon was the blueprint of God's mind; it was a mirror held up to the divinely created cosmos. It was, in the words of Frank Manuel, "the most important embodiment of a future extra-mundane reality, a blueprint of heaven; to ascertain every last fact about it was one of the highest forms of knowledge, for here was the ultimate truth of God's kingdom expressed in physical terms."[2]

Isaac Newton was fascinated by the temple all his life. He made an extended effort to work out the exact measurements of the sacred cubit by which it had been built.[3] He did not include a history of the Jews in his *Chronology of Ancient Kingdoms Amended* but instead filled five pages in the middle of the book with the floor plan, as if that said everything there was to say about God's chosen people. He and his colleagues thought that if you could know precisely the measurement of every pillar and every tile and every nook and cranny in God's sacred temple then you might be able to tease from it the answer to every question in the universe, for it itself was the full measure of the universe.

In the early Renaissance there began to appear in Europe scale models not only of the Temple of Solomon but of the Temple of Jerusalem and the Tabernacle, for they too were mirrors of God's mind (the Temple of Jerusalem having been built in accordance with the design of the Temple of Solomon but also modified a little later, with details from Ezekiel's dream of the temple). Entrepreneurial scholars trundled these models across England, giving lectures for the edification of the masses and for very small sums; in 1726, the mathematician and divine William Whiston had a craftsman build scale models of the Tabernacle of Moses and of "the Temple of *Jerusalem,* serving to explain *Solomon's, Zorobabe's, Herod's,* and *Ezekiel's* Temples; and had Lectures upon that at *London, Bristol, Bath* and *Tunbridge Wells.*"[4] Much earlier, in February 1675, the celebrated 1:100 models of the Temple of Solomon and the Tabernacle of Moses, complete with every last moving and unmoving part and constructed by the famed Rabbi Jacob Judah Leon, were brought from Amsterdam and put on display, every day except Saturdays and Sundays, beside the tree near the Great Synagogue in Duke's Place in London.

Every morning, the local rabbi crossed the lawn from the nearby synagogue and answered questions about this magnificent artifact.

And on the first day he answered questions from the king and queen of England, who had been among the very first to come and visit. Every day thereafter, amid the barking of dogs and the shrieking of children and the muffled clattering of carriages passing by on the muddy streets, he fielded questions that were usually put by well-dressed dandies and were meant, not to edify the asker, but to entertain the asker's lady companion, and the rabbi bore up without condescension in the face of these questions that were bitter proof of the ignorance of the gentiles in the face of the holiness of Rabbi Jacob Judah and that of Leon's model of the sacred temple.

It is very likely that, one chilly February day, Isaac Newton came to examine the scale model of the Temple of Solomon.* He was thirty-three; nobody knew him, or that he was anything else than just another lean and hungry young man; and so nobody would have paid any attention to him as he gazed with a sort of preternatural intensity into the bowels of the scale models of the Temple of Solomon.

In 1675, Isaac Newton had already begun work on his celebrated exegesis of biblical prophecy, *Observations upon the Prophecies of Daniel, and the Apocalypse of St. John*. And it's likely that, as early as 1675, Newton had reached the startling conclusion, shared by very few of his fellow countrymen, that the entire prophecy of the Book of Revelation—thundering horsemen, cascading stars, woman clothed in the sun, all of them—unfolded within the Temple of Jerusalem. That is why, scrutinizing Rabbi Leon's brilliantly wrought scale model of the Temple of Solomon on that chilly February day, ignoring the poorly informed sallies of the foppish young, Newton might well have superimposed, in his mind's eye, the Temple of Jerusalem on the Temple of Solomon. And if he did so, he very likely saw, scaled-down and Lilliputian-sized like

---

*Newton visited London just five times before he moved down permanently in 1696. One of these early visits was in February–March 1675. Of this visit, we know only that Newton attended two meetings of the Royal Society. It's unimaginable that he would not have been informed by a colleague of the presence of Rabbi Leon's celebrated scale model and that he would not have gone to see it—though there is no record of this.

the temple, making his way with dignity and forthrightness across the miniaturized floor of the model of the Temple of Solomon-Jerusalem, the tiny but eager figure of John of Patmos.

Isaac Newton's method of interpreting the Book of Revelation was so intricate, so complex, so revolutionary, that its readers might be forgiven for taking it in slowly, bit by bit—a little here, a little there.

It was based on the assumption that the Temple of Jerusalem was indeed the blueprint of God's mind; that it was a replica of the universe; and that therefore John could find within it, however cosmic or outré, every single prophetic hieroglyph or figure that he used in the Book of Revelation.

But these images could not be seen directly. John's task was to derive from them the physical layout of the temple, its vessels and its paraphernalia and the ceremonies that unfolded within the temple. The original cosmic delineation of the tabernacle had been made by Moses in the desert; seers had made additions ever since; and John himself was a seer who knew more than had ever been written down.

In *Antiquities of the Jews,* Josephus describes what Moses did, telling us of the traditional symbolism of the Temple of Jerusalem: the three parts of the sanctuary were

> every one made in way of imitation and representation of the universe. When Moses distinguished the tabernacle into three parts, and allowed two of them to the priests, as a place accessible and common, he denoted the land and the sea, these being of general access to all; but he set apart the third division for God, because heaven is inaccessible to men. And when he ordered twelve loaves to be set on the table, he denoted the year, as distinguished into so many months. By branching out the candlestick into seventy parts, he secretly intimated the Decani, or seventy divisions of the planets; and as to the seven lamps upon the candlesticks, they referred to the course of the planets, of which that is the number. The veils,

too, which were composed of four things, they declared the four elements.[5]

Here is an example of what (or so Newton believed) John does all through the Book of Revelation: passage 10:1 reads, "I saw another mighty angel . . . his face blazed like the sun, his legs like pillars of fire, and he had a little book open in his hand. He planted his right foot on the sea and his left foot on the land, and then shouted with a loud voice like the roar of a lion." How did John come by this image? He has seen, in the temple, a priest place one foot in the brazen laver or "sea of glass"—a large bowl filled with water beside the altar—and the other foot on the ground. In the lexicon of prophetic figures that John knew and that Newton had rediscovered, the *earth* symbolizing the Greek Empire, and the *sea* symbolizing the Latin Empire. But these basic images are drawn from the temple, so that the priest with one foot in the laver and the other on the ground becomes, in the visions, an angel with one foot in the sea and one foot on the earth, which, taken altogether, signifies an event that will concern both the Latin and Greek Empires. And since the high priest of the ceremonies is reading from the Torah (or the Book of Law) this little book becomes the "little book" that the angel shows John.

As Matt Goldish writes, Newton in his *Observations*

> explains with the aid of his prophetic lexicon how the images presented are in fact representations of the prophet's experience in the Temple. [Newton explains] move by move, Saint John's entrance into the Temple precinct, his viewing of the various ceremonial objects, and his observations of the ceremonies of the daily worship, the Day of Atonement and the Feast of Tabernacles. Every aspect of the Temple—its physical layout, vessels and ceremonies—thus becomes critical to the unraveling of the secrets held in the Apocalypse. . . .
>
> The scheme of Revelation, in Newton's conception, is that of the

prophet viewing events in the physical Temple, but reporting them in a cryptic or allegorical form which was designed to hint at the major events of apocalyptic history. Thus, the vision functions for him at three levels: the concrete physical scene of the prophet walking in the Temple; the cryptic manner in which this activity is described in Revelation; and the prophecy that this cryptic description is attempting to convey to those who understand.[6]

But there was such a kinship between the tabernacle and the two temples—all three are the warp and woof of the mind of God—all three have operated on some realm of being too mystical to be explained or understood or even imagined but which are the shape of God's mind— that all three eternally vibrate with the other, and if need be John can summon into the Tempe of Jerusalem resonances of the Temple of Solomon and the tabernacle; the boundaries of all three are blurred. The Temple of Jerusalem bursts its bonds, and events and objects skirting it also become the earthly correlatives of major events in the future history of mankind. For example, sometimes the Tabernacle of Moses with the twelve tribes of Israel encamped around it seem to be superimposed on the Temple of Jerusalem, and the twelve tribes become the root, the anchor point, the essential substance of events in the future history of mankind to which John will give hieroglyphic form 1,500 years later.

What became clear to Newton as he pored over the Book of Revelation was that the brilliant and despairing prophet John, incarcerated at Patmos, has re-created in his imagination, bit by bit, stone by stone, the entire structure of the Temple of Jerusalem (with the tabernacle and the Temple of Solomon flickering in and out). And within this structure, and from this structure, he has created the entire Book of Revelation.

It makes sense that he should have done this. It's the measure of how traumatized he was by the destruction of the Temple of Jerusalem, which was far and away the worst of the many atrocities he knew had been inflicted on the Jews or the Christians, or that he had experienced. He was inconsolable in his grief over the Temple of Solomon. But he was a creative genius, and there is a way in which grieving creative geniuses can restore their psychic balance. They can re-create that which they have lost, and in this case John re-created in his mind's eye the Temple of Jerusalem.

John built a temple in his mind—and, lo and behold, God walked in. Let's recall that not in any synagogue, but only in the Holy of Holies in the sanctuary of the temple, could God reside on Earth; the destruction of God's home on Earth was the reason many pious Jews grieved so deeply for the loss of the temple; and, on some lofty level of creative genius, John welcomed God back into that temple in his mind that would become the narrative vehicle of the Book of Revelation.

The Book of Revelation was for Newton an immensely intricate and complex artifact. We're used to thinking of John's Apocalypse as having been composed in a single intense burst of inspiration. Newton did not believe this. John of Patmos was a divinely inspired prophet, to be sure. But he was no eye-rolling ecstatic. Rather, he was an artist in the classical mode, wholly in command of his creation—divinely inspired by but not intimidated by God—calm in the presence of the sublime and prophesying, as one seventeenth-century divine wrote, "from the *still voice* of a great humility, a sound mind, and a heart reconciled to himself and all the world."[7]

The true prophet wasn't without passion, but he was coolly scientific—almost like Isaac Newton! John constructed the Book of Revelation carefully, piece by piece, rationally, deliberately, like Praxiteles sculpted parts of the Parthenon in ancient Athens or Michelangelo painted the ceiling of the Sistine Chapel at the Vatican in Rome or Johann Sebastian Bach composed the Mass in B minor in the tranquillity of the Collegium Musicum in Leipsig—maybe, perhaps, a little like Isaac Newton composed the *Principia Mathematica*!

Still, we get the impression when we read the Book of Revelation that it is a "channeled" text; that John didn't exactly write it but, rather, that God transmitted it to John through Jesus Christ. As such, it poses the same problem as all channeled texts: how is it that something that is ineffable, numinous, not graspable by the senses—something coming from a higher level of reality—can possibly be communicated to the hard-as-rocks physical time-space continuum in which we live? *Les Misérables* author Victor Hugo's spirit guides told him this wasn't something that could really be done: "There is no alphabet of the uncreated, there is no grammar of heaven. You don't learn Divine like you learn Hebrew. . . . Angels are not Professors of Divine Language. . . . All that which is uncreated is unnamed, the speech of celestial language is bedazzlement, to express oneself is to be resplendent, clarity of speech is luminosity."[8]

James Merrill (1926–95), the Pulitzer Prize–winning American poet whose masterpiece *The Changing Light at Sandover* (1981) is built out of channeled texts, explained that such texts are by nature highly subjective: "The powers they [the spirits] represent are real—as, say, gravity, is 'real'—but they'd be invisible, inconceivable, if they'd never passed through our heads and clothed themselves out of the costume box they'd found there. *How* they appear depends on us, on the imaginer."[9]

Isaac Newton reveals yet another unknown facet—that of an expert on channeling—when he describes what he calls the "Preamble to the Prophetic Visions," which is Revelation 5:6–7. This preamble explains the mechanics of John's prophesying: God gives Christ a scroll with seven seals. Christ, loosening the seals one by one, reads the scroll. The essence of the scroll is transmitted to John—but what Christ reads is not what John writes. Newton writes movingly that the contents of the scroll are "of so transcendent excellency that they were fit to be communicated to none but the Lamb. . . . You are to conceive that the Lamb opened the book for his own perusal only & that the concomitant visions which appeared to Saint John were but general & dark emblems of what was particularly & perspicuously revealed to the Lamb in this book."[10]

The contents of Revelation are merely "certain visions which Saint John saw concomitant to the opening of the seals." Christ reads, and transmits raw creative energy to John, who, seizing *"the motions of some and voices of others"*[11] from the costume box of the Temple of Jerusalem through which the Book of Revelation is unfolding, clothes that energy in Apocalyptic visions.*

But Isaac Newton isn't interested in explaining channeling. His concern is to demonstrate that God wrote the "Preamble to the Prophetic Visions" to demonstrate the true nature of the relationship between God and Christ. That true nature is that Christ is not equal, but subordinate, to God. The proof of this? Christ doesn't know the future history of the world until he reads it in the scroll God has given him. Christ doesn't already possess the knowledge of the future history of mankind that God possesses; he is *not* equal to God.

The God we meet in Revelation is the God of Arius and Newton, not the God of Athanasius. Otherwise, asks Newton, "Why were we told of this book [the scroll] if it contained a revelation for the Lamb only, & not for us?" Why did God bring up the subject at all? Newton answers that

> it was done in prosecution of the main design of the Apocalypse [John's Book of Revelation], which was to describe & obviate the Great Apostasy. That Apostasy was to begin by corrupting the truth about the relation of the Son to the Father in putting them equal [as in the Doctrine of the Trinity], & therefore God began this prophecy with a demonstration of the true relation: showing the Son's subordination, & that by an essential character: his having the knowledge of futurities only so far as the father communicates it to him. And lest you should think he had this knowledge given him from all eternity, the book was represented in the hand of God alone sealed at first.†[12]

This equating of Christ with God is at the heart of what Newton calls the Great Apostasy. According to *Webster's* an *apostasy* is the "renun-

---

*Did John resurrect the Temple of Solomon in his head to serve as a memory palace—a vast *ars memoria* mnemonic tool—in which to store his developing Book of Revelation in anticipation of one day being able to set it down? Those interested in following this thought through should consult *The Art of Memory* by Frances A. Yates.

†This passage is from Newton's "Untitled Treatise on Revelation" (section 1.4). The reader wishing for clarification will find a 3,200-word excerpt from the "Untitled Treatise," including this passage, in appendix D.

ciation or abandonment of a former loyalty (as to a religion)." As has been mentioned in an earlier chapter, the Great Apostasy really got under way in 325, when the Council of Nicaea adopted an early form of the doctrine of the Trinity. It took on more power when it was ratified by the Council of Constantinople as the Nicene Creed in 380. So Newton is telling us that (according to the Word of God as expressed in the preamble in Revelation 5:6–7) the Book of Revelation is not only the future history of the Roman Empire and Christianity, but it is also, over and above that, the future history of the developing evil of the Great Apostasy, which will become a very dangerous idolatry because, by making Christ the equal to God, it will encourage the worship of Christ at the expense of the worship of God. And nothing is more dangerous for the soul of man, Newton seems to believe, than the failure to worship God.

To the modern reader, this seems a little exaggerated. Why not worship Jesus and God at the same time? Or why worship either? Surely Newton is taking the doctrine of the Trinity too seriously—or not. Let us recall that, according to Newton's most intimate colleagues, the great mathematician possessed an intuitive faculty, almost a "seeing eye . . . that deep original instinct which peers through the surface of words and things—the vision which sees dimly but surely the other side of the brick wall or which follows the hunt two fields before the throng."[13] Let us bear with him through this chapter and the next, and the next, and see if a doctrine revered today by the Roman Catholic Church is actually one of the most heinous spiritual crimes in the history of mankind.*

---

*Newton's discussion of the "Preamble to the Prophetic Visions" and the Great Apostasy raises some provocative issues: First, most modern scholars believe Athanasius was "the first, as far as we know, to place the Book of Revelation in his version of the New Testament canon" (Pagels, *Revelations,* 135). But if Newton's interpretation of the preamble is correct, wouldn't the brilliant and vicious Arius-hater Athanasius have sniffed this out and absolutely not suggested that Revelation be a part of his Ur-canon of the New Testament? Second, how could John of Patmos, when he wrote Revelation at the end of the first century AD, possibly have known that a Trinitarian/anti-Trinitarian schism would develop in the church by the fourth century? In John's time, the early apostolic church gave no indication that any such schism would ever develop.

But the Book of Revelation is also the future history of the decline and fall of the Roman Empire, in that Christianity developed within its confines, and the fortunes and failures of Christianity were intertwined with those of the empire, the one continually affecting the other, as Newton will vigorously strive to demonstrate in his deciphering of the Book of Revelation.

So Newton's interpretation of Revelation is part of his grand scheme of mapping the progressive corruption of the soul of man; in so being, it intersects with Newton's letters to John Locke on twenty-seven corruptions of scripture and Newton's formidable exposé of Athanasius, the man who single-handedly, and in Newton's view fraudulently, kept the doctrine of the Trinity alive and enabled it to triumph, to, as Newton saw it, the everlasting sorrow of mankind.

As the Book of Revelation opens John finds himself in the Temple of Jerusalem. A high priest enters, lights the lamps, mounts the altar, and reads from the Torah. A religious ceremony has begun. But John's mind is already operating on two levels. Newton explains:

> On the first day of that month, in the morning, the High-Priest dressed the lamps: and in allusion hereunto, this Prophecy begins with a vision of one like the Son of man in the High-Priest's habit, appearing as it were in the midst of the seven golden candlesticks, or over against the midst of them, dressing the lamps, which appeared like a rod of seven stars in his right hand: and this dressing was performed by sending seven Epistles to the Angels or Bishops of the seven Churches of Asia, which in the primitive times illuminated the Temple or Church Catholick.[14]

The Book of Revelation will foretell the future history of mankind—and the mundane religious ceremonies John will witness in the temple provide a terrestrial foundation for that exalted purpose. Newton writes: "For the Temple & ceremonies of the law were types and shadows of things to come.... The allusions to the Law [in Revelation] are partly in describing the Christian worship & partly in predicting things future. The first is done by allusions to the Jewish daily worship, the second by allusions to the Feast of the seventh month."[15] The feast of the seventh month has become the feast of Hannukah, which is wholly bound up with the future history of every Jew.

Now John stands before the temple altar, which is decorated with cherubim, and immediately finds himself standing before the throne of God, of which he says, "in the midst of the throne, and round about the throne, were four beasts full of eyes before and behind. And the first beast *was* like a lion, and the second beast like a calf, and the third beast had a face as a man, and the fourth beast was like a flying eagle" (Rev. 4:6–7). John also tells us, "And round about the throne were four and twenty seats & upon these seats I saw four and twenty Elders sitting clothed in white raiment; & they had on their heads crowns of gold."

And then the lamb—Jesus Christ—appears and, as has already been described, from his throne God hands Jesus Christ a scroll sealed with seven seals. Christ loosens the seals one by one; visions explode in the head of John, who is standing nearby; and the prophet of Patmos commences writing the Book of Revelation.

Such, as set forth above, is the narrative pattern of Revelation. And it should be obvious from the above that this literary masterpiece of John of Patmos breathes a God-orientation, a numinosity, a plenitude and sumptuousness of religiosity, piety, and awe in the face of sublimity that we men and women living in a twenty-first century, almost completely denuded of God and numinosity, may find it difficult to understand.

There is a twentieth-century equivalent of the Book of Revelation, and we will include it in our discussion of John's Apocalypse because, this may help us understand the Book of Revelation.

In 1899, the British-Polish author Joseph Conrad wrote a novel, *Heart of Darkness,* that is the Book of Revelation without a trace of God or a scintilla of numinosity, so much so that it becomes a demonic, or anti-, Book of Revelation. The narrator, Marlow, comes to Brussels in Belgium to sign a contract with La Société Anonyme Belge pour le Commerce du Haut-Congo (The Belgian Company for Trade in the Upper Congo), a company that will send him to the Congo, in Africa, then a Belgian colony, to pilot a steamboat up the Congo River.

This corporate headquarters—which is a front for a rapacious Belgian government operation pillaging the Congo for all it is worth under the guise of bringing Christianity to it—is an anti-temple, a demonic temple, a temple to Mammon, or money. All of the action in *Heart of Darkness* emanates from this anti-temple insofar as La Société Anonyme Belge controls everything that goes on in the Congo.

Marlow enters a large, almost empty room and sees "a deal table in the middle, plain chairs all around the walls, on one end a large shining map marked with all the colors of the rainbow." Conrad means us to compare this with the sumptuous numinosity of Revelation 4:3–4: "There was a rainbow round about the throne, in sight unto an Emerald. And round about the throne *were* four and twenty seats: and upon the seats I saw four and twenty elders sitting, clothed in white raiment; and they had on their heads crowns of gold." In this executive headquarters of La Société Anonyme Belge in Brussels, God on his throne is a "pale plumpness in a frock-coat" seated behind a heavy desk. He gives Marlow not a scroll foretelling the future, but a bare contract listing Marlow's future duties in the Congo. That Marlow's destiny—his future, his fate—is bound up with this contract is indicated by the presence of two ladies continuously knitting black wool, like the Fates of antiquity continuously wove the fate of mankind.[16]

That *Heart of Darkness* is a bitterly ironic novel mirroring Saint John's Apocalypse is borne out by its having been adapted by Francis Ford Coppola as the 1970 blockbuster film *Apocalypse Now*. Vietnam is Coppola's Congo; his overriding theme is the imperialistic aspirations of the United States that bring blight and destruction to Vietnam, while one of Conrad's themes is the imperialistic aspirations of Belgium, which bring atrocities, plundering, and destruction to the Congo. In both cases the imperialists are deluding themselves or patently lying; "'defending democracy' was the American equivalent of the 'civilizing work'" of the European imperialists in the Age of Empire.[17]

Marlow makes an explicit allusion to the Roman occupation of Britain (which serves to link the reader to the Roman imperialism that is, according to Isaac Newton, a major thread in Revelation's future history of mankind): "And this [England] also has been [like the Congo] one of the dark places of the earth. I was thinking of very old times, when the Romans first came here, nineteen hundred years ago—the other day." While John of Patmos is a fiery prophet flushed with the presence of the divine, Conrad gives Marlow as storyteller just the barest hint of spirituality, describing the storyteller as "lifting one arm from the elbow, the palm of the hand outwards, so that with his legs folded before him he had the pose of a Buddha."[18]

Revelation is a part of the history of the corruption of the soul of man; it tells the story of that Great Apostasy that will end with the Apocalypse, beyond which lies a thousand years of peace and, ultimately, the redemption of mankind. *Heart of Darkness* is also about the corruption of the soul of man, but this dark tragedy offers no hope of redemption, and at the end we gaze wordlessly into the heart of darkness that is the fully corrupted soul of its evil genius, Kurtz.

In *Heart of Darkness,* Marlow boards a ship and begins his journey down the coast of Africa to the Congo. In the Book of Revelation, Christ loosens the first seal of the scroll, and John begins his journey through the future history of mankind, which is also a stroll through the Temple of Jerusalem. We'll recall the "four beasts full of eyes before and behind" around God's throne about whom John says that "the first beast *was* like a lion, and the second beast like a calf, and the third beast had a face as a man, and the fourth beast was like a flying eagle" (Rev. 4:6–7). Newton takes this information and connects it with the twelve tribes of Israel encamped around the four sides of the tabernacle 1,500 years before John wrote. These tribes are the raw tabernacle- and temple-related data—the "costume boxes"—out of which John will mold into prophetic hieroglyphic form the divine visions now bursting one after the other from beneath the first four seals. Drawing on rabbinical as well as biblical sources, Newton tells us:

> on the east side [of the Tabernacle] were three tribes under the standard of Judah, on the west were three tribes under the standard of Ephraim, on the south were three tribes under the standard of Reuben, and on the north were three tribes under the standard of Dan, Numb. ii. And the standard of Judah was a Lion, that of Ephraim an Ox, that of Reuben a Man, and that of Dan an Eagle, as the Jews affirm. Whence were framed [centuries later, on the sides of the Temple of Jerusalem's altar] the hieroglyphics of Cherubims and Seraphims, to represent the people of Israel. A Cherubim had one body with four faces, the faces of a Lion, an Ox, a Man and an Eagle.[19]

Newton interprets "the faces of a Lion, an Ox, a Man and an Eagle" as pointing to the heraldic animals of the tribes of Judah (a lion), Ephraim (a calf), Reuben (a man), and Dan (an eagle). From another rabbinic source, he identifies the heraldic colors of the four tribes as white, red, black, and pale. He factors in on what side of the tabernacle each tribe is encamped. From this he decides that the Four Horsemen of the Apocalypse are four Roman emperors and their dynasties. From all this we can draw up a table as shown on page 145.

Newton writes: "The visions of the opening of these seals relate only to the civil affairs of the heathen Roman Empire. So long the apostolic traditions prevailed, and preserved the Church in its purity: and therefore the affairs of the Church do not enter into these first four seals."[20]

These four emperors and their dynasties, whose reigns stretch from about AD 9 to 297, do not come into conflict with Christianity, which remains untouched by apostasy and "apostolic"; that is, "what the apostles taught." Newton piles on further proofs that the Four Horsemen of the Apocalypse are who he says they are. He tells us that the differing

FOUR HORSEMEN OF THE APOCALYPSE

Seal/ Horseman	Tribe	Heraldic Color	Heraldic Animal	Emperor/ Dynasty
First Seal/ First Horseman	The tribe of Judah encamped to the east of the tabernacle/ temple	white	a lion	Vespasian and his family entering Rome from the east
Second Seal/ Second Horseman	The tribe of Ephraim encamped to the west of the tabernacle/ temple	red	a calf or ox	Trajan and his Spanish family entering Rome from the west
Third Seal/ Third Horseman	The tribe of Reuben encamped to the south of the tabernacle/ temple	black	a man	Severus, Antoninus, and the African dynasty entering Rome from the south
Fourth Seal/ Fourth Horseman	The tribe of Dan encamped to the north of the tabernacle/ temple	pale	an eagle	Decius, Gallus, and the northern emperors entering Rome from the north

colors of the four horses tell us how much blood the emperors have spilled, and whose.

The winner of this bloodletting sweepstakes is Trajan (53–117), the second horseman, whose horse is red (Rev. 6:4). This isn't surprising, since Trajan and his dynasty span the Jewish-Roman War of 66–70 and the Jewish revolt of 113–117 that saw a million and a half Jewish men, women, and children killed. This tribe's heraldic animal is the ox, sometimes the calf, and these were the beasts that were most often sacrificed, and therefore the beasts that gave the most blood to God.

The third horseman (Rev. 6:6) rides a black horse; this color, says Newton, denotes the *type* of men killed by Emperor Septimus Severus and his dynasty, who reigned from 145 to 211. Newton says Severus and his armies usually killed only distinguished people in battle, for instance, generals or politicians, who deserved to be properly mourned, and black was the proper color of mourning.

The period of the fourth horseman, who horse is pale, commences, says Newton, during the time of Maximinus the Thracian (ca. 173–238) and represents death that came upon both rulers and people in abundance for the thirty-three-year period beginning with the Thracian: "The deaths here were not to be by bloodshed only because . . . [starvation and plague were also widespread, and came to the great and the common alike;] therefore this horse is neither red nor in mourning as in those seals, but pale, a color which equally expresses all sorts of deaths & the deaths of all sorts of persons."[21]

The first horseman, whose horse is white, is traditionally identified as Jesus Christ, and Newton accepts this identification along with his identification as Vespasian and his sons, who, Newton apparently thought, shed the least blood of all.

Let's examine in detail just one of these Four Horsemen of the Apocalypse and marvel—or perhaps feel uncomfortable, for sometimes what Newton says feels contrived—at the wealth of detail he brings to his exegesis to these four, perhaps the best-known, hieroglyphs in John's Book of Revelation. Christ opens the third seal. John is once more admonished to "Come and See!" He sees a black horse, "and he that sat on him had a pair of balances in his hand." Newton says the scales allude not to famine but to Septimus's early vocation as a judge. This emperor, he writes, "had a natural affection to judicature from a child, was so expert a Lawyer that at the age of 32 years the Emperor Marcus designed him Prætor & that more than among the candidates, after he came to the Empire heard causes daily all the morning, was very severe against criminals."[22]

Severus wins the award for the most bizarrely surrealist passage in Revelation's account of the Four Horsemen of the Apocalypse: "A measure of wheat for a penny, and three measures of barley for a penny; and see thou hurt not the oil and the wine" (Rev. 6:6).

Newton discusses this exhaustively, citing numerous sources and explaining that a penny was "the daily wages of the soldiers, & of other laborers . . . a measure containing so much as was allowed for the maintenance of a poor man for a day." Severus increased this figure, says Newton, his reign being "therefore remarkable for the increase of the Roman provisions above what other Emperors had done . . . there followed great tranquility & plenty."[23]

Historians tell us that Severus was indeed very concerned with the proper distribution of grain, barley, wine, and oil to the poor and increased quotas all around. Gibbon writes that his "expensive taste for building, magnificent shows, and, above all, a constant and liberal distribution of corn and provisions, were the surest means of captivating the affection of the Roman people. . . . Severus celebrated the secular games with extraordinary magnificence, and he left in the public granaries a provision of corn for seven years, at the rate of 75,000 *modii*, or about 2,500 quarters, per day."[24]

The fourth seal, says Newton, is linked to the tribes of Dan, whose color is "pale," whose heraldic beast is the eagle, and who were encamped to the north of the tabernacle. Newton comments conclusively: "Lastly to all this the standard of this Rider is very agreeable, being an eagle, a Bird of prey which feeds on carcasses."[25]

In *Heart of Darkness*, the "four horsemen" are the imperialist powers of Europe clandestinely jockeying for control of the dark continent of Africa. Conrad conveys their presence with just a few words. Marlow, in the pay of the Belgians, sees, when partway down the coast of Africa, a single French man-of-war anchored just offshore and shelling the jungle and remarks, "There wasn't even a shed there. . . . It appears the French had one of their wars going on thereabouts. . . . In the empty immensity of earth, sky, and water, there she was, incomprehensible, fitting into a continent."²⁶ (The film *The African Queen* vividly records the presence of Germany in Africa during World War I; and, at the same time as Conrad was writing *Heart of Darkness* [1899], the British and the Dutch Boers were locked in a bloody battle for pieces of South Africa.)

Two things should have become apparent to the reader by now: First, that Newton's proofs for his interpretations of John's prophetic hieroglyphs and figures often seem to be contrived, and sometimes outrageously contrived. And, second, that it is absolutely impossible, in the short space of the two chapters in this book dedicated to Newton's interpretation of the Book of Revelation, to deal with any more than a tiny bit of Saint John's Apocalypse. So, in our discussion, we will try to choose passages that, first of all, show us exactly how Newton does what he does (so the reader will be able to do it on his or her own) and, secondly, that bear directly on Newton's prophecy that the Apocalypse will come in the year 2060.

When, in Revelation, Christ loosens the fifth seal, John sees "under the altar the souls of them that were slain for the word of God, and for the testimony which they held" (Rev. 6:9–10). Newton says this bitter prophetic hieroglyph emblemizes Diocletian's persecution of the Christians, so one of the souls under the altar is Archbishop Peter of Alexandria, whose death plays a poignant role in Newton's excoriation of Athanasius. Here, Newton's *Observations* on Revelation suddenly intersects with the "Paradoxical Questions Concerning the Morals & Actions of Athanasius"; we are approaching a point of high tension in Newton's "History of the Corruption of the Soul of Man," where a number of events will pile up to trigger the first stage of the Great Apostasy. (The temple-related raw data that John, prompted by explosions of inspiration from Christ, draws on to create this prophetic figure derives from the ancient practice of sacrificing animals in the outer courtyard of Solomon's Temple; the prophet has jumped far back in time to secure what he needs for this image.)

Meanwhile, in *Heart of Darkness,* Conrad gives us a harshly ironic picture of the Congo natives undergoing persecution at the hands of their imperialist masters. Here the exploitation is all the more brutal in that it is passive; the Congolese are displaced, starved, and left to die. Marlow tells us that

> black shapes crouched, lay, sat between the trees, leaning against the trunks, clinging to the earth, half coming out, half effaced within the dim light, in all the attitudes of pain, abandonment, and despair. . . . They were dying slowly—it was very clear. . . . Near the same tree two more bundles of acute angles sat with their legs drawn up. One, with his chin propped on his knees, stared at nothing in an intolerable and appalling manner. His brother phantom rested its forehead as if overcome with a great weariness; and all about others were scattered in every pose of contorted collapse, as in some picture of a massacre or a pestilence.[27]

Now Christ loosens the sixth seal, and we seem to be in a whole new universe, one of catastrophe on a cosmic level: the sun turns black, stars plummet to earth, blazing mountains are tossed into the sea, abominations crawl out of the earth. However, if we are expecting to find behind these garish hieroglyphs interstellar warfare in the mode of Doris Lessing's "space-fiction" novels (*Shikasta, The Making of Representative Seven,* and so on) or interdimensional conflict after the fashion of Stephen Spielberg's *Indiana Jones and the Kingdom of the Crystal Skull,* we will be in very disappointed. Newton is pursuing his agenda of seeking an "analogy between the world natural and . . . a world politic,"[28] and he ruthlessly reduces whatever seems transcendent or paranormal or fantastical or from another reality to the unrelentingly mundane. John sees "a tremendous earthquake, the sun turned dark like coarse black cloth; and the full moon was red as blood. The stars of the sky fell upon the earth" (Rev. 6:12).

The hieroglyphs of the sixth seal foretells what looms behind and follows upon the Council of Nicaea, which epochal synod Newton also describes, with his own follow-up, in "Paradoxical Questions" (see chapter 4, "Bloodbath in a Boghouse: Murder in the Fourth Century, Part 1): This is the disintegration of the pagan religions of the Roman Empire as Constantine makes Christianity its official religion. The sun is the prophetic figure for a ruler; the moon, for "the body of the common people, considered as the King's wife"; the stars, for "subordinate Princes and great men" (or, when the ruler-sun is Christ, for bishops);[29] and earthquakes, for wars and similarly major disruptions. Revelation 6:14 tells us that "every mountain and island were moved out of their places," and Newton tells us that mountains and islands are "the cities of the earth and sea politic," so this is about shifts in political entities. Revelation 6:15 says many distinguished men "hid themselves in the dens and in the rocks of the mountains." Newton says dens and rocks are temples and that this passage prophesies "the shutting up of idols in their Temples, or burying them in the ruins thereof."[30] David Castillejo says the "motions of some and voices of others" from which John fashions these prophetic figures are, for the sun, "the bright flame of the fire of the Altar (or by the face of the Son of man) shining through this flame like the Sun in his strength," and, for the moon, "the burning coals upon the altar convex above & flat below like an half moon." We're being asked to imagine that at one point, while the high priest is reading from the Torah, the flame of the altar flickers and almost goes out and the burning coals are momentarily reduced to glowing embers.[31]

Now we have a hiatus: the angels, standing at the four corners of the Earth, hold the winds back. Then the "sealing" of the 144,000 "of all the tribes of the children of Israel" takes place. Newton says this prophetic hieroglyph draws its imagery from the twelve tribes (144,000 Israelites) encamped around the tabernacle; it is also linked, through the Feast of Tabernacles, to Jews who have repented during the High Holy Days and whose names God as a consequence is setting down in a positive mode in his as-yet-unsealed Book of Life.

Heart of Darkness gives us a black parody of the sealing. The Congolese natives, whose only crime is that they are pagan Africans in the wrong place at the wrong time, are not sealed; they are shackled together with iron collars. In their own universe they are perfectly innocent, and their shackling is a kind of demonic sealing. Marlow tells us that

> a slight clinking behind me made me turn my head. Six black men advanced in a file toiling up the path. They walked erect and slow, balancing small baskets full of earth on their heads, and the clink kept time with their footsteps. Black rags were wound round their loins and the short ends behind waggled to and fro like tails. I could see every rib, the joints of their limbs were like knots in a rope, each had an iron collar on his neck and all were connected together with a chain whose bights swung between them, rhythmically clinking.[32]

Iliffe tells us that, for Newton, the spiritual disaster of the Great Apostasy, "accomplished by making Athanasian Trinitarianism the official religion across the Roman Empire in 380, was described by the opening of the seventh seal."[33] This making of the doctrine of the Trinity the cornerstone dogma of the Catholic Church was only the second of the three phases of the settling-in of the Great Apostasy, as we will see.

CHAPTER EIGHT

APOCALYPSE 2060?

Let us pause for a moment in our headlong race with Isaac Newton to the Apocalypse via Revelation and look at a curious treatise of his, unknown for three hundred years, that discusses an unusual, disastrous event he believed would take place sometime during the End Times. This was the *diluvium ignis,* or "flood of fire."

If we've come to believe, by the time this chapter ends, that Newton really did prove the Apocalypse will arrive in 2060, then we'll be unnerved by this disastrous event predicted by many of the millenarians of the seventeenth century. The *diluvium ignis* sounds like the culminating stage of what we call global warming, when, say some of our scientists, parts of the Earth's crust will burst into flame.*

The flood of fire was first predicted in 2 Peter 3:10. The text was probably written in the middle of the first century AD. The passage runs: "But the day of the Lord will come as a thief in the night; in the which the heavens shall pass away with a great noise, and the elements shall melt with fervent heat, the earth also and the works that are therein shall be burned up." This event is generally predicted for

---

*In July 2016, NASA and the National Oceanic and Atmospheric Administration (NOAA) reported that June 2016 was the hottest June ever recorded. Global average temperatures in June were 0.9°C (1.6°F) hotter than the average for the twentieth century. These temperatures broke the previous record, set in 2015, by 0.02°C (0.036°F). Arctic sea ice now covers 40 percent less of the Earth than it did in the 1980s, NASA added.

Judgment Day (right after Armageddon); it was first prophesied some thirty years before John of Patmos wrote Revelation.

Isaac Newton writes, in "The Synchronisms of the Three Parts of the Prophetic Interpretation," that "it is a received opinion that this judgment shall be accompanied with a conflagration of the world; & some hearing that in the future world the Wolf shall lie down with the Lamb & all beasts shall become gentle & harmless & the Earth become fuller of rivers & more fruitful . . . have conceived that an amendment of the whole frame of nature shall ensue that conflagration."[1] Newton means that those who believe the world will be transformed into a place of harmony and beauty believe there must be a global conflagration beforehand ("an amendment of the whole frame of nature"), because only a total cleansing of the Earth's surface will properly prepare for this new world to arrive.

Newton doesn't believe this global conflagration will necessarily happen and cautions that those who read the Bible in this way are mistaking mystical for real-life, historical, language: "these fancies have been occasioned by understanding in a vulgar & literal sense what the Prophets writ in their own mystical language. For the conflagration of the world in their language signifies the consumption of Kingdoms by war, as you may see in Moses, where God thus describes the desolation of Israel."[2] Thus the conflagration is entirely political in nature.

This being said, Newton *does* seem to believe there will be a smallish, localized conflagration, a *diluvium ignis,* around Judgment Day. He writes, "but in the day of judgment there is also a literal conflagration of the world politique in the lake of fire & to those that are cast into it a conflagration also of the world natural, the heaven & earth where they are being on fire & the elements melting with fervent heat."[3]

This conflagration will not burn up much territory.

> And whilst the Apostle Peter tells us that none but the wicked shall suffer in this conflagration & that this is a time of refreshing to the Godly I cannot take it for a conflagration of any considerable part of

this globe whereby the rest of the habitable world may be annoyed. And if the world natural be not burnt up there is no ground for such a renovation thereof as they suppose: The glorious Sun & Moon, multiplied rivers & copious vegetables of the new world are its Kings & people, the peaceable & harmless Beasts its peaceable Kingdoms, & the new Jerusalem that spiritual building in Sion whereof the Chief corner stone is Christ & the rest of the stones & gold are the saints.[4]

Early church fathers like Origen and Eusebius suspected the passage in Peter was a forgery. Some first- and second-century thinkers thought the predicted disaster had already occurred; perhaps, they ventured, Peter was foreseeing the fiery destruction of the Temple of Jerusalem. Many believed the flood of fire would rain down right after the Jews returned to the Holy Land and became converts to Christianity.[5] (See chapter 9, "The Conversion of the Jews.")

The belief of a great many English millenarian thinkers of the seventeenth century was that God would aim the *diluvium ignis* straight at the heart of the Roman Catholic Church. Trying to determine the physical cause of a flood of fire—global warming was an inconceivable idea at the time—researchers decided it would have to be a volcanic eruption from the bowels of the Earth. The most volcanically volatile region in Europe, they decided, was central Italy (notwithstanding that Europe's most active volcano, Mount Etna, was some little distance away, on the east coast of Sicily). And the most potentially dangerous area of all was beneath Rome's Vatican City. So there were many thinkers, especially Protestants, who saw the raison d'être of the biblical *diluvium ignis* as the incineration of the Antichrist diabolically inhabiting Saint Peter's Basilica![6]

Some believed the *diluvium ignis* wouldn't arrive until near the end of the thousand years of peace. To Newton's way of thinking, the Apocalypse, though a dreadful time (it included Armageddon and multiple plagues), was the storm before the calm. At the end of this bloody interregnum, Satan would be bottled up in the bottomless pit and a thousand years of peace would commence. Only a sainted minority would survive; our world would be strange and new, but still intact. At the end of this semi-blissful period of one millennium (during which Newton hoped to wield some influence as a saint), there would be a final clash between Christ and Satan, Christ would triumph, and our world would morph into something wholly different and unknown at the same time as a tiny remnant of mankind transitioned to another world. (See chapter 10, "With Noah on the Mountaintop.")

Does John's Book of Revelation give any hint of the arrival on our planet of a *diluvium ignis* (which flood of fire sounds strangely like the final paroxysm of global warming as predicted by some for our Earth)? Let us return to a necessarily impressionistic discussion of Newton's interpretation of the Book of Revelation and see if Newton has anything to say about this.

We have arrived at the opening of the seventh seal. From beneath this seal there successively emerged seven trumpets a-blasting and, in rapid succession after, seven vials (or, in some translations, "bowls") a-pouring. Accompanying the noisy blasts of the seven trumpets was the din of seven thunderings.

Such was the interpretation of all the exegetes of the seventeenth century—all except Isaac Newton. Newton defied everyone by stating that the seven vials emerged concurrently with the seven trumpets; that is, that the seven trumpets (with the seven thunderings) and the seven vials described the same event from two different angles: they trumpeted and poured together.

John, receiving the raw stuff of visions in a steady stream from Christ, molds the prophetic hieroglyphs of the trumpets and the vials from the trumpet blasts and the exquisite scents wafting from the vials of the ceremony of the Feast of Tabernacles unfolding within the temple. These exquisite scents undergo a drastic change in Newton's deciphering of the hieroglyphs: they become horrible poisons. In Newton's ever-nuanced and subtle interpretation, they have a double purpose: some just give you the plague, while others give you the ability to give others the plague.* The future history of the world that the trumpets and vials tell is the battles of the Romans against the barbarians; Newton says that often during the Festival of Tabernacles the high priest read not from the Torah but from Jewish history; there are many wars in Jewish history, and John makes use of this history when he concocts the prophetic hieroglyphs that emerge from the seven seals.

So, the yoked trumpet blasts and vial pourings emerge one after another. John creates, encoded in hieroglyphs, vivid images of bloody, desperate struggles between Rome and the Goths, or the Huns, or the Vandals, or the Alans, or the Franks, and so on. The adoption in 380 of Athanasian Trinitarianism as the official religion across the Roman Empire (which marked the beginning of the second phase of the Great Apostasy) has begun to operate like a corrosive acid, weakening the moral fiber of the Romans, because, by distracting their attention with Jesus Christ, it dilutes the power of man's exclusive and necessary love for God.

As has been said, all other Revelation exegetes, including Newton's tutor Joseph Mede, saw the seven vials as emerging from the seventh seal long after the seven trumpet blasts. This was a reasonable assumption, since the vials come long well after the trumpets in the text of Revelation, the first group in passages 8:7–12; 9:1–13; 11:14; and 12:1, 3; and the second in 16:2–17. These exegetes interpreted the prophetic hieroglyphs of the seven vials as foretelling the struggles of Martin Luther and the Protestant Reformation against the papacy.

But Newton dismissed this interpretation, because he virtually dismissed the Protestant Reformation. He thought it was a mere bump in the road in the face of the oncoming juggernaut of the now fully consolidated Great Apostasy. Iliffe writes that, for Newton, "the demonic power of the Papacy was so great that Martin Luther's revolution had

---

*Newton writes: "The pouring out of a Vial is taken in a double sense, signifying some times the execution of a plague on that thing whereon it is poured, & sometimes the incitement & invigoration of that thing, as it were by a contagious virtue of the medicament, to execute the plague on another thing." (Newton, "Untitled Treatise on Revelation," section 1.6.)

no dampening effect whatsoever on the papacy." He adds that with his "astonishing and utterly original analysis," Newton "totally inverted what orthodox Christians of all persuasions took to be the heroes and villains of history. His colleagues had reserved a special place for the six vials as a specific account of the trials of Protestantism. Newton composed a detailed and extensive analysis of the way in which Catholics—whom he termed 'sorcerers' and 'magicians'—fulfilled the conditions of the sixth trumpet and vial."[7] The trumpet-led battles with the barbarians, the potions in the vials, had merely exacerbated the Catholic Church's relentless corruption of the soul of man.

The first trumpet blast hurls hail and fire mingled with blood on the earth (Rev. 8:7). The first vial (or bowl) empties "loathsome and malignant ulcers" on the enemies of Christ (Rev. 16:2). These seemingly cosmic hieroglyphs, says Newton, actually

> direct us to the invasions which broke forth immediately after the death of Theodosius. For during the reign of that Emperor the Empire flourished very much bearing up against the endeavors of all foreign enemies, & enjoying a more then usual tranquility. There were indeed between the wars of Maximus & Eugenius some attempts upon Gallia by the Franks, but these were but short & unsuccessful & may be compared rather to gentle breathings then winds. But so soon as Theodosius was dead, Ruffin, to whom Theodosius left the tuition of Arcadius thinking to get the Empire to himself, called in all the nations of the North to trouble the Roman waters.[8]

For more details, including Newton's explication of the various symbols, the reader is directed to section 1.6 of the "Untitled Treatise on Revelation," on the Newton Project website.

Let's jump ahead to the fourth trumpet blast and its concurrent vial pouring. Revelation 8:12 reads: "The fourth angel blew his trumpet, and a third part of the sun, a third part of the moon and a third part of the stars were struck. A third part of the light of each of them was darkened, so that light by day and light by night were both diminished by a third part." We are, of course, not in a cosmic realm but in the thoroughly mundane domain of kings, kingdom, and princes. The kingdoms of western Europe are darkened and will be for some time. Newton says the passage refers to Belisarius (505–565). Flavius Belisarius was a Byzantine Empire general who played a decisive role in emperor of the Eastern Roman Empire Justinian's attempt to reconquer much of the former Western Roman Empire, which had been lost to the barbarians less than a century before. What lies within this hieroglyph, says Newton, is the story of how Belisarius "conquered the *Vandals,* invaded *Italy AD 535,* and made war upon the *Ostrogoths* in *Dalmatia, Liburnia, Venetia, Lombardy, Tuscany,* and other regions northward from *Rome,* twenty years together. In this war many cities were taken and retaken." There was a colossal loss of both military and civilian life; when the Ostrogoths retook Milan from the Romans, they murdered 300,000 males, from the young to the old, and "sent the women captives to their allies the *Burgundians."* Rome itself, and other great cities, were taken and retaken several times, and Rome, practically without governance, fell into great decay. In AD 552, "after a war of seventeen years, the kingdom of the *Ostrogoths* fell." A remnant of the Ostrogoths, supported by a German army, fought on for three or four more years. This was followed by the war of the *Heruli,* which, says Newton, quoting authorities, "slew all *Italy.*"[9]

But what about the fourth vial, whose contents, when deciphered, should complement the story of the fourth trumpet? Revelation 16:8 tells us that "the fourth angel emptied his vial over the sun, and the sun was given power to scorch men in its fiery blaze. Then men were terribly burned in the heat, and they blasphemed the name of God who has control over these afflictions; but they neither repented nor gave him glory."

Newton says his interpretation of the fourth trumpet is

> notably confirmed by the correspondent Vial; the tenor of this is that, it was poured upon the Sun & power was given him to scorch

men with fire & men were scorched with great heat & blasphemed God &c, that is, that the pouring out of this Vial was an incitement of the supreme terrestrial potentate to torment men with war & men were tormented with vehement war & blasphemed God. And thus it happened. For the Greek Emperor (the supreme terrestrial potentate) [Justinian] was the cause of the wars of this Trumpet by sending his armies [led by the formidable Belisarius] into Italy in pursuance of his claim to those regions.[10]

It's as if a sun-induced fever has seized Justinian and compelled him to unleash the extremes of warfare. The brightness of western Europe is reduced by one-third during these years of war; faith is lost and God is ignored; and the world is swept more and more powerfully toward the third and cumulative stage of the Great Apostasy.

In *Heart of Darkness,* Revelation 8:12 and 16:8 are reflected in the growing darkness of the soul of man as expressed by the murder, terror, and depravity confronting Marlow as he pilots his steamboat up the Congo River. At the end of the trip is the object of their search, the maverick white hunter Kurtz, whose soul is already dark; it is, in fact, the "heart of darkness." As Marlow proceeds, with the foliage ever thickening on the riverbanks, a hail of native spears sprays the steamboat from both sides; the helmsman is killed in front of Marlow's eyes. A harlequinesque figure hails the steamboat from the bank; Marlow puts in, and this bizarrely dressed naive young European turns out to be in the spell of the evil genius Kurtz; he is the anti–John the Baptist to the counter-Christ of this once-promising leader who has gone wholly over to the dark side. Guided by the acolyte, they arrive at Kurtz's encampment; and as they draw nearer strange round objects atop stakes in the front of his house resolve themselves into shrunken heads. Nature seems to droop and crawl and be a part of the steady corruption of the human soul that Marlow senses all around him.

In the simultaneous blast of the fifth trumpet and pouring out of the fifth vial, John of Patmos has foretold, says Newton, the birth and rise of Islam.

When the fifth trumpet sounds (Rev. 9:1), a star falls to earth and an angel opens a bottomless pit from which arises smoke so thick it darkens the sun and moon. Out of the smoke come locusts as strong as scorpions, who have the power to harm every living creature except those with the mark of God on their foreheads. Newton says these plagues of locusts are armies, and these armies are the warriors of Muhammad. Revelation 16:10 says the fifth angel emptied his vial (the plague of darkness) upon the throne of the animal; Newton says the throne of the animal is Islam.

Ron Iliffe explains:

This was the rise of Islam, to be dated from when Mohammed found his vocation as a prophet in 609 CE (as Newton dated it). His flight from Mecca to Medina in 622 was the opening of the pit, but the fifth trumpet and vial lasted from 635 to 936. . . . The extended torment referred to the fact that Muslims had repeatedly laid siege to Constantinople without being able to take it.[11]

Newton tells us something about the temple components out of which John molds the prophetic hieroglyphs that encode the story of Islam's birth. Newton says that "the bottomless pit or lower parts of the earth called Hades & hell" was based on "the sink which ran down into the earth from the great Altar & was covered with a stone to open & shut," adding that, "the opening of the bottomless pit . . . denotes the letting out of a false religion: the smoke which came out of the pit, signifying the multitude which embraced that religion."[12]

Now an angel hands John a "little book" and tells him to eat it (Rev. 10:1). The angel tells him to prophesy anew according to this book; it's generally considered that God has given John access to the backs of the pages of the scroll that Christ had read to him.

We are nearing the point where the Great Apostasy will begin to take hold, and it is at this point that two women enter the Book of Revelation.

We know that, for Newton, a woman is always a church. In Revelation 12:6, a woman clothed in the sun gives birth. Her baby is snatched away. A beast rises up out of the water and attacks her. She escapes. It is not clear where her newborn son is. Most exegetes, including Newton, believed the woman clothed in the sun is the Christian Church in its purest form, which is now, with the ascendancy of Trinitarianism, being exiled, sent to be "nourished in the wilderness" for the duration of the supremacy of the Roman Catholic Church, the papacy, or, for Newton, as Iliffe sums it up, "that impious monster bristling with stakes, relics, indulgences, saint-worship and miracle-soliciting."[13]

The other woman is the Whore of Babylon riding the Beast, which beast is the Antichrist, or the Roman emperor Nero. It was thought that Nero had never died, and in fact he lives on in the Revelation narrative through various "incarnations." He is Rome; he is the Rome that becomes entangled, through Constantine, with Christianity; he is the Roman Empire harshly compromised by the growing power of the church. Finally, Rome is subsumed into the Roman Catholic Church, which becomes, for Newton and most non-Catholic exegetes, the Whore of Babylon riding on the back of the Beast (Rev. 17:3). As that, she is the demonic opposite of the woman clothed with the sun, who represents the true Christianity of Jesus's apostles, who taught love to God and our neighbors. She is the pope's kept whore, the totally corrupt Roman Catholic Church that Newton and his colleagues despise with a virulence that ceases to surprise postmodern man.

We also encounter two women in Joseph Conrad's *Heart of Darkness*. One is Kurtz's black pagan mistress, who has been utterly corrupted by Kurtz himself; she is the Whore of Babylon riding the Beast who is Kurtz/the Antichrist—except that this woman comes from the polar opposite of a great city like Babylon; namely, the bare jungle. The text hints of "the unspeakable acts" she and Kurtz have committed together;

she appears, lost, mourning, and savage, on the shore as Marlow's boat leaves with the dying Kurtz on board. Marlow describes her thus:

> She walked with measured steps, draped in striped and fringed clothes, treading the earth proudly, with a slight jingle and flash of barbarous ornaments. She carried her head high; her hair was done in the shape of a helmet; she had brass leggings to the knee, brass wire gauntlets to the elbow, a crimson spot on her tawny cheek, innumerable necklaces of glass beads on her neck; bizarre things, charms, gifts of witch-men, that hung about her, glittered and trembled at every step.... She was savage and superb, wild-eyed and magnificent; there was something ominous and stately in her deliberate progress.[14]

The counterpart to this black pagan goddess is Kurtz's fiancée, whom he long ago left behind in Belgium. She is the "woman clothed in the sun"; her purity is such that she cannot know that her beloved Kurtz could ever have become what he became. Marlow, visiting her on his return to Brussels, doesn't try to enlighten her. Hers is the spotlessness that cannot see dirt. Kurtz's last words were, "The horror! The horror!" Marlow tells her that those last words were her name.[15]

For Newton, the resurrection of the Two Witnesses and the recommencement of the preaching come at the same time as the fall of Babylon, the corrupt church—that is, the Roman Catholic Church—and perhaps all other churches that fail to preach the necessity of worshipping God, and God alone. Newton believed that, as the 1,260-year period of the Great Apostasy comes to an end, we would see "many begin to run to and fro as knowledge of the church begins to be preached in all nations by the two witnesses ascending up to heaven in a cloud."[16] The fall of Babylon and the preaching of the true Gospel end with the blasting of the seventh trumpet, which heralds Armageddon, the coming of Christ, the resurrection, the judgment, and the establishment of the Kingdom of God on Earth. Newton also links the return of the Jews (see chapter 9, "The Conversion of the Jews") with the sounding of the seventh trumpet of Revelation.

Newton mulled over several "commencement dates." One was AD 538, when the emperor of the eastern remnant of the Roman Empire decreed that the bishop of Rome should be the head of all the Christian churches. This decree became effective with the defeat of the Arian (i.e., anti-Trinitarian) powers that had ruled much of the western empire (including Italy). Adding 1,260 years to 538 brings us to 1798, the year when Napoleon's general Berthier took the pope prisoner and proclaimed the end of all political rule of the Roman Catholic Church. Other commencement dates Newton toyed with included AD 476, then considered the year of the fall of the Roman Empire, which put the end of the world at 1736; and AD 609, roughly the year the Roman emperor Phocas gave Pope Boniface IV the right to institute image worship in churches, which put the end of the world at 1869.

"The two-horned Beast slowly rises up out if the earth and the ten-horned Beast receives the Dragon's throne in the west. . . ." We have now entered that part of Newton's exegeses that fairly bristles with horns. This is because Newton is about to lean heavily on the Book of Daniel to supplement his journey of discovery through the Book of Revelation. We find the same horns, with the same meanings, in Revelation, but in the Book of Daniel they are the star performers. Horns stand for kingdoms, usually political and sometimes spiritual. Newton believed the beast in Daniel's fourth vision was the prophetic hieroglyph for Rome.

This was the beast with "huge iron teeth" and ten horns, of whom Newton wrote, "Daniel's fourth Beast, which is the same with Saint John's [beast with ten horns], signifies properly the western nations of the Roman Empire, & those alone."[17]

The ten horns were the ten barbarian kingdoms that harried and sometimes occupied Rome: the Vandals and Alans in Spain and

Africa, the Suevians in Spain, the Visigoths, the Alans in Gallia, the Burgundians, the Franks, the Britons, the Huns, the Lombards, and Ravenna. As Daniel watched (Dan. 8:20, 24), three of the horns were "plucked out by the roots" and a "little" horn sprouted in their place. This little horn quickly became "greater than his fellows."

Newton argued that this eleventh horn was the Roman Catholic Church. It differed from the other horns in claiming to be a "universal bishopric; Daniel 25 says it will "alter the seasons and the Law." This horn had "a mouth full of boasting" (Dan. 7:8), which Newton thought symbolized Pope Leo III's haughtiness. Newton wrote: "With his mouth he gives laws to kings and nations as an Oracle; and pretends to Infallibility, and that his dictates are binding to the whole world."[18]

(In *Heart of Darkness,* there are no horns save those the Africans are wearing decoratively; but there are tusks—elephant tusks. The action revolves around them. The Europeans lust after them, killing, maiming, and enslaving to have these tusks; they are beautiful and valuable in themselves, but man's rapacity makes them become deliverers of death. The lust to have them brings the colonial powers into Europe, makes these powers clash—and they end up acting as badly as the battling Romans and barbarians symbolized by the horns of the Book of Daniel and John's Book of Revelation.)

Newton had been tempted to choose 774 as the commencement date, which would have put the Apocalypse at 2034. It was in 774 that Pope Adrian I first gained "temporal dominion" for the Vatican by acquiring three earthly principalities. Newton says these acquisitions were the three uprooted horns of the fourth beast, which stood for Ravenna, Lombardy, and Rome. The pope acquired this political power with the help of the Franks and their king Charlemagne.

It is at this time that the Catholic rites of the canonization of saints and of transubstantiation (the belief that the bread and wine offered in the celebration of the sacrament of the Eucharist turn into the actual blood and body of Jesus Christ during the service) took hold. For

Newton this is the vilest sort of idolatry, and an ominous intensification of the Great Apostasy. But Newton finally decided that the year the papacy achieved political dominion over the Earth—the year when the Great Apostasy became established in full control of the religious life of mankind—was 800.

This was the true commencement year.

One event, both real and symbolic, captured the moment. It took place in the old Basilica of Saint Peter's on Vatican Hill in Rome on Christmas Day 800. On that day, an old, trembling man, with red swollen eyes and a floppy tongue that made him hard to understand, got down on his knees before a fifty-six-year-old, black-bearded soldier built like a sumo wrestler (and between five and seven feet tall, depending on who you talked to) and pledged allegiance to him.

The kneeling figure was Pope Leo III (750–816), who, eight months before, had been dragged off his horse by two assassins who tried to squeeze out his eyes and cut off his tongue. Leo was thrown into prison but escaped two months later. The pope's brush with death—not the first—had made him more determined than ever to protect the Holy See by forging a revolutionary alliance with the greatest political figure of his day.

Before he had knelt, Leo III had placed a crown on the head of the Frankish king Charlemagne (742–814) and named him Holy Roman Emperor. Now, as the pope struggled to his feet, Charlemagne bowed to him and called him "My brother."

It was a power-sharing agreement. The pope would share power with Charlemagne over the countries the newly crowned Holy Roman Emperor ruled: France and Germany. Charlemagne would add to his dominions the lands Christianity had held sway over almost since the time of Constantine: Rome, Britain, the Holy Lands, and parts of Africa and Asia.

The first Holy Roman Emperor would soon convert the Saxons and Huns on his conquered lands to Catholicism, thereby significantly magnifying the power of the Roman Church.

Here is Newton's description of the coronation:

> Soon after, upon Christmas-day, the people of Rome, who had hitherto elected their Bishop, and reckoned that they and their Senate inherited the rights of the ancient Senate and people of Rome, voted Charles their Emperor, and subjected themselves to him in such manner as the old Roman Empire and their Senate were subjected to the old Roman Emperors. The Pope crowned him, and anointed him with holy oil, and worshiped him on his knees after the manner of adoring the old Roman Emperors.[19]

We'll recall that "beasts" are kingdoms; the "seven-headed beast restored" of Daniel may be the seven kingdoms (two, France and Germany, belonging to Charlemagne, and five, Rome, Britain, the Holy Lands, and parts of Africa and Asia, belonging to the pope) that came together as one on Christmas Day 800. Daniel 7:25 says the "little horn"—for Newton, the papacy—will rule for a "time, two times and half a time." ("And they shall be given into his hand for a time, two times, & half a time.") Newton took this phrase to signify the mystical/prophetic time span of 1,260 years, whose acquaintance we have already made. It is by adding this mystical number of 1,260 to 800 that Newton will arrive at a date for the Apocalypse of AD 2060.

Before we get into commencement dates, however, we must take very quick note of other hieroglyphical actions fulminating, as if to emphasize its importance, around the commencement date of 800.

"The outer court trodden down by the gentiles" takes its hieroglyphic trappings from the outer court of Solomon's Temple, whose walls were torn down by the Babylonians and never repaired, being left open for non-Jewish visitors. Here the court seems to refer to the figurative breach torn in the side of the Roman Catholic Church to let in Charlemagne and his minions. "The witnesses on earth prophesying in sackcloth" are those who preach right Christian doctrine despite the pressure on them from the false, idolatrous, Trinitarian preachers in the Vatican and the court of Charlemagne (which have become, disastrously for Newton, almost a single entity. Newton writes:

> And so by the prophesying of the two witnesses we are to understand their promulgating & spreading the laws & word of God according to the utmost power they are able to speak with. For this their prophesying is opposed to the speaking or prophesying of the false prophet, & so signifies here not foretelling future things by immediate revelation, but the preaching to the world the word of God according to the right interpretation & meaning of it & propagating it in God's name with all the authority they are able; as on the other hand the false Prophet propagates false interpretations & other lies in the name of his God, pretending the authority of his God, as the true prophets do that of theirs; to make his sentences pass for law & gain to himself a law making power in matters of religion.[20]

To return to the subject of the "commencement date": It had been Joseph Mede who had first set forth the "time" equals a "day" equals a "year" formula; Snobelen lists those biblical passages that influenced him:

- Numbers 14:34: "After the number of days in which ye searched the land, even forty days, each day for a year, shall ye bear your iniquities, even forty years, and ye shall know my breach of promise";
- Ezekiel 4:6: "And when thou hast accomplished them, lie again on thy right side, and thou shalt bear the iniquity of the house of Judah forty days: I have appointed thee each day for a year";
- Daniel 2:7: "and I heard him swear by him who lives forever that it would be for a time, two times, & half a time."

To reiterate: because a "day" stood for a "year," then "time, two times, & half a time" equals 360 years (reckoning a year as 360 days according to the old standard) + (2 × 360 =) 720 years + (½ × 360 =) 180 years, totaling 1,260 years.

But, if 1,260 days stood for 1,260 years, why the figure "1,260"—that is, why "time, two times, & half a time"—in the first place?

Because, said Isaac Newton, 1,260 days (three-and-one-half years) was roughly the life span of a "short-lived Beast." And "short-lived beasts" was the prophetic hieroglyph in the Book of Revelation for "lived [sic] Kingdoms."[21]

This period of 1,260 years was thus the period when the Great Apostasy, represented by the Roman Catholic Church, would rule supreme over the souls of humankind. This was the period when the Catholic Church would, in Newton's view, have completely abandoned Christianity's true teachings.

Newton had calculated the time of the Apocalypse using a number of other commencement dates, and we do not know how seriously he took his prediction that the world would end in 2060. In one manuscript he seems to suggest that he made up the prediction so as to shut everybody else up. He wrote:

> This I mention not to assert when the time of the end shall be, but to put a stop to the rash conjectures of fanciful men who are frequently predicting the time of the end, & by doing so bring the sacred prophesies into discredit as often as their predictions fail. Christ comes as a thief in the night, & it is not for us to know the times & seasons which God hath put into his own breast.[22]

None of Newton's contemporaries used his commencement date of AD 800. Joseph Mede reckoned the commencement date for the 1,260-year period of total apostasy was the year of the fall of the Roman Empire—476. Adding 1,260 years to this date, he came up with an End Time of 1736—nine years after Newton's death.

For Newton, this 1,260-year period was represented in the Apocalypse as the 1,260 days during which the woman clothed in the sun (the apostolic church) is nourished in the wilderness (Rev. 12:6), and the Great Whore rides the Beast that is the triumphant Roman Catholic Church (Rev. 17:3). Toward the end, the Beast of the Bottomless Pit kills the Two Witnesses (Rev. 11:7), but they are resurrected after having lain dead for three and a half days. The 1,260 days when "the outer court of the Temple is hidden underfoot by the Gentiles" (Rev. 11:2–3) encompasses this same time period.

And when will that 1,260 time period end? Possibly, as we have seen, and according to Newton, in the year 2060.

The Stoics of antiquity believed in the periodic renewal of the universe. The millenarians of Newton's time believed that an "eternal sabbath" would follow Judgment Day.

Steven Snobelen explains, and we mentioned at the beginning of this chapter, that Newton did not believe the world would literally end in the year 2060. There would be a great battle, Armageddon, and then Christ and the saints would intervene to establish the worldwide thousand-year Kingdom of God. It would be a time of peace and prosperity; in his *Observations upon the Prophecies of Daniel*, Newton quotes Micah 4:3: "[The people will] beat their swords into plowshares and their spears into pruning hooks," and during this time "nations shall not lift up a sword against nation, neither shall they learn war any more."[23]

Manuel tells us that into a few terse phrases from the Apocalypse, Newton

> compressed a wealth of scriptural evidence for his belief that the world was moving inexorably toward a cataclysm, a great conflagration, to be followed by a yet undefined form of renewal.... I quote the whole passage (Rev. 20:10): "And the devil that deceived them was cast into a lake of fire and brimstone, where the beast and false prophet are, and shall be tormented day and night for ever and ever" ... [Newton] merely jotted down the phrase, "Days and nights after the Judgment." ... Tormenting the wicked for ever and ever

is quite comprehensible and sufficient unto itself, and the prophet could have been expected to stop at that. But when John inserted the words "day and night," which are seemingly superfluous and in excess, he surely meant to inform us of something—in this instance that the succession of days and nights would still be marked after Judgment Day. And that presupposed a new heaven and a new earth without which such a succession would be meaningless. Thus John in Revelation was communicating an important fact about the future history of the physical universe which later became part of one version of Newton's cycloid cosmological theory.[24]

Let us quote Newton more fully in this regard. He is almost lyrical when he intones:

A new heaven & new earth, New Jerusalem comes down from heaven prepared as a Bride adorned for her husband. The marriage supper. God dwells with men wipes away all tears from their eyes, gives them of ye fountain of living water & creates all things new saying. It is done. The glory & felicity of the New Jerusalem is represented by a building of Gold & Gems enlightened by the glory of God & ye Lamb & watered by ye river of Paradise on ye banks of wch grows the tree of life. Into this city the kings of the earth do bring their glory & that of the nations & the saints reign for ever & ever.[25]

As a youth, Newton believed in the succession of worlds. He based his belief on arguments from theology. The second book of Peter 3:8 reads: "Nevertheless we, according to his promise, look for new heavens and a new earth, wherein dwelleth righteousness." Newton took this prophecy in a literal, physical sense; he had checked the text in the Vulgate, "*in which text,*" he wrote, "*an emphasis upon the word WE is not countenanced by the original.*"[26]

Revelation 21:1 reads: "And I saw a new heaven and a new earth: for the first heaven and the first earth were passed away; and there was no more sea." Isaiah 65:17 reads: "For, behold, I create new heavens and a new earth: for the former shall not be remembered, nor come into mind," and Isaiah 66:22 added, "For as the new heavens and the new earth, which I will make, shall remain before me, saith the Lord, so shall your seed and your name remain."

Newton found evidence for his belief in Revelation 20:10: "And the devil that deceived them was cast into the lake of fire and brimstone, where the beast and false prophet are, and shall be tormented day and night for ever and ever." Newton jotted down the phrase "Days and nights after the Judgment."[27] The days and nights during which the beast and the false prophets were being tormented must have taken place on a world, since only worlds have day and night. And, since the old world had been destroyed, there must be a new world upon which this tormenting could take place.

So, we must not despair. Even if we take seriously Newton's assertion that the Apocalypse will come in 2160—an assertion that he doesn't seem to have taken entirely seriously himself—we don't have to face annihilation but merely Armageddon and, afterward (if we are among the chosen few), a strange and silent thousand years of peace. After that, we will transition to a new world; that is, if we're among the chosen few.

In chapter 10, we'll discuss Noah, whose existence—and whose prowess as the patriarch captain of the ark—were articles of faith for Newton and his friends. In that chapter, we'll become acquainted (to some degree, through Newton's own writings) with the concept of the "remnant," the fortunate few who are charged with the custodianship of the essence of mankind's knowledge—in fact mankind's very soul—during certain times of transition, both metaphorically and really, from one world to the next. (Isaac Newton thought he was a member of that remnant.)

Before that, though—in the next chapter—we'll look at another set of prophecies and Newton's interpretation of them: those prophecies in the Book of Daniel bearing on the return of the Jews to the Holy Land.

CHAPTER NINE

THE CONVERSION OF THE JEWS

Had we but world enough, and time,
This coyness, Lady, were no crime
. . . I would
Love you ten years before the Flood,
And you should, if you please, refuse
Till the conversion of the Jews.

ANDREW MARVELL,
To His Coy Mistress (CIRCA 1650)

"The air over Jerusalem is saturated with prayers and dreams," wrote the poet Yehuda Amichai,[1] and these yearnings of the human soul are nowhere more vibrant than above the Temple Mount in East Jerusalem. Here the three great monotheistic religions of the world—Islam, Judaism, and Christianity—stand in uneasy confrontation with each other. On one side, the Dome of the Rock shelters a huge, rough-hewn piece of limestone marking the site where the temples of Solomon and Jerusalem once stood. From its craggy surface, Muhammad soared skyward on his Night Journey to Mecca, Jacob saw a ladder stretching to heaven, and Abraham came near to sacrificing his son to God. Jesus delivered his last sermon from the temple's sun-baked outer court. The Ark of the Covenant, missing since Babylonia destroyed the Temple of

Solomon, lies hidden in a cave beneath the dome. Such are the best-known legends!

Christianity's rule over Jerusalem ended in 638 when the Muslim armies of Caliph Omar Umayyad overran the Holy City.* In 691 the caliph's son, also Omar, constructed an edifice that, elaborated over the decades and intended as an "expression of the superiority of Islam," became the octagonal golden-domed shrine that is the Dome of the Rock. In 715, the Muslims extended their architectural power to the other side of the Mount, where Caliph al-Walid built the Al-Aqsa Mosque, still considered to be the most beautiful mosque in the world. All this transformed the Temple Mount into the holiest site in the world for Jews and Christians and the third holiest site (after Medina and Mecca) for Muslims.

In 1099, after a bloody siege and wholesale slaughter that left invaders and defenders up to their ankles in blood, the Crusaders crushed Jerusalem. Al-Aqsa became the palace of the Crusader kings, then, in 1118, the headquarters of the Knights Templar. In 1187, the Kurdish Muslim sultan Saladin drove the Crusaders out of Jerusalem as ruthlessly as the Crusaders had stormed their way in a century before.

For the next 730 years, with almost no exceptions, only Muslims were permitted to tour the Temple Mount and worship at the Dome of the Rock and Al-Aqsa. That policy didn't change when the Ottoman Turks seized control of Jerusalem in 1516. Even the most eminent non-Muslim foreigners were forbidden entry to the Mount. In 1832, when the celebrated poet Alphonse de Lamartine (regarded as the William Wordsworth of France and destined to become the president of France) arrived in Jerusalem and asked if he could visit the dome and Al-Aqsa, the Turkish governor replied with exquisite backhandedness, saying:

---

*There were two exceptions. In AD 361–363, the Roman emperor Julian the Apostate, bent on eradicating Christianity, allowed the Jews to begin rebuilding the temple. This ended when Julian was assassinated by one of his soldiers, who was probably a Christian. In 614, the Persians captured Jerusalem and handed control of the city to their allies the Jews. The Persians revoked the agreement in 617.

> If you require it, all shall be open to you, but I shall expose myself to the risk of grievously irritating the Mussulmans of the city; they are still ignorant, and believe that the presence of a Christian within the precincts of the mosque would be perilous to them, because a prophecy says, that whatever a Christian may ask of God in the interior of Al-Aqsa he shall obtain; and they have no doubt that the petition of a Christian to his God would be the extermination of Mussulmans and the ruin of their religion.[2]

The Turks remained masters of Palestine until 1918 when, just before the end of World War I, they were defeated by the Allies. The Promised Land was now a British protectorate. But Britain, seeing that the Muslims were as hostile toward the Jews and Christians as ever, decided not to lift the ban on non-Muslim visits to the Mount.

Astonishlngly, the state of Israel was created in 1948. But East Jerusalem was still part of Jordan and the sacred sites on the Mount remained closed to the non-Muslim world. Then, in 1967, came the lightning bolt of the Six Days' War, when Israel crushed four Arab armies and took control of East Jerusalem. Israeli defense minister Moshe Dayan, fearing a clash of religions and the rekindling of the war, left the administration of the Mount to the Palestinians. Christians and Jews could tour the Temple Mount, but they were not permitted to worship in the dome or in the mosque.

The Muslims increasingly refused to admit there had ever been a Judeo-Christian presence on the Temple Mount, with some declaring that a Jewish temple may have existed, but not in Jerusalem. Yasser Arafat and other leading Palestinians took up the refrain, insisting that "I will not allow it to be written of me that I have . . . confirmed the existence of the so-called Temple beneath the Mount."[3]

Incessant quarreling over spiritual precedence on the Mount lit the fire of the first intifada, or first Palestinian uprising; it burned from 1987 to 1991. On September 28, 2000, a visit to the Mount by the then opposition leader Ariel Sharon sparked off the second intifada, which raged until 2005; a thousand people were killed in 2001 alone. Ever since, even as Israel has fought bloody wars with Hamas, bitter debate over who has precedence on the Temple Mount has not ended. In September 2015, TV viewers around the world were greeted by the sight of stone-throwing Palestinian youths being chased around and between the gorgeous pillars within the Al-Aqsa Mosque by tear-gas-hurling Israeli police and howling attack dogs. As of this writing, sporadic street warfare still flares up between Palestinian youth and the Israeli police, while Palestinian authorities insist with mounting fury that the Jewish presence on the Temple Mount is just another Israeli settlement that must be eradicated.

Such are the political forces that were unleashed in 1967 when the Temple Mount was liberated by the Israelis.

Mystical forces were unleashed as well. Millennia ago, when the specter of Jewish dispersal from the Holy Land was becoming a reality, the Hebrew prophets declaimed that one day God would call the Jews back to Jerusalem, that they would rebuild the temple and the Messiah would come, and that in the Holy War that followed evil would be vanquished. Over the centuries, the Christians added their own gloss to these prophecies, declaring that when the Jews arrived back in Israel they would convert to Christianity, and Christ, not the Messiah, would arrive and that this would trigger the Holy War of Armageddon and then one thousand years of peace and a final conflict that would end with the triumph of good and the arrival of the New Jerusalem.

The outcome of the Six Days' War thrilled fundamentalist Christians around the world, the creation of the state of Israel was as wildly wonderful as it was wholly unexpected; suddenly it seemed possible that the Apocalypse scenario prophesied for Jerusalem could begin to unfold. And now the Israelis had liberated the sacred plateau in East Jerusalem! Fundamentalist Christian leaders all over the world leaped to their podiums, particularly in America. In the 1970 global bestseller *The Late Great Planet Earth,* preacher Hal Lindsey reminded his readers that the creation of the state of Israel in 1948 had been the first manifestation of that return of the Jews which, as predicted in scripture, would pave the way to the Last Days of the world. But the capture of the Temple Mount in 1967 had broken everything wide open and made possible the complete fulfillment of scripture: now the temple would be rebuilt, Christ would come a second time, and the final battle with the Antichrist would be engaged. The Middle East, Lindsey declared, was "the Fuse of Armageddon!"[4]

Of course it was true, he added, that the presence of the Islamic dome and the Al-Aqsa Mosque on the Mount presented an obstacle, but that would be overcome, because prophecy demanded that the temple be raised again.

It has been half a century now since the Mount was liberated. The sleeping giant of the Apocalypse has been awakened, but his power of movement is severely impeded by the sectarian storms that rage around the Mount, and when he strains against these forces he only produces dreamlike micro-parodies of what was supposed to be. Clandestine cells of radicalized Jews, Christians, and Arabs plot how to storm the Mount, take the temple, and provoke the Apocalypse all by themselves. They roam the narrow streets of East Jerusalem like restless anarchists from a Dostoevsky novel, protesting, preaching, and harassing the authorities. Sometimes Christians and Jews cooperate (the Christians try to convert the Jews and the Jews want American money); other times they squabble. Rumor has it that they sneak into the caves beneath the dome at night to search for the Ark of the Covenant, which they will use as a shield against impiety on the day they storm the temple; or that they are raising rare red heifers whose blood, mixed with ashes, will protect them when they invade the sacred precincts of the Mount. Most alarming of all are the cells of radicalized fundamentalist Muslims, who—and again this is largely rumor—are plotting to capture the Temple Mount and, when their own messiah, the Mahdi, comes, use it as a base of operations from which to launch a jihad against all the Jews in the world. As 2017 dawned, the Temple Mount fought for its soul amid the storm clouds scudding across Middle Eastern skies.

Four hundred years ago in England, Isaac Newton heard the distant keening of the Temple Mount yearning for fulfillment. He heard it not directly but through the prophetic books of Daniel and Revelation and in the impassioned prophecies of Hebrew prophets like Ezekiel and Isaiah. The Apocalypse scenario of the great millenarian thinkers of the seventeenth century was not unlike (though far more learned!) than that of Hal Lindsey in *The Late Great Planet Earth*—with one extraordinary addition. Not only did Newton think the process couldn't begin until God called the Jews back to Israel from the four corners of the Earth to which they had been scattered, which could happen only when the Jews had settled in every single country in the world. He also thought it wouldn't happen until all the Jews, once back in Israel, or shortly before, had been converted to Christianity en masse. Then the liberation of the Temple Mount would strip away "the mystery of the restoration of all things that the Prophets had spoken about," and he wrote of "the final return of the Jews [from] captivity & their conquering the nations of the four Monarchies & setting up a peaceable righteous & flourishing Kingdom at the day of judgment."[5]

The notion that the Jews will convert en masse to Christianity seems quaint and old-fashioned to us today, and so it appeared to some of Newton's contemporaries. In the poem *To His Coy Mistress*, which begins this chapter, Andrew Marvell (1621–1678) uses the phrase "the conversion of the Jews" to indicate an event that will take place very far in the future if it takes place at all—in other words, an extremely improbable event.

If the idea of the mass conversion of the Jews seems strange to us today, many find it even stranger that so many in Newton's time ardently desired the end of the world even if that meant the end of themselves. Professor Diarmaid MacCulloch explains:

> It is not surprising that so many have sought the Last Days. The writing and telling of history is bedeviled by two human neuroses:

horror at the desperate shapelessness and seeming lack of pattern in events, and regret for a lost golden age, a moment of happiness when all was well. Put these together and you have an urge to create elaborate patterns to make sense of things and to create a situation where the golden age is just waiting to spring to life again.[6]

But the notion that the Jews would convert en masse to Christianity had somehow become woven into that pattern, and the religion scholars of Newton's time not only wrung every last drop of meaning out of biblical prophecy, but they also resorted to every last strategy they could think of to find out when God would, or if he already had, summon the Jews back to Israel.

William Oughted, vicar of Albury (1574–1660) and the probable inventor of the slide rule, used a "calculation coincidence with the diluvian period" to come up with the prediction that an extraordinary event, which he thought was the conversion of the Jews, would take place in 1658.[7] The eminent prelate-scholar Johann Heinrich Alsted (1588–1638), notorious for the sheer weirdness of his learning, examined with a fine-tooth comb the personal hieroglyphs of Giordano Bruno and immersed himself in the musical triangles of Raymond Lull, and did a great deal more, until, adding it all together, he decided that 1694 was the date of the conversion of the Jews and the beginning of "the divinely ordained End Times."[8] In 1715, chronologist Arthur Bedford decided that the demise of Cromwell's Republic and the restoration of church music (the Commonwealth having banned the latter along with Christmas, maypole dancing, and the theater) "had a providential purpose and would play a role in the conversion of the Jews to Christianity."[9]

But the greatest angler in the murky waters of biblical prophecy about the Jews was Sir Isaac Newton—perhaps. Oliver Cromwell (1599–1658), lord protector of the Commonwealth, came a close second. This wasn't because Cromwell was a scholar of biblical prophecy and a seeker after End Times dates himself, though he was. It was because he had acted politically in a way that may actually have brought the time of the return of the Jews a little closer—or at least that's how Cromwell's contemporaries saw it.

Here's what happened: In the summer of 1656, the mystically minded Rabbi Mahasseh ben Israel (1604–1657), an internationally acclaimed scholar of religion and since age eighteen a leader of the Jews in Amsterdam, paid a visit to Cromwell in London. There were no Jews in England at the time. They had been expelled en masse by Edward I in 1290. Mahasseh wanted to annul the expulsion, and for that reason he had asked for a series of meetings with the lord protector.

Cromwell agreed to the meeting because the Commonwealth needed money, and he knew how rich the Amsterdam Jews were and how good they were at making money for others. He was prepared to listen to Mahasseh.

The rabbi arrived. The meetings began. They discussed money. Then Mahasseh startled Cromwell by giving him another, wholly original, reason for letting the Jews back into England. He quoted Deuteronomy 28:64 and Daniel 12:7. Both passages stated that Judgment Day would not come until the scattering of the Jews was complete "from one end of the earth even unto the other"; only then would God call the Jews back to the Holy Land to set in motion the events that would lead to the End Times. Mahasseh knew the gentiles desired this; but if there were no Jews in England then the Jews had not settled in every country in the world, and God could not call them back to Israel.[10]

Cromwell listened. This argument impressed him. What impressed him more was the wealth of the Jews and the financial problems of his government. (It's likely the payment of a large sum of money to Cromwell from the Jewish bankers of Amsterdam had much to do with making these meetings possible.) Cromwell agreed to let the Jews back into England.

But Parliament wouldn't agree. Then, after some investigations, it turned out that Edward I had never really passed a law banning the

Jews from England. Cromwell decided to act as if Rabbi Mahasseh ben Israel's request had actually been granted by his government. The Jews began, slowly and inconspicuously, to slip into England. A cemetery and a synagogue were established in London. A Naturalization Bill for Jews was passed in 1753. In 1868, Benjamin Disraeli, a Jew, was elected prime minister of England.

At no time during this period did the Jews suddenly begin to return to the Holy Land; for whatever reasons, God held back his call. But Newton and his colleagues never ceased to search for signs that the recall of the Jews was under way.

Newton held the Jews in the highest esteem. He believed they really were God's chosen people, "unique among the nations of the world and the special recipients of God's grace."[11] Newton's library contained more than two hundred books related to Judaism, including, for starters, a Hebrew Torah, the Kabbalah, and multiple works by Maimonides. As has been mentioned earlier, the great ancient Semitic-languages scholar Abraham Yahuda, scouring Newton's papers on theology, detected in Newton a distinctively Jewish cast of character. He saw the mathematician as a "Judaic monotheist of the school of Maimonides,"[12] a man whose writings breathed forth the implicit belief that "Jehovah is the unique God."[13] Yahuda read Newton's theological writings in 1936 when Hitler was stripping German Jews of their citizenship and forbidding them to marry non-Jews. He confessed to his wife that "in these times of crisis and ordeal he [Newton] exercises a calming and reassuring influence upon me." Scholar Sarah Dry explains that Yahuda saw "the redemptive potential of Newton's papers for the Jews, who could benefit from Newton's sympathy with their faith at a particularly vulnerable moment. . . . [She says that, for Yahuda, the papers] contained truths that could survive 'destructions and isolations' of the sort hinted at by Daniel."[14]

Newton believed Abraham's covenant with God (Gen. 12:1–3) was a mark of the special favor the Creator had bestowed on the Jewish people. God had promised them that Canaan would be theirs forever

(Gen. 13:14–17)—numerous biblical passages attested to this*—and Newton believed a promise from God was the equivalent of divine law and therefore eternal and immutable.

Some objected that the Jews had already returned once, from the Babylonian Captivity, and that therefore God's prophecy had been fulfilled; there could be no second return. Newton replied that God had promised the Jews they would return forever but that the return from Babylonia had not been forever; the Diaspora, the first signs of which could be seen at the time of the Babylonian Captivity, would expand into an immense Earth-spanning exodus once the Jews were defeated by the Romans in AD 70.

Newton pointed out that God had promised the Jews they would return to *all* of their lands, but that that, too, hadn't happened when they returned from Babylon. "Israel possessed neither the land of the Edomites nor that of the other nations at the time of their return," clarifies Steven Snobelen.[15]

If the Jews were God's chosen people, unique among nations, then why had he allowed them to be scattered to the winds? Not because they bore any responsibility for the crucifixion and death of Christ, insisted Newton. The Jews had brought the Diaspora on themselves by failing to "understand prophecy"; they had refused to recognize that Jesus was the prophesied Messiah. They had instead put their faith in Daniel's Prophecy of the Seventy Weeks, which they thought provided them with divine assurance that their Messiah would arrive at the right time. Tragically, this apparent assurance had emboldened them to go to war with the Romans.†

---

*The rebuilding of Jerusalem and the waste places was predicted in Micah 7:11, Amos 9:11, 14; Ezekiel 20, 26, 33, 35, 36, 38; and Isaiah 54:3, 11, 12; 55:12; 61:4; 65:18, 21, 22. There are allusions to the return from captivity of the Jews, and the coming of the Messiah and his kingdom, in Daniel 7; Revelation 19; Malachi 24; Joel 3; Ezekiel 36, 37; Isaiah, 60, 62, 63, 65, 66, and many other places of scripture. (See Snobelen, "Mystery of the Restitution," 101.)

†Gustave Flaubert provides us with a succinct description of the Jewish Messiah in his story "Herodias": "The Jews gave the name of Messiah to a liberator who would give them

Here's how Newton puts the Jewish-Roman problem: "This Prophecy [of the Seventy Weeks] which had for some time put the Oriental nations in continual expectation of a temporal Potentate out of Judaea, and which the Jews understood of the Messiah with that confidence of temporal domination as to rebel against the Romans and begin that war which caused their ruin."[16]

When the Jews failed to recognize that Jesus was their Messiah, "God began to reject them from being his people, & to call the Gentiles without obliging them to observe the law of Moses, & soon after caused the Jewish worship to cease & the Jews to be dispersed into all nations . . . as at this day so that at present they are no body politique or people but a scattered servile race of men without any government of their own."[17]

This was not a patronizing statement. Newton had nothing but praise for how well the Jews, despite not having created an independent Jewish political entity anywhere in the world, had maintained their identity through 1,700 years of exile. He marveled at how they "in a wonderful manner continue numerous & distinct from all other nations: which cannot be said of any other captivated [captive] nation whatever, & therefore is the work of providence."[18]

Newton regarded Judaism as the purest religion. The Jews had never tried to describe their God; they had never set up graven images of Him; they had never been idolatrous. It may be that Newton regarded the enclaves of Jews buried away in the far-flung metropolises of the world as shining beacons of monotheistic light, dispelling the blackness of Trinitarianism with the purity of their beams. In the context of his history of the corruption of the soul of man, they were steadfast flashes of light gleaming out of the gathering darkness of the Great Apostasy.

(cont. from p. 181) possession of all goods and dominion over all peoples. Some even maintained that two Messiahs were to be looked for. The first would be vanquished by Gog and Magog, demons of the north, but the other would exterminate the Prince of Evil, and for centuries they had been expecting him to arrive any minute. . . . The Messiah would be a son of David and not of a carpenter. Then, too, he would uphold the Law, whereas this Nazarene attacked it." (Flaubert, "Herodias," 115.)

But the fact was that the Jews had erred grievously in failing to understand biblical prophecy that told them that Jesus was their Messiah, and they had suffered accordingly. And now we find ourselves in the presence of some of Newton's most mysterious pronouncements. They are pronouncements that frighten us a little. The great mathematician put the greatest store by the proper interpretation of prophecy. He believed that, since the prophets always spoke the word of God, to interpret prophecy correctly was "no matter of indifferency but a duty of the greatest moment."[19] Moreover, if the Christians failed to interpret prophecy correctly, how much harsher must their punishment be if they as Christians, "who have had ample instruction in how properly to understand Scripture, are not able to recognize the signs of the Antichrist and the true apocalyptic future?"[20]

As Newton puts it:

> For certainly it must be as dangerous and as easy an error for Christians to adhere to Antichrist as it was for the Jews to reject Christ. And therefore it is as much our duty to endeavor to be able to know him that we may avoid him, as it was theirs to know Christ that they might follow him.... Therefore beware that thou be not found wanting in this trial. For if thou beest, the authority of these scriptures will as little excuse thee as the obscurity of our Saviour's Parables excused the Jews.[21]

Only the pure of heart can interpret prophecy correctly, declares Newton. And they will be mocked if they do, because, Newton says enigmatically, "It is the wisdom of God that his Church should appear despicable to the world." Moreover, those who speak truth about prophecy will have to endure the "reproaches of the world,"[22] and, the greater the truth, the greater the reproach. As we near the End Times, says Newton, we will experience a "quickening" of perception; this is a statement that could easily come from the lips of a twenty-first-century

New Age aficionado. Newton concludes that we will understand prophecy "still better at the return of the Jews from their long captivity predicted by Moses & the Prophets."[23]

These are among Newton's severest warnings to mankind. But what does he mean? What are these biblical prophecies that we must not fail to understand? Everything seems to be bound up with the time of the return of the Jews to the Holy Land. Newton scoured the Book of Daniel for answers to these questions, but he never really tells us what he discovered. Instead, he gives us a commencement date and a specific prophetic time span.

This brings us back to the question of the prophetic numbers. Where exactly did they come from? It was essential to make proper use of these mysterious numbers, and in chapter 8 we saw Mede and Newton working with the prophetic formula "a time, time-and-a-half, and two times" (equals 1,260 days equals 1,260 years). We'll recall Newton's assertion that "the ancient sages had more sorts of days weeks months & years than one. They had their vulgar years & their great years, vulgar months & great months, weeks of days & weeks of years, natural days & days mystical. And therefore where the things prophesied of are mystical it's proper to understand their times in a mystical sense."[24]

We will find new mystical time spans in Daniel: "1,290 days," "1,335 days," "2,300 days." It was obvious to Newton that these days were years: "Tis altogether impossible that within the 1290 & 1335 days in Daniel 12, which extend from the setting up the abomination Daniel 11.31 to the end of the prophesy & to the resurrection of the dead, [that so many historic events could have been packed in] unless those days be mystical [i.e., years and not days]."[25]

The mystical formulae of "a time, time-and-a-half, and two times" and its variants seem to be as much constants of the world of biblical prophecy as Einstein's relativity equation, $E = mc^2$, or Wolfgang Pauli's fine structure constant, 137, are constants of the physical world. But, again, where do these mystical numbers come from? For starters: How did the Hebrew prophets get in the habit of calling a prophesied year a day?

In *New Theory of the Earth* (1696), Newton acolyte William Whiston traces the day-equals-year formula back to the first days of our planet. Whiston is trying to reconcile the findings of astronomy with the Book of Genesis's assertion that the world was created in seven days. He decides that when the world was created a day really did equal a year, because the earth hadn't yet begun to rotate. Because, according to Whiston's pre-Darwinian worldview, mankind was created in a state of perfection on on the sixth day, then we were aware that our planet did not rotate. This we remembered, and in some mysterious way it became, when our planet was spinning at full speed, the mystical formula of a day's equaling a year.[26] This notion appears in various expressions in some of the writings of ancient poets and thinkers. Empedocles (ca. 490–430 BC), for one, wrote, "When Mankind Sprang originally from the Earth, the Length of the Day, by reason of the slowness of the Sun's [perceived] Motion [the Earth's motion], was equal to ten of our present Months."[27]

Did Pythagorean number theory play any role in the formulation of these mystical numbers? We first encounter "time, time-and-a-half, and two times" in the visions of the prophet Ezekiel. For centuries the rumor ran through the Middle East that Ezekiel, using the pseudonym of "Nazaratus Assyrius," was a teacher of Pythagoras. This rumor persisted until it was determined, not long ago, that Ezekiel lived in 622–570 BC and Pythagoras from 582–507 BC; the life spans of the two overlap by only twelve years. But finally no one has been able to establish a connection between formulae such as "time-and-a-half and two times" equals 1,260 years and revolutionary sixth-century BC Pythagorean theories that involve triangles, magic numbers, irrational numbers, vibrating strings, and the like.*

---

*The legend of a relationship between Ezekiel and Pythagoras is possibly a modern one and was fueled by our new awareness of what we call the Axial Period—the time, mostly confined to the sixth century BC, when an axis of supreme genius zigzagged across the known world. The Orphics, Zarathustra, Gautama, the authors of the Upanishads,

▲

So how did Newton use the mystical numbers in the Book of Daniel? What conclusions did they enable him to reach about the return of the Jews to the Holy Land?

Scholar David Castillejo tells that

> the political events that he [Newton] will be interpreting sweep right over Newton's own head and into our future; but he himself never mentions a date beyond his own life-time, citing the biblical warning against false interpretations. He merely devotes his attention to fixing the start of a prophesied period. Thus he will discuss in great detail when the reign of the Whore of Babylon began, but he will not mention its end. So the reader is left to perform the second half of the operation by [e.g.,] adding 1,260 years onto the opening date.[28]

Let's see how Newton deals with Daniel's prophecy (12:11): "And from the time that the continual burnt offering is taken away, and the abomination that makes desolate is set up, there shall be a thousand two hundred and ninety days." Most of Newton's contemporaries thought "the abomination that makes desolate" took place in the year

---

(cont. from p. 185) Jeremiah, and Ezekiel—all were all up and about at roughly the same moment in time. We don't really know what caused this sudden efflorescence, right across the map, of titanic human genius; perhaps the human mind, evolving through many millennia, had reached a kind of critical mass and took a sudden leap forward. Scholar Steven Sittenreich writes: "Sometimes these towering waves of genius clashed. But the extraordinary achievements of Pythagoras, though partly derived from his travels in the Middle East, weren't necessarily anything with which a Hebrew prophet could connect. Nowhere in the Book of Ezekiel or anywhere in the whole range of the Hebrew/Aramaic Bible is there the slightest touch of Pythagoreanism. The Greek and the Hebrew worldviews were two contrasting modes of vision, and the Pythagorean Weltanshauung found its way into Judeo-Christian literature only by way of the Stoics, who posited a Logos as Mediator between a distant Creator and mankind; this doctrine shines forth here and there in the Gospel of Saint John." (Steven Sittenreich, personal communication with the author, April 2014.)

168 BC, when the Hellenistic king of Syria and also ruler of Judaea Antiochus IV Epiphanes (215–164 BC) forced the Jews to sacrifice swine on the altar of the Temple of Jerusalem. Antiochus also tried to destroy all copies of the Torah and suppress the Jewish festivals, and he murdered the legitimate high priest, Onias III (2 Mac. 4:30–38; Dan. 9:26), selling the priesthood to Onias's brother Jason (2 Mac. 4:8).

Newton, on the other hand, thought Daniel's "abomination that makes desolate" was a reference to the year AD 609 when Pope Boniface IV signed a decree granting every church in Christendom the right to set and worship images of Christ.

This was idolatry of the highest order, and Newton despised it. So, unique among his colleagues, he advises the reader to add the prophetic time span of 1,290 days, or years, to AD 609. If we do so, he says, we will arrive at the year when "God shall have accomplished to scatter the power of his holy people"—that is, the year when the Jews will finally have been dispersed to every country in the world—"and He will then summon His Chosen People back to the Promised Land."[29]

When we add 609 to 1,290 we get 1899. What could possibly have happened in that year that has any bearing on the return of the Jews to the Promised Land?

Theodor Herzl (1860–1904) was a handsome and debonair Hungarian Jew who up until 1896 had no interest in Judaism. He made his living writing sketches, criticism, and lightweight frothy plays and spent his nights hobnobbing with the wealthy gentiles of Budapest high society.

In 1896, Herzl suddenly became a new man. He was purposeful and powerful. He published a book, *Der Judenstaat* (*The Jewish State*), that was dedicated to "turning the messianic currents of longing for a return to Zion into a political force."[30] Zion is the Hebrew name for the Temple Mount; Herzl was advocating that a sustained political effort be made to create a "national home," eventually a state, for the Jewish people in Palestine.

Herzl's book was the talk of the town, but it was disliked by

European Jewry who feared that the precarious position they now held in gentile society would be endangered if they rallied to the author's call. But Herzl pushed on forcefully; he was charismatic to the highest degree, "cast," writes one biographer, "in the heroic mold, with a prophetic face, an aristocratic poise, and a manner which enthralled his followers and impressed his intellectual opponents."[31] In the face of intense opposition, the fervent young crusader convened the first Zionist Congress in Basel, Switzerland, in 1897.

The congress was well attended, but few practical proposals and less money were forthcoming. Still, this first Zionist Congress was the first step on an extremely rocky road, beset by every sort of pitfall, that would lead to the foundation of the state of Israel a half century later. Enthusiasts of biblical prophecy who take Isaac Newton's formulations seriously believe that 1,290 years added to AD 609 and equaling 1899 refers to the First Zionist Congress; they brush off the fact that the congress took place in 1897, accounting for the discrepancy by citing a difference between solar years and lunar years.

Newton does better, though, when he tackles Daniel 9:25, which reads, "From the going forth of the commandment to restore and to build Jerusalem unto the Messiah the Prince *shall* be seven weeks." Once again he uses a commencement date different from that of most of his colleagues, who believed the prophetic time span of seven weeks (equals 49 days, which equals 49 years) eventuated in the year Christ was born. Newton, however, thought the culminating year was that of the Second Coming of Christ. He counsels us to add the 49 years to 1,290 years added to AD 609.

This takes us to the year 1948.

Nineteen forty-eight was hardly the year of the Second Coming of Christ, who hasn't arrived yet. But it is the year when, on May 14, due to an extraordinary confluence of political events almost as rare as the alignment of five planets, the United Nations voted to establish the state of Israel in Palestine. The Jews, after wandering the world for two millennia, had found a home again.

And, if 1948 wasn't the year when Christ came a second time, the foundation of the state of Israel was the one indispensable step toward making that happen. For Christ could only return when the temple was rebuilt, and that could only happen once Jerusalem truly belonged to the Jews again.

Newton added a second observation to his interpretation of Daniel 9:25, and this one, when we think about it, truly gives us pause. Newton writes: "Since the commandment to return and to build Jerusalem, precedes the Messiah the Prince 49 years, it may perhaps come forth not from the Jews themselves, but from some other kingdom friendly to them, and precede their return from captivity, and give occasion to it."[32] Newton doesn't identify this "other kingdom friendly to them." What country might it be?

Other Zionist congresses followed Herzl's first congress. In between, the Hungarian leader spent nearly all his time personally lobbying the leaders of countries in Europe and the Middle East; he was looking for a tract of land on which the Zionists could settle. In 1903, Britain's colonial secretary Joseph Chamberlain offered Herzl a huge chunk of land in Uganda, British East Africa. "It is hot on the coast," the colonial secretary enthused, "but as you travel inland, the climate improves and is splendid even for Europeans. You can raise sugar and coffee there."[33]

Herzl was grateful but not enthused. He reluctantly put this offer before the Sixth Zionist Congress of 1903. It caused so much bitter dissension that it had to be quietly dropped.

On November 3, 1917, the British made a far more substantial offer. World War I was almost over; Britain knew it would soon control Palestine. The British Foreign Office sent a letter to the Zionists that began, "His Majesty's government view with favor the establishment in Palestine of a national home for the Jewish people, and will use their best endeavors to facilitate the achievement of this object."[34]

This was the Balfour Declaration, on the basis of which 150,000 Jews were able to immigrate to Palestine between 1918 and 1936. The British reneged on the declaration in 1939 to mollify the

Arabs, who wanted to bar them from access to their oil fields in the Middle East. But, by this time, a strong enough Jewish presence existed in Palestine that the Zionist dream was able to stay alive until 1948.

There's no doubt Britain played an indispensable role in the establishment of Israel. Was it the "other kingdom friendly to them" of which Newton speaks?

Newton also gave much thought to Daniel 12:1: "Blessed is he who waits and comes to the thousand three hundred and thirty-five years." He suggested that we add the prophetic time span of 1,335 days/years to the commencement date of AD 609.

When we do, we arrive at the year 1944.

Whatever did 1944 have to do with the return of the Jews to the Holy Land?

Newton scholar David Castillejo has an answer.[35] He points out that June 6, 1944—D-Day—was Day One of the liberation of Europe. That was the day when 176,000 Allied troops, ferried across the English Channel in an armada of 6,000 landing craft, ships, and other vessels, stormed ashore at a dozen French coastal towns and launched a massive counteroffensive against Nazi Germany. Months of desperate fighting followed. The Allied armies ground through France, Belgium, and the Netherlands. Finally, they plunged into Germany.

Here the Allies encountered Hitler's concentration camps. This was the first the world learned of Hitler's savage plan to annihilate world Jewry, a plan that had already seen the deaths of six million Jews. The Nazis were defeated in the summer of 1945. Revulsion at the Holocaust mounted. A feeling developed that the world must somehow compensate the Jews. The worldwide Zionist leadership, hesitant during the war, committed itself to the establishment of a Jewish state in Palestine. The refoundation of a state of Israel, inconceivable for two millennia, wouldn't have happened if a colossal cross-Channel expedition, launched in 1944, hadn't liberated the captive countries of Europe and exposed the full extent of Hitler's crimes against the Jews.

▲

Newton frequently changed his mind about when the return of the Jews might occur. Still using the Book of Daniel, he pointed out the path of some dates that lie very far in our future.

In Daniel 8:13, a saint asks, "How long shall be the vision concerning the daily sacrifice, and the transgression of desolation, to give both the sanctuary and the host to be trodden under foot? In Daniel 8:14, the saint answers, "Unto two thousand and three hundred days; then shall the sanctuary be cleansed."

Newton decided that this 2,300-day (or 2,300-year) prophetic time span was the time span of the Jewish Diaspora. But he had difficulty settling on a commencement date, because he had to decide among the year the Romans destroyed the Temple of Jerusalem (AD 70), the year when they built their own temple to Jupiter Olympus on the burned-out site of the Temple of Jerusalem (AD 132), and the year of the third Jewish-Roman War—the Bar Kochba revolt of AD 135–36—which ended in the bloody defeat of the Jews. Adding the prophetic time span of 2,300 years to these commencement years gives us dates for the "cleansing of the sanctuary," said Newton—the time when the Jews return to Jerusalem and the End Times begin—AD 2,300; AD 2,430; and AD 2,433–34, respectively.[36]

That's 354 to 420 years in the future, some twelve generations from now—far enough away to make even the most eager of biblical prophecy enthusiasts lose a little interest!

Newton sometimes seemed to feel that a period of prosperous governance in Palestine would take place between the time of the return of the Jews and the advent of the End Times—that the Jews, once returned, would make Palestine a wealthy state and a powerful military force for a period of time before the Apocalypse. Newton's source was Ezekiel, chapters 38 and 39, where, Newton wrote:

> He [Ezekiel] represents how the Jews after their return from captivity dwell safely and quietly upon the mountains of Israel in unwalled towns without either gates or bars to defend them until they are

grown very rich in Cattle and gold and silver and goods and Gog of the land of Magog stirs up the nations round about, Persia and Arabia and Africa and the northern nations of Asia and Europe against them to take a spoil, and God destroys all that great army, that the nations may from thenceforth know that the Jews went formerly into captivity for their sins but now since their return are become invincible by their holiness.[37]

Some see in Ezekiel 38 and 39 a foretelling of the Israeli War of 1948 or the Six Days' War of 1967. Newton himself made only one concrete prediction about a restored Jewish state—one that sounds curiously like a half quatrain from Nostradamus. The future nation of the Jews would not come about, predicted Newton, "Til Egypt Has a Greek king / And Turkey was no more"[38]

Newton was right about Turkey; the return of the Jews to Israel was out of the question until the British chased the Turks from Palestine in 1918. But the chances that Egypt will ever have a Greek king are so impossibly remote as to push the time of the return of the Jews very, very far indeed into the future.

It is amazing to think that Isaac Newton's public reflections on the fate of the Jewish people as foretold in biblical prophecy actually helped the Jews return to Israel.

This was because those speculations strongly influenced his theological disciples like William Whiston. Newton's strange and provocative interpretations of Daniel merged with a prophetic tradition that, writes Professor Stephen Snobelen, "helped create during the nineteenth and twentieth centuries the religious and political climates that paved the way for the resettlement of Jews in Palestine."[39]

So Newton's intense meditation on the Jews ultimately transformed itself into a contribution to their well-being. Thus Newton becomes, along with so much else, virtually a Zionist and certainly an enabler and abettor of the state of Israel.

CHAPTER TEN

With Noah on the Mountaintop

On July 1, 2004, the hybrid spacecraft *Cassini-Huygens* slipped into orbit around the planet Saturn. During its second orbit, the three-ton probe, launched by NASA in 1997, separated into its two component parts. *Cassini* continued on its way around Saturn while *Huygens* sped off toward Titan, Saturn's largest moon. *Huygens* landed on Titan on January 14, 2005, and radioed data back to Earth by way of *Cassini* for ninety minutes.

The hybrid craft scored a double first. *Cassini* was the first space vehicle to orbit Saturn, and *Huygens* was the first space probe to land on a celestial body beyond the orbit of Mars.

If the shade of Isaac Newton had been speeding through interplanetary space and chanced to come upon Saturn when these events were taking place, he would have had good reason to feel satisfied. He had been the first person to calculate the orbital velocities of Saturn and Titan and work out the interactions of their gravitational fields. He had invented the mathematics and physics that enabled NASA to fling *Cassini-Huygens* across two billion miles of space into orbit around Saturn. He had known personally the two scientists after which *Cassini-Huygens* was named: Jean-Dominique Cassini (1625–1712), the French Italian astronomer who discovered four of Saturn's moons, and Christiaan

Huygens (1629–1695), the Dutch mathematician who discovered Titan and first realized that the bulges on Saturn's sides must be rings.

What would have held Newton's attention most as he hovered watching the antics of *Cassini* and *Huygens*, was the huge ringed world that hung suspended in space before his eyes. This wasn't because of Saturn's great beauty or its scientific potential. It's because Newton would have seen in his mind's eye, as if superimposed upon the yellow-orange disk of the ringed planet, the face of Noah, the patriarch of the ark, who had brought all the species of the world through the Flood and had saved mankind.

The planet Saturn was named for the god Saturn, and Newton was convinced that the god Saturn was derived from the memory of Noah, who with his family had been a sort of template of the gods. And Noah was a personage of consuming interest for Newton, for he had brought with him on the ark priceless knowledge of the antediluvian world. Moreover, Noah was one of those rare beings whom God produces only once every few thousand years: a great leader to lead a remnant of mankind from a dying world to a new.

Newton had given much thought to all this in his lifetime, as we are about to see, and a glimpse of the planet Saturn would have brought it all back.

On almost the same day, July 1, 2004, as *Cassini-Huygens* swept into orbit around Saturn, another expedition, completely different but perhaps equally visionary, was being called off back on Earth. A twelve-man team of American Turkish climbers had assembled at the foot of Mount Ararat to scale the sacred mountain and look for Noah's ark. But Ararat lies in disputed Kurdish territory, and in 2004 tensions ran high between Turks and Kurds. Iran is ten miles away and Iraq, where a vicious civil war was raging, not much farther away than that. The Turkish government, fearing the area was swarming with spies, mercenaries, and terrorists, had called off the expedition at the last moment.

The binational team, sponsored by the Christian Shamrock/Trinity

Corporation of Honolulu, Hawaii, must have been disappointed. The previous summer had been the hottest in two hundred years, and melting snows on the mountaintop had revealed a black rectangular shape. Aerial photos suggested this was the wreckage of a boat the size of an ocean liner. To the expedition's Christian organizers, it sounded like Noah's ark.

Now Christian/Shamrock would never know. Theirs would be the last expedition to prepare to climb Mount Ararat in search of the lost ark. Today, political storm clouds roil even more menacingly around the slopes of the 17,820-foot saddle-backed mountain as Syrian refugees pour into Turkey and a savage terrorist army, ISIS, seeks to establish a caliphate throughout the Middle East. Iraq simmers and boils, and Iran broods in isolation. Turkey's ban against climbing Mount Ararat has become a permanent one.

Perhaps it doesn't matter that there will be no more trips up the mountain. Few today, evangelical Christians excepted, believe in Noah's ark. This was far from the case in Isaac Newton's time. In seventeenth-century Europe, Noah and the ark and the Flood were articles of religious faith. The story of the Flood, important because it told the story of humanity's second fall into sin and scarcity (the expulsion of Adam and Eve from the Garden of Eden was the the first), was the subject of endless speculation. How big was the ark? How did it hold so many animals? How many generations of Noah's family did it take to repopulate the world? Was that even possible? What were the not just moral but physical causes of the Flood?

The greatest minds of the time grappled with these problems. Comet chaser Edmund Halley (1656–1742) speculated that a passing comet might have caused the Earth to wobble on its axis enough to make the oceans slop up over the land for forty days.[1] Jesuit linguist-scholar Athanasius Kircher (1602–1680) used Galileo's studies of floating bodies to try to figure out how the six-hundred-year-old Noah had been able to pack two of every living species into one big box-shaped boat.[2] Swedish scientist Olaus Rudbeck (1630–1702), at age twenty-two the codiscoverer of the lymphatic system, used details from the Bible to calculate how many children were born in the generations between Jacob and Pinehas, then used that information to calculate the population of the world one century after the Flood: 44,288. Rudbeck also reconstructed the circuitous route taken by Magog (hence "Goth"), son of Japhet and grandson of Noah, when he led his progeny from the Middle East to the land that would one day be called Sweden.[3]

All of this held the attention of Isaac Newton, but, more than most of his colleagues, he was fascinated by that other, special, cargo, having nothing to do with animals, that he believed Noah had brought with him on board the ark. This was the intellectual and spiritual riches of the world before the Flood that had been passed along, though suffering much corruption, until it came to Noah. And Newton wanted to know everything he could about this man, Noah, who had been ordained by God to transport such a cargo.

First, though: Was there ever a flood at all? Was there ever a Noah? Was there ever—could there still be—an ark atop Mount Ararat?

The story of Noah and the ark is a part of the spiritual heritage of all three major monotheistic religions of the west, Christianity, Judaism, and Islam.

The Book of Genesis tells us that God, despairing that humankind will ever be cured of its wickedness, decides to drown the human race. He singles out Noah as the one righteous man in the world and tells him to build a huge, box-shaped, pitch-lined ark and fill it with two of all the fauna in the cosmos. Noah does so, and he, his family, and the male and female of every species ride out the catastrophe in their ark for forty days and forty nights. The waters subside; Noah sends out a raven and two doves. The second dove doesn't return; he decides there must be dry land, and soon he makes landfall. Noah and his family disembark; the patriarch builds an altar and makes a burnt offering to the Lord. God blesses Noah, makes a covenant with him, and promises not to punish mankind again. A rainbow appears in the sky. Noah and his family set out on the monumental task whose historical reality was never doubted in Newton's time: the repopulating of the world. This is the story told in Genesis 6:9 to 9:17; there are allusions to the flood in Ezekiel 14:14, 20; Isaiah 24:5, 18; 54:9; Psalm 29:10; Job 22:15; and elsewhere, and in some sections of the Koran.

We've known since the latter part of the nineteenth century (from excavations made at Ashurbanipal's library in Nineva) that the Old Testament story of Noah and the flood was borrowed from a four-thousand-year-old Sumerian epic called *Gilgamesh*. In this poem, the god-hero Gilgamesh, tormented by questions of why we die and whether death can be avoided, travels to the underworld to talk to Utnapishtim, the Sumerian Noah, to whom the gods have given immortality because he rescued mankind from the flood.

Utnapishtim tells Gilgamesh the flood story. The gods decided to drown mankind. One of them, Enlil, warned Utnapishtim and told him to build a huge ark in seven days. Utnapishtim filled it with "the seed of all living creatures"[4] and all the animals on his estate. The flood lasted a week; it took another week for Utnapishtim to make landfall; then humanity began again. The gods reproached themselves for causing the flood and promised never to do it again; then they created wolves and tigers to harass man perpetually!

Josephus says the third-century BC Chaldean priest-historian Berossus, "following the most ancient records, gave an account, like Moses, of the flood and the destruction in it of humankind, and the ark in which Noah, the founder of our race, was saved when it was carried onto the peaks of the Armenian mountainside."[5] Berossus's Noachic hero is called Xisuthrus.

Newton accumulated a large amount of material about Noah. He tells us the flood story of the Chaldean historian Abydenus (circa 200 BC), who claimed to have had access to primordial texts. Noah is the Chaldean king Sisithrus, who is warned of the flood by Saturn, the same classical god whom Newton believed was derived from Noah. The flood lasts for three days. Once Sisithrus has completed his mission he is swept up into heaven by the gods—just as, according to the Apocryphal literature, Noah's grandfather Enoch was swept up into heaven.*

Glittering fragments of the Noah and Utnapishtim legends tumble kaleidoscopically through the Sisithrus version. Are all three versions fiction? In 1965, researchers taking inventory in the basement of London's British Museum stumbled upon two cuneiform tablets that mentioned an extensive flood in the Babylonian city of Sippar in 1640 to 1626 BC. The tablets told the story of how a water god, Enki, warned a priest-king, Ziusudra, that God was planning to drown the human race; the priest-king built a boat and with all members of his family survived.

There really was a Ziusudra. His name appears in an early column of the Sumerian king list; he ruled over the Babylonian city of Shuruppak in about 2900 BC. There is evidence that there really was a flood at Shuruppak at about this time.

But it was a small flood. Moreover, archaeologist Sir Leonard Woolley, excavating in the Sumerian city of Ur, found evidence that this mini-deluge took place between 4000 and 3500 BC. Nonetheless, writes author Paul Johnson, "the savior-figure of Ziusudra, presented in the Bible as Noah, thus provides the first independent confirmation of the actual existence of a Biblical personage."[6]

There seems to be plenty of literary evidence suggesting that, in

---

*"Sisithrus was king [of Chaldea] and Saturn predicted to him that there would be a massive downpour of rain on the fifteenth day of the month Desius, and ordered that everything that was connected with books should be hidden and put away at Heliopolis in the country of the Sippari. In obedience to the command of God, he immediately set off to sail towards Armenia [with birds, reptiles, and horses], and during this voyage he was overtaken by the sudden fulfillment of the prediction. But on the third day when the storm had begun to abate, he sent out birds to explore, in case they might see land that had emerged anywhere and was standing up above the waves. And when all they found was a measureless extent of water, and nowhere at all appeared where they might take refuge, they flew back again to Sisithrus, and others after them did the same. But when he did the same a third time, and his prayers were answered (for the birds returned with their wings covered in mud), he was immediately removed by the power of the Gods from the society and eyes of men; but the vessel came to land in Armenia, and became a source of amulets for the natives which they made from its beams and wore suspended from their necks." (King, *Finding Atlantis*, 61–62.)

ancient times, there were numerous floods that were at least local. Journalist Charles Berlitz writes, typically "there are over 600 variations of this legend among the ancient nations and tribes and the story has been told through the millennia in all quarters of the globe."[7]

But these ancient "nations" may have been extremely small, and the ancient floods may have taken place at very different times. Whatever the case may be, no scientist has come forward in modern times to present evidence that there was a global flood, and none seems to be interested in researching the subject.

No one in Newton's time doubted there had been a flood, or that the remains of the ark rested on Mount Ararat. Newton quotes Berossos: "Some part of this vessel [the ark] is preserved to this day in Armenia by the mountain of the Cordyaei, from which some inhabitants of the place are accustomed to wear bits of bitumen scraped from it as a talisman of the same thing."[8]

Newton also quotes the Greek philosopher-historian Nicolas of Damascus (born ca. 64 BC), who is mostly remembered as having tutored the children of Antony and Cleopatra: "There is above Minyas [a great mountain in Armenia which] they call Baris, to which, as the story goes, many fled for refuge at the time of the deluge and were saved; and a certain man borne on an ark landed on top of the mountain, and the remains of the timbers were preserved for a long time."[9]

From the early fourth century AD on, Jews, Christians, and Muslims alike were convinced that God had laid down an interdiction, to be lifted only on Judgment Day, against climbing Mount Ararat and visiting the ark. And in fact nothing was heard of the sacred mountain for a thousand years, not until 1357, when a travel guide appeared, called *Travels of Sir John Mandeville,* asserting that Noah's ark indeed rested on Mount Ararat (which was seven miles high, the guide said), and it could be seen from very far away when the weather was clear. Moreover:

> some men saith that they have been there and put their fingers in the hole where the devil went out when Noah said a blessing. But

a man may not well go thither upon that hill for snow that lieth always upon that hill winter and summer. For there cometh never man since Noah was, but a monk that through the grace of God went thither and brought a plank, that is yet at the abbey at the hill's foot.[10]

No one has ever seen that plank.

Nothing was heard of an ark on Mount Ararat for another five hundred years. And then, in 1829, the mountain was climbed, and Ararat began to be demystified. A German natural philosophy professor had taken a quick look around and then returned. He hadn't found the ark—but he hadn't been punished by God. Was the divine injunction against climbing Ararat no longer in force? So it seemed for eleven years. Then, in 1840, Mount Ararat unexpectedly erupted. The little town of Ahora at the mountain's base was totally destroyed. A gaping hole marked the spot where the monastery of Saint Jacob's had once stood. The German natural philosophy professor had prayed in this monastery before setting off on his expedition.

Divine wrath did not speak out again on the mountain. In 1845, a second German professor ascended its slopes but did not find the ark. He returned unharmed. In 1856, a team of British ex-army officers scaled Ararat and also came away empty-handed. Their Kurdish guides descended the mountain with the conviction that British aplomb and a stiff upper lip had broken the divine interdiction of the holy mountain once and for all.[11]

The fifty ascents of Ararat that had followed the ascent of the two Germans had their share of, variously, adventure, joy, injuries, romance, awful weather, incompetence, and fraud, but there was nary a trace of the ark. A rumor persists to this day that the czar of Russia sent a military expedition up Ararat during World War I. Photographs of the ark's interior, brought back to the czar, were still being passed around by townspeople in eastern Turkey in the 1960s, but none of these photos has survived. A Frenchman, Fernand Navarra, climbed the mountain several times in the 1950s, each time bringing back a piece of the ark; each time the piece turned out to be good only for firewood. In 1957, a number of Turkish air force pilots claimed to have spotted a boat-shaped mass lying near the base of Ararat. As far as is known, the Turkish government did not follow up on these reports. Later on, Soviet complaints that the holy mountain was being climbed by American spies forced the Turkish government to seal it off to foreigners. The embargo ended, and very soon the expeditions of the moon-walking American astronaut colonel James Irwin aroused new interest in Ararat and Noah's ark. All that this evangelical Christian space explorer got for his enthusiastic efforts, however, was a fall down the mountainside that nearly killed him.[12] None of these expeditions brought back evidence that the ark existed.

Christianity, Islam, and Judaism each poured an ocean of apocryphal literature into the Noah story. Islamic sources say Noah dug up Adam's body and carried it on board, laying it down between the sleeping quarters of the men and those of the women as a sort of chastity belt. Saint Hippolytus (died AD 235) says Noah packed frankincense, myrrh, and gold—the gifts of the Magi to the baby Jesus—into the hold. Thirteenth-century AD Islamic scholar Abdallah ibn 'Umar al-Baidawi asserts that the name of a prophet was written on every plank in the ark.

There was a unicorn on board the ark—perhaps. The rabbis tell three unicorn stories: (1) Noah brought a few small specimens on board; (2) a single unicorn was brought on board, but it was so big its excrement clogged the Jordan River and only its head could be gotten inside; and (3) the unicorn was so big that only the tip of its nose could be pulled on board. The Talmud says the unicorn was so gigantic that none of it could be gotten on board, and it had to be lashed to the hull.[13] A Ukrainian folktale bestows on the unicorn a reckless grandeur: "All the beasts obeyed Noah when he admitted them to the ark. All but the unicorn. Confident of his strength, he boasted, 'I shall swim.'"[14]

There were hitchhikers including the giant king Og, whom Noah let ride on the roof. On board was Methuselah's niece, who is nameless, but who at a very great age dictated the history of the antediluvian world to Noah's grandchildren.

All three of the great monotheistic religions agree there was a light in the ark that never went out. The Koran (sura 29:13–14) says Noah's boat carried two precious stones that, shining like the sun, told the passengers when to pray and if it was day or night. Scholars say the Hebrew word *tsohar*, once thought to mean a window of the ark, actually means "a brightness," "a brilliance," "the light of the noonday sun." Christian sources describe the light as a "shining crystal" and a "power" that illuminated the "entire vessel for the duration of the flood voyage."[15]

It's likely Isaac Newton would have dismissed most of these apocryphal claims as nonsense. But he would certainly have agreed that there was a light on board the ark that never went out. Only, for Newton, that light was neither mechanical nor magical: it was a "consecrated flame" that had once belonged in an antediluvian temple, or *prytaneum*.*

Newton's belief that there was a consecrated flame on board the ark came from his reading of Genesis. He quotes Genesis 7:2–3 to the effect that along with a male and female of every species the ark contained "seven pairs of all clean animals, the male and female" and "seven pairs of the birds of the air also, male and female." Newton explains that "Noah, when he went into the ark [at the beginning], provided for sacrifices by taking in with him a greater number of Clean Beasts and clean fowls than of unclean ones. . . . For so soon as Noah came out of the ark, he built an altar & offered burnt offerings of every clean Beast & every clean fowl unto the Lord."[16]

When clean beasts and clean fowl were sacrificed they had to be burned by a consecrated flame. From this Newton deduced that Noah

---

*A consecrated flame is one that cannot be allowed to go out. It is not to be relit. In early Rome, if a consecrated flame went out it could be rekindled only by the rays of the sun focused through concave mirrors (Plutarch, *The Lives of the Ancient Greeks and Romans*, 82). A modern-day equivalent of the consecrated flame is the "eternal flame," or torch, of the Olympic Games, which is not allowed to go out during the games or the four-year period between the games.

had brought with him an eternally burning flame from an antediluvian temple. Newton writes: "no doubt for the same end he [Noah] took in [to the ark] with him also the sacred fire with which he was to offer them.... Therefore, 'tis reasonable to believe that they sacrificed also with a consecrated fire ... and accounted it as irreligious to sacrifice with strange or profane fire as to sacrifice an unclean Beast."[17]

That ever-blazing flame in the hold of the ark turned it into a floating temple. For forty days and forty nights, it was the only temple in the world. Now the ark was a sort of "blueprint of God's mind"; it was God's home on Earth; and from the flame God maintained contact with mankind. (Utnapishtim's ark was similar; a *Gilgamesh* translator writes that its dimensions "suggest something more like a ziggurat [tower topped by a temple] or temple than a ship.")[18] Not Newton himself, but later scholars, have made much of the dimensions of the ark as set forth in the Book of Genesis: 611.62 feet long, 85.24 feet wide, 51.56 feet in height between keel and top deck, and 18,231.58 tons in weight (when empty).[19] They believe that through a reconstruction of the shape of the ark, of its every contour and nook and cranny, they can arrive at knowledge of the world before the Flood. Some modern-day Noachic Flood mavens see the ark as a floating computer chip that is encoded with all knowledge of the antediluvian world of the *prisca sapientia*.

Our story has now taken us to the moment on Mount Ararat when Noah unpacks the consecrated flame from the ark and uses it for the animal offering to the Lord.* This ceremony was of tremendous importance, because it cleared the way for a conversation with God. Certainly it was essential that, at this moment atop the mountain, man and God

---

*Animal sacrifice seems primitive to us today, but it was what was required at the time. Scholar Emil Fackenheim tells us that such a sacrifice, rather than being "barbaric," was a form of prayer, prayer being "the sacrifice of the mouth." Making a burnt offering was the "holding open" of oneself that made possible an "incursion of the Divine." (A second interpretation is that the sacrificed animal bears the sins of the sacrificing human, and that when the animal is dead the suppliant has purified himself before the Lord, and therefore may speak.) (Fackenheim, *What Is Judaism?* 183–84.)

talk. The human race had failed miserably, precipitating the Flood, Noah's agonizing voyage, and this meeting. But God had his share in the blame. He had bungled the making of the human race, and he had admitted this. (Thousands of years later, the sect of the Gnostics would tirelessly poke fun at this allegedly omniscient Creator who at the same time had acknowledged the "failure of his creation.")[20]

A circle of worshippers around a consecrated flame: The eight members of Noah's family would have sat or knelt in a circle around the altar atop which burned the consecrated flame. Just these two elements, a consecrated flame and a circle of worshippers, turned it into a temple, one practicing what Newton called the "religion of the Prytaneum."

Newton believed the religion of the "first nations of the world" was the "religion of the Prytaneum," and he held that the consecrated flame in the center of the circle symbolized the sun. This simple religious ceremony was a remembrance of the heliocentric universe and proof that the earliest men and women knew that the Earth went around the sun. The religion of the prytaneum was an "astronomical theology"; however simple in form it seemed, it was a fusion of all science and all theology.

At this moment in time mankind (or this remnant of it), atop Mount Ararat, still practiced the simple, twofold religion that it had been practicing since the beginning of time. Newton defines it in "Of the Church": "The moral part of all religion is comprehended in these two precepts: Thou shalt love the Lord thy God with all thy heart & with all thy soul & with all thy mind. This is the first & great commandment & the next is like unto it. Thou shalt love thy neighbor as thy self. These are the laws of nature"—and have the same authority as, for example, the law of universal gravitation. "These are that part of religion which ever was & ever will be binding to all nations, being of an eternal immutable nature because grounded upon immutable reason."[21]

At the conclusion of the burnt offering, God makes a covenant with Noah. The Deity seems to have felt that the lack of an explicit instruction manual for the practice of the religion of the prytaneum was one reason for mankind's second fall. So God gives Noah that manual: the seven Noachic commandments. As Newton writes in the "Irenicum":

> All nations were originally [from the Flood on] of the Religion comprehended in the Precepts of the sons of Noah, the chief of which were to have one God, & not to alienate his worship, nor profane his name; to abstain from murder, theft, fornication, & all injuries; not to feed on the flesh or drink the blood of a living animal, but to be merciful even to bruit beasts; & to set up Courts of justice in all cities & societies for putting these laws in execution.[22]

Centuries later, God would introduce three more commandments; they would appear, along with the first seven, on the two tablets that Moses brought down with him from Mount Sinai. On Mount Ararat, God has given to Noah an Ur- or proto-Judaism.

Scholars today are almost certain now that Newton embraced the Stoic concept of cycles or vast rhythms affecting the totality of things. Plato taught that the physical universe and everything in it is organic and mortal, moving through growth to their decay and final destruction by fire or water. A priest tells Solon in the *Timaeus,* "There have been, and will be again, many destructions of mankind arising out of many causes; the greatest have been brought about by the agencies of fire and water.... [There is a tradition of] a great conflagration of things upon the earth, which occurs after long intervals [of time]...."[23]

The notion of a succession of worlds haunts us still. The physicist Stephen Hawking told the audience at the BBC's annual Reith lectures that "although the chance of a disaster to planet Earth in a given year may be quite low, it adds up over time, and becomes a near certainty in the next thousand or ten thousand years.... By that time we should have spread out into space, and to other stars, so a disaster on Earth would not mean the end of the human race.[24]

The notion has been cropping up in the creative arts with increasing frequency. In the Matthew McConaughey film *Interstellar* (2014), the premise is that "Earth's future has been riddled by disasters, famines, and droughts. There is only one way to ensure mankind's survival: interstellar travel. A newly discovered wormhole in the far reaches of the solar system allows a team of astronauts to go where no man has gone before, to a planet that may have the right environment to sustain human life."[25]

When Worlds Collide and *After Worlds Collide* captivated readers in the early thirties. Here is the plot: "A runaway planet hurtles toward the earth. As it draws near, massive tidal waves, earthquakes, and volcanic eruptions wrack our planet, devastating continents, drowning cities, and wiping out millions. In central North America, a team of scientists race to build a spacecraft powerful enough to escape the doomed earth. Their greatest threat, they soon discover, comes not from the skies but from other humans."[26] But they survive their fellow men. And, since it so happens that a twin Earthlike planet accompanies the planet that is on target to strike the Earth, the escapees, now aboard their spaceship, are able to land on that planet and save the human race. (In his *In the Days of the Comet*, H. G. Wells had nicely finessed the problem. Mankind does not transition from one planet to another. Instead, a green gas from a passing comet transforms men's minds, erasing negative thoughts and making mankind entirely good. Our species will transform the planet in accordance with its thoughts, and thus the transition will occur within our planet itself.)

Newton believed that emigrating from our world to a new one would be the last act before the curtain of the Apocalypse crashed down. He writes:

> then doth this present world perish by fire & the heavens pass away with a great noise, & the elements (the Gold, Silver, precious stones, wood, hay, stubble, as Paul expresseth it) melt with fervent heat, & then the Saints according to God's promise look for new heavens & a new earth wherein dwelleth righteousness. For as Noah was saved

out of the waters, so in the judgment by fire a remnant may be preserved to replenish the earth a second time.[27]

The operative word is "remnant." Newton sometimes uses it in the sense of an individual, one of "a few scattered persons which God had chosen, such as without being led by interest, education, or humane authorities, can set themselves sincerely & earnestly to search after truth."[28] The superlative human being could be a great prophet or the bringer of a wholly new body of knowledge to mankind. "Ezekiel, his [Daniel's] contemporary, in the nineteenth year of Nebuchadnezzar, spake thus of him to the King of Tyre: Behold, saith he, thou art wiser than Daniel, there is no secret that they can hide from thee, Ezek. xxviii. 3. And the same Ezekiel, *in* another place, joins Daniel with Noah and Job, as most high in the favour of God, Ezek. Xiv. 14, 16, 18, 20."[29]

To ferry a remnant of mankind across the river of space required a very great leader, and Newton was fascinated by Noah because he regarded him as being such a leader. Did Newton feel this about himself? Westfall writes: "There is no doubt that Newton placed himself among the select few. Some of his descriptions of the remnant have the poignancy of personal experience. . . . Isolated in his chambers from the hedonism and triviality of Restoration Cambridge, Newton may have wondered if he was another Elijah, like the first almost the only true believer left."[30] A Dead Sea scroll unearthed near Qumran in 1947 would have reassured Newton of the greatness of Noah and of his status as one of the remnant. First translated in 1967, called the Genesis Apocryphon, it tells the story of the first day of Noah's life.

We don't know who wrote the scroll, but the fictional narrator is sometimes given as Noah's father, Lamech. We'e told that Noah's nature was encoded in his name, that name meaning "rest." The Genesis Apocryphon quotes Genesis 5:29, when Lamech says, "This *name* shall comfort us concerning our work and toil of our hands, because of the ground which the Lord hath cursed." Lamech believed the word "rest" meant "respite from labor."

Old Testament scholars Aryeh Amihay and Daniel A. Machiela believe the word actually means "remnant" or "leftover," because Noah, in bringing the ark through the flood, made himself "simultaneously a savior and a survivor, or remnant." They quote Ezekiel 14:12–20, which states that Noah is someone "who would be saved from any plague or calamity that God would bring on the earth." They cite the apocryphal Book of Enoch, which asserts that Noah "will survive the flood and 'be your remnant, from whom you will find rest.'"[31]

Noah is, then, the great hero who will save mankind. One of the hallmarks of the hero is that he or she displays great strength at a very early age. As a baby Hercules strangled two snakes in his crib. Frontiersman Davie Crockett "kilt him a bear when he was only three." Britain's King Arthur pulled the sword Excalibur out of a stone, a feat no one else could accomplish, when he was just eleven. Jesus Christ bested the temple priests in religious debate when he was twelve.

Noah outdoes all of these great heroes. He is born able to speak. His first conversation, on the first day of his life, is with God. Amihay and Machiela comment:

> Unlike other animals, who learn to stand just hours after their birth, the human newborn takes hours to master the human traits of standing upright, walking, and speaking. In our texts, however, Noah is an astounding exception. His instantaneous speech marks him as outstanding, and the fact that his first words are addressed to the Lord of righteousness marks him as a wise and prophetic individual.[32]

Noah is born with a face that radiates light. His newborn body is "whiter than snow and redder than a rose; his hair was all white and like white wool and curly."[33] Amihay and Machiela believe that red and white were the second-century BC equivalent of black and white, and that these extremes of coloration of Noah's body express "the notion of good and evil residing together"[34] within him. At birth Noah already encompasses the extremes of human experience.

He is so amazing that his father, Lamech, can't believe this is really his biological son. He accuses Batenosh, his wife, of having conceived Noah by a "watcher," one of the two hundred angels or "sons of God" who (Gen. 6:1–4; 1 Enoch) descended to Earth to couple with the "daughters of men."*

With Batenosh's anguished reply to Lamech, we seem to feel on our cheeks the hot breath of a sexuality four thousand years old. Lamech narrates: "Then Batenosh my wife spoke with me very harshly, and wept / And she said, 'O my brother and my husband, you yourself should [believe me], remembering my pleasure . . . / in the heat of the moment, and my panting breath! Now I am telling you everything truthfully.'"[35]

But Lamech doesn't believe her and rushes off to consult with his father, Methuselah (the same Methuselah who will live to be 969 years old). Methuselah doesn't believe her either and rushes off to consult with his own father, Enoch. Professors Amihay and Machiela explain that Enoch "quells all fear regarding Noah and foretells the child's key role in the post-deluge reestablishment of righteousness upon the earth. . . . Enoch predicts one of the activities that Noah will undertake: 'He is the one who will divide the entire earth' [between his three sons and their progeny]."[36]

Enoch gives Lamech a birthday present for Noah (1 Enoch 108:1): a special book of knowledge that includes "the secret of the *Ibbur*." Some say this is the secret of how to put together a combined solar and lunar

---

*The authors of the Genesis Apocryphon don't say that having sexual relations with a watcher necessarily produced a "bad" child. Watchers sometimes fathered good children and angels sometimes fathered bad ones. According to the Apocryphon, when a higher being (it's not clear what this means) had relations with Zophanima, Noah's uncle Nir's wife, she delivered Melchizedek, a hybrid child of the highest virtue. (The birth was, however, darkened by tragedy. All through Zophanima's pregnancy, her husband accused her of being unfaithful. And she *had* been—but the father of her child had cast a spell over her so that she remembered nothing. Moments before Melchizedek was born, she was overcome by doubt and her husband's accusations and died of grief [Stone, Amihay, and Hillel, *Noah and His Book(s)*, 72].) When the angel Samuel slept with Eve, she gave birth to the notorious Cain; but for all his future treachery Cain was almost as precocious as Noah and went out and picked flowers for his mother when he was three days old.

calendar"; others that it's the secret of how "Great Souls" can temporarily merge with and light up worthy mortals in the furtherance of a great end.[37]

A prodigy of prodigies, Noah must have been a young man of overwhelming charisma and an old man of prodigious presence. Nonetheless, the Flood trip seems to have exhausted him, and once in the new world he has that episode of drunkenness about which so much has been written.

But Noah gets his second wind and advances down the slopes of Mount Ararat with his family to undertake the extraordinary task of repopulating the world. There were few thinkers in the seventeenth century who did not believe that Noah had actually done this, and a good many of them tried to figure out the various permutations and combinations of sons and grandsons and great-grandsons that would do the trick. (See chapter 12, "Deconstructing Time.")

And then Noah dies—perhaps. From here on the Noah story goes off in two different directions. The first would have interested Newton not at all; the second would have engaged him deeply. In the first story, God confers immortality upon Noah. This may be an echo of the *Gilgamesh* flood story, where the gods reward the Sumerian Noah Utnapastim by making him immortal.

But Noah doesn't take easily to his immortality. He wanders through the world like the Wandering Jew, forlorn and alone, perhaps noting with growing sadness as the centuries go by that mankind is backsliding once again, building up to a new crisis of sin and painful redemption. (Some scholars believe he is the mysterious white-bearded old man who appears briefly in the "Pardoner's Tale" of Geoffrey Chaucer's *The Canterbury Tales*.*)

In the second version of Noah's post-ark existence he becomes

---

*See Peter G. Beidler, "Noah and the Old Man in the Pardoner's Tale," *Chaucer Review* 15, no. 3 (Winter 1981): 250–54. It's doubtful that Newton took much if any interest in this aspect of Noah.

more than immortal; he beomes the father of the gods and, later on, he becomes a planet.

When Noah and his family came down from the mountain to begin the task of repopulating the world, we can only assume that they were surrounded by the din and splattering mud of thousands of galloping animals while myriad flocks of screeching birds swept overhead and crawling creatures slithered between their legs as the living cargo that Noah had brought with him on the ark enthusiastically began its own work of repopulation.

But we mustn't make fun of this first family of refugees, because, for Newton and his contemporaries, to witness the descent of Noah and his family from the heights of Ararat was to witness the birth of the gods. This first family of the post-Flood era was a kind of template, the proto-pantheon or Ur-pantheon of all the pantheons of gods that would one day come to populate the universe of polytheism.

Richard Westfall explains:

> Common characteristics distinguished the corresponding gods of all ancient peoples. All peoples worshiped one god whom they took to be the ancestor of the rest. They described him as an old and morose man and associated him with time and with the sea. Clearly, Noah furnished the original model of the god called (among other names) Saturn and Janus. Like Noah, Saturn had three sons. Every people had a god whom they depicted as a mature man, the god they held most in honor. They had translated Ham into Zeus, Jupiter, Hammon, and others. All worshiped a voluptuous woman variously named Aphrodite, Venus, Astarte, et alia, originally a daughter of Ham. The histories of the gods of one people frequently became confused with those of another, and people invented fables which confounded the origins of the gods by claiming the gods of others for their own.[38]

We find ourselves in the realm of euhemerism. Euhemerus (331–251 BC) was a Greek mythographer, probably born in Sicily, who, one day while sojourning as a guest at the palace of King Cassander of Macedonia, had a eureka idea. He had been thinking about Alexander the Great (356–324 BC), who had conquered most of the known world by the time he was thirty and thereafter had insisted on being worshipped as a god. So prodigious were Alexander's accomplishments that his soldiers acquiesced. Soon many of his subjects were worshipping him as a god. This didn't end with Alexander's death.

But of course Alexander the Great was not a god, mused Euhemerus. Then his eyes wandered over the many marble statues of gods like Aphrodite and Hermes and Zeus that bedecked the niches of the walls of Cassander's palace and the idea occurred to him: Were they too never gods? Were they merely mortal men and women who were deified because they had done good service to their country?

This idea wasn't new in the fourth century BC. But Euhemerus decided to take possession of it. He wrote a book, *The Sacred Scripture,* in which the narrator stumbles on an unknown temple on an imaginary island called Panchaea. He enters and sees before him a golden pillar inscribed with a registry of the births and deaths of the gods, including Zeus, who is listed as king of Crete. The narrator has discovered that the gods were once mere human beings. He leaves Panchaea to take this knowledge to the world.

Most readers thought *The Sacred Scripture* was fact, not fiction. Euhemerus had managed to popularize his idea, and it was dubbed *euhemerism.* Two thousand years later, it was the lens through which Newton and his colleagues viewed both mythical personages and mythical places.*

Newton assembled numerous references to Noah as the god Saturn.

---

*Newton applied euhemerism to mythical places in this way: He believed the river Styx of Greek mythology, across which the boatman Charon ferries the souls of the dead, was based on the River Nile (the Egyptians had made the Nile sacred to Osiris). Newton quotes Diodorus, who quotes Homer: "They come to the waves of Ocean, and the Leucadian rock / And to the gates of the sun, where dreams rule, a wandering race, / and further on they reach the grassy green meadows, which are frequented / by the shades, mere images of men, lacking life."

After quoting this [says Newton], he [Diodorus] tells us that by "Ocean" here the Nile

He wrote:

> Saturn because of his great age is made the God of time. He was accounted the author of husbandry [cultivation of plants or livestock/farming/agriculture] and in token thereof carries a scythe. Drunkenness was attributed to him and in memory thereof the Saturnalia were instituted. He was painted by the Egyptians with eyes before and behind [an allusion to Noah's looking backward to the antediluvian age and forward to the post-Flood age] and reputed the justest of men and the father of truth. And in all these respects he agrees accurately with Noah.[39]

And in another text:

> ... The Egyptians passed down the tradition that the most ancient of the Gods reigned for a space of one thousand two hundred years and the later ones not less than three hundred years. And such longevity only Noah with his sons and grandsons achieved. Saturn and Rhea with the other Gods of that time are said by Philosophers and Poets to have sprung from Ocean. That is why the Egyptians also depicted their Gods in a boat on the waters.[†40]

(cont. from p. 212) is meant, and by "the gates of the sun" Heliopolis, and by "the grassy green meadows" of the dead are meant the pastures by the Acherusian marsh near Memphis. For, he says, many grand Egyptian funerals were conducted there; they transport the corpses across the river and the Acherusian marsh and lay them in crypts situated there. Hence also the Acheron and Charon of the Greeks. (Newton, "Miscellaneous Draft Portions of 'Theologiæ Gentilis Origines Philosophicæ.'")

†Do the names of Noah's family hint at the gods and goddesses they purportedly became? We know that "Ham" means "hot" in Hebrew, and the name of Noah's eldest son may also have affinities with "Héammu" (as in Hammurapi), a west Semitic sun god; that "Shem," the name of Noah's middle son, is perhaps associated with the Sumerian "Kengir" (sometimes "S[h]umer"), the Sumerian term for southern Mesopotamia (which Shem is supposed to have settled); and that "Japheth" may be etymologically linked with the Greek "Iapetos," one of the Titans. But these connections are so obscure as to prove nothing. The author is indebted to Michael Avioz, Department of Bible, Bar-Ilan University, Ramat-Gan, Israel, for much of this information.

Not only did Noah become the god Saturn, but he also became the god Janus, and Janus and Saturn sometimes became one and the same god. Janus was the Roman god of beginnings and endings, arrivals and departures, transitions, gates, doors, doorways, and passages. *Janua* is the Latin word for "door," and *janitor* is the Latin word for "doorkeeper" (hence our word *janitor*). Open a door and observe the two doorknobs: one faces out and the other faces in. Newton believed the peoples of the ancient world associated this backward- and forward-looking god with Noah standing on Mount Ararat. He noted that "a coin was struck in Italy at one time with the double face of Janus on one side and a ship on the other. These things clearly refer to the flood."[41]

Moreover, Janus

> was depicted by the Egyptians with eyes before and behind, as if he had seen both before the flood and after.... Saturn was the God of time; and in the Orphic writings he is called "father of all" and "ruler of created things" and his wife Rhea is "Mother of gods and mortal men."... Janus too was God of the year and of time and ... he is called the Sower of things and the source of the Gods; and he was depicted with two faces. All this can be understood only of Noah, a man long-lived beyond all men, and father of all mortals.[42]

It would be several generations before the pantheons of gods evolve from Noah's family. First, Noah's sons had to spread out from Mount Ararat, each bearing a portion of the sacred fire and each founding a different nation. They then established the religion of the prytaneum in their nations and practiced the seven Noachic precepts.

Noah's son Shem had originally founded Assyria, then been driven out by his nephew Nimrod. He had ruled areas bordering Egypt. Shem's son Cush founded Egypt, and continued his grandfather's worship of the Noachic God whom Newton described as "all eye, all ear, all brain, all arm, all power to perceive, to understand, and to act; but in a manner not at all human, in a manner not at all corporeal."

Not long after Cush founded Egypt, an event took place that looms large in Newton's history of the corruption of the soul of man suddenly and that registers a seismic shock in the religious consciousness of mankind. This was the birth of that form of idolatry called polytheism. It seems that it's difficult for mankind to worship something that it can't see or hear or smell or taste or touch. The ancient Egyptians began to take the symbols of religion for the substance. Newton writes: "The frame of the heavens consisting of Sun, Moon and Stars being represented in the Prytanea as the real temple of the Deity, men were led by degrees to pay a veneration to these sensible objects, and began at length to worship them as the visible seals of divinity."[43]

Worship began to include more and more of the elements of the physical universe: "And because the sacred fire was a[n arch] type of the Sun and all the elements are part of that universe which is the temple of God, they soon began to have these also in veneration. For 'tis agreed that idolatry began in the worship of the heavenly bodies and elements."[44]

Newton scholar Kenneth Knoespel explains:

Men discard an absolute faith in God for a "veneration" of the secondary effects by which his wisdom can be apprehended, thereby confusing form and content, the ideal and the material, the timeless and the corrupt. Worshiping the mere representations of divine order—the Sun, stars, and planets—turns men and women away from techno-scientific knowledge and true faith and makes them subject to self-willed delusions.[45]

It wasn't long before the next step in the process of corruption occurred. Newton writes: "Worshiping the sun, the known planets, and the four elements, mankind began to honor the memory of his most illustrious ancestors by naming the planets after them. Finally, mankind, believing that the souls of his ancestors had transmigrated to these planets, began to worship them as gods."[46]

Man had created a set of gods whom he worshipped. The memory of the one true God remained within him, but polytheism was in the ascendant. Instead of worshipping God, we began to worship our ancestors as gods. Wanting to give them a visible habitation, we decided their souls had transmigrated to heavenly bodies, and we named those heavenly bodies after them.

And so, in the "Theologiae Gentiles Origines Philosophicae," Newton declares that all ancient peoples worshipped the same twelve gods if under different names, the originals of these twelve gods being Noah, his children, and his grandchildren. If we compare the pantheons of the twelve gods of the Greeks, the Romans, and the Egyptians, we see that

> the 12 Gods were all of a kindred, parents and children, brothers and sisters, husbands and wives to one another, and divers had one common mother, Cybele. They lived all at the same time, which is called the age of the Gods. . . . And in their age the brothers and sisters for want of further choice became husbands and wives. All which characters agree best to the times next the Flood.[47]

Newton linked the first four post-Flood generations with the Gold, Silver, Bronze, and Iron ages of high antiquity, declaring that "the Saturn therefore who reigned in the Golden age and his son Jupiter who reigned in the Silver one can be no other than Noah and his son Ham. For Ham himself was the warrior Mars."[48]

Moreover, says Newton:

> Every nation deifying their own kings applied. . . . the name of Jupiter to him whom they had most in honor, as the Arabians [originally the Chaldeans] to their common father Chus, the Assyrians to their common father Nimrod . . . and the father of their Jupiter every nation called Saturn and one of his sons Hercules or Mars.[49]

Newton piles complexity on complexity:

The Planets and the Elements which are signified by the names of Gods were enumerated by the Egyptians in this order: a Saturn, Jupiter, Mars, Venus, Mercury, Sun, Moon, Fire, Air, Water, Earth (Terra). The Earth (Tellus) which is represented/produced/foreshadowed by the four Elements is the Fifth essence and completes the number twelve. The whole of Philosophy is comprehended in these twelve, provided that the stars indicate Astronomy, and the four Elements the rest of Physiology.[50]

Professor Knoespel explains that, "according to Newton, Noah's children and grandchildren became absorbed within other mytho-histories, or pre-histories, of antiquity through a process in which they first became localized as historical figures and then memorialized as stars and planets."[51]

Here we must leave our discussion of Noah. But, if Newton and his contemporaries are to be believed, Noah has not left us.

The next time you go to a New Year's Eve party, you can experience him in two of his guises. Father Time—the little old man with the pointed white beard and the long scythe—is a racial memory of Noah, and it is to this figure, who metamorphoses from old man to babe in swaddling clothes at the stroke of midnight, that we raise our glasses. (In the midstroke of his metamorphosis, he is also the god Janus looking backward and forward in time.)

In this act of raising our glasses, we all take on a bit of Noah. Newton and his colleagues believed that the god Bacchus—the great celebrator of the grape—derived his existence from Noah. On New Year's Eve, we re-create Noah's notorious act of drunkenness (and perhaps redeem it) and recognize God's gift of the grape to mankind.

CHAPTER ELEVEN

IN THE DAYS OF THE COMET

When Halley's comet flashed across the sky in 1910, thousands of people were terrified they would be killed by the poison gas in its tail. Comet pills, comet insurance, and hard hats to protect against the rain of fire went on sale. Some clergy advised their parishioners to store their valuables in the church to keep them safe; then the clerics made off with the goods. A report out of Oklahoma said a sheriff had rescued a local virgin who was about to be sacrificed to the comet.[1] A German astronomer insisted that "a single thread of a spiderweb would pose more danger to a charging elephant than would Halley's comet."[2] But nobody listened; on the nights that the comet was closest, thousands tried to drown their fears in frenetic "comet parties" held in the major metropolises of the world.

But there was no poisonous gas, and the worst that happened was that Mark Twain, who was born in the year of Halley's comet 1834, predicted he would pass out of this world in the year of Halley's comet 1910, and he turned out to be right.

When the legendary comet next paid a visit, in 1986, everything had changed. The comet was far enough away (39 million miles, as opposed to 13.3 million in 1910), and the general public was sufficiently educated, that there was no panic. This time a gleaming armada of spaceships—two joint Soviet-French ventures, two Japanese spacecraft, and the European Space Agency's *Giotto* space probe—traveled out to meet the celebrated wanderer of the skies. They prodded it, X-rayed it, palpated it, listened to it, analyzed it, photographed it, and watched as, stripped of many of its secrets, it sped away toward its rendezvous with the sun.

Back on Earth, comet pills had gone on sale again. This time they were manufactured by a single agency, the Grand Rapids, Michigan, Public Museum, and were made of yogurt-covered sunflower seeds bearing the consumer-protection label: "Museum Surgeon General has determined that worrying about comets can be hazardous to your health."[3]

For millennia, comets instilled fear in the hearts of humankind. They appeared without warning; they lit up the sky with fire; nobody knew what they were. A widespread belief persisted that they were direct acts of God portending calamity to mankind. Often they seemed to single out distinguished individuals: In Shakespeare's *Julius Caesar*, a comet streaks across the sky after Caesar's assassination, bearing out the prophetic words of his wife, Calpurnia: "When beggars die there are no comets seen; / The heavens themselves blaze forth the death of princes" (2.2.30–32).

The comet that would come to be called Halley's comet streaked across the sky in 1066 and was believed to foretell England's defeat by William the Conqueror at the Battle of Hastings. In 1456, a huge comet filled Christendom with terror. Pope Calixtus III (1378–1458) thought it both heralded and caused the fall of Constantinople, and the frightened pope added a prayer to the Ave Maria: "Lord save us from the devil, the Turks, and the comet." There is a legend that Calixtus III actually excommunicated the comet as a dangerous heretic.*[4]

---

*The astronomer Fred L. Whipple writes: "The story that Pope Calixtus III actually excommunicated the comet of 1456 (an apparition of Halley's Comet) is a hoax. The Pope was clearly worried, however. He ordered public prayers for deliverance from the comet and from the enemies of Christianity. (Whipple, *The Mystery of Comets*, 13.)

Thomas Aquinas (1225–1274) wrote that comets were one of the fifteen signs heralding Judgment Day.[5] Martin Luther (1483–1546) warned that "the heavens write that the comet may arise from natural causes, but God creates not one that does not foretoken a sure calamity."[6] Comets were said to foretell the death of King Frederick of Sicily in 1264 and that of Pope Urban IV in 1327. In 1531, Louise of Savoy (mother of King Francis I), seeing from her sickbed a comet in the sky, observed somberly: "Behold an omen which is not given to one of low degree. God sends it as a warning to us. Let us prepare to meet death." Three days later, she was dead.[7]

It wasn't until the latter half of the seventeenth century that the men and equipment needed to unlock the secrets of the comet began to be put in place. Isaac Newton would be the prime mover. Just as he had cleansed the New Testament of Trinitarian fraud, so he would lift the comet out of the mire of myth and superstition. Aristotle thought comets were merely atmospheric phenomena, something like sheet lightning; Newton would restore the comet to its rightful place in the heavens, demonstrating that it pursued a majestic path around the sun and teaching mankind how to measure that path.

He would do more: he would intimate that the universe is divine sacrament as well as physical phenomenon and that God has given the comet a special role in the destiny of mankind.

In London in 1675, King Charles II established the Royal Greenwich Observatory on hilly terrain south of the Thames River. Initially, its purpose was to reduce shipwrecks by helping ships determine longitude while at sea. A Scot, John Flamsteed (1646–1719), half crippled by teenage pleurisy, neurasthenic, defensive to the point of paranoia—but brilliant and totally dedicated to astronomy—was named Britain's first Astronomer Royal. The salary the king granted Flamsteed was so small that he was forced to buy most of the equipment himself with the money he earned as an Anglican preacher.

Greenwich Observatory is world-famous today. Tourists crowding through its rooms are surprised to discover (they'd never thought about it) that the line that separates the Western Hemisphere from the Eastern Hemisphere runs right between their feet, on the floor of the observatory, where it is marked off as the prime meridian. The line also marks Greenwich Mean Time, the anchor point of all the time zones in the world. So famous did Greenwich Observatory become that in 1895, two anarchists, defying it as a symbol of establishment oppression, detonated a bomb beside its walls. One anarchist died in the blast; the observatory was unharmed.*

The observatory was completely operational when, in November 1680, a vast new comet swam into the skies above Europe. Flamsteed followed its progress with elation, writing to a friend: "I believe scarce a larger hath ever been [seen]."[8] Astronomer Pierre Lemonier wrote, "It issued with a frightful velocity from the depths of space" and by the end of November "seemed falling directly into the sun."[9] Flamsteed followed this comet, called the Great Comet of 1680 (and also Kirch Comet, after its discoverer) night after night, whenever the London fog and rain let up, until it disappeared into the precincts of the sun.

Then, in mid-December, another comet appeared, streaming away from the other side of the sun. Or was it another comet? John Flamsteed was unique among astronomers of the time in suggesting that this second streaking ball of fire was the Great Comet of 1680 all over again. Flamsteed theorized that a combination of magnetism and Cartesian "vortices" had caused the comet to ricochet off the face of the sun and start on its way back toward Earth.

At Cambridge, Isaac Newton, already famous for his work on optics, had been watching the new comet with furious intensity. Or, rather, the two new comets, since Newton didn't accept Flamsteed's single-comet theory. When the "second" comet made its closest approach to Earth on December 27, Newton sketched it as the length of four full moons set side by side, it stretched across seventy degrees

---
*Joseph Conrad's 1907 novel *The Secret Agent* is based on this incident.

(5 percent) of the night sky. In his drawing, the comet hovers above King's College Chapel, almost parallel to the roof and only a little longer, with its head jutting out over one end and its tail slanting up just a little beyond the other.[10]

Let's hold that drawing of comet and chapel in our minds, for we will see that its hieroglyph-like contents had a significance for Isaac Newton of which even he was likely not yet quite aware.

The great mathematician continued to observe this "second" comet, first with the naked eye (sometimes aided by a monocle, since he was nearsighted), then with a three-foot telescope, and finally with the seven-foot telescope that is on display today in Newton's rooms in Cambridge. Toward the end of March, the comet faded away among the stars.

Gone with the comet was Newton's belief that it was a second comet. He had decided Flamsteed was right—though Newton believed the Great Comet of 1680 had circumnavigated the sun before starting back toward Earth.

It was an unlucky day for John Flamsteed when it turned out he was right and Newton was wrong. Frank Manuel observes: "One rarely proved Newton wrong. If one did, retribution, though it might be long delayed, ultimately followed."[11] The relationship between Flamsteed and Newton, though it had only just begun, was already souring.*

By Christmas 1681, Newton had concluded that all comets circle the sun, just as planets do. He set out to demonstrate this mathematically, succeeding in time to insert his calculations into the *Principia Mathematica* in 1687.

---

*The relationship would grow worse over the years as Newton increasingly demanded raw astronomical data from Flamsteed. The Astronomer Royal couldn't supply it fast enough to suit Newton, and the latter grew steadily more imperious. Toward the end of Flamsteed's life both men were hurling invective at each other, not the least because Newton, working with Edmund Halley, virtually stole Flamsteed's new comprehensive star map from him, publishing it without the Astronomer Royal's permission. (An amended version of Flamsteed's landmark text was published five years after his death by his widow.)

This was a prodigious achievement. Professor Dobbs writes that

> the taming of the comets, making them more or less domesticated members of the solar system, was not the least of Newton's achievements in the *Principia*. Such a notion [that comets regularly orbit the sun] was almost unheard of at the time. Comets had always seemed radically alien; they were erratic and ephemeral bodies that portended no good, and had traditionally been taken as "signs" and portents of disaster—not disaster itself, for none had ever been known to crash into anything.[12]

But this was only the half of what Newton had to say about comets. We've become aware of the other since the release of the great bulk of Newton's nonscientific papers in 1937. Dobbs explains that, with Newton's work, "the newly domesticated comets were promoted in status from signs to agents of destruction."[13] Newton not only believed comets portended disaster; he believed they *brought* disaster. And this was at the behest of God.

The great visionary poet William Blake (1757–1827) wrote, "'What,' it will be question'd, 'When the Sun rises, do you not see a round disk of fire somewhat like a Guinea?' O no, no, I see an Innumerable company of the Heavenly host crying, 'Holy, Holy, Holy is the Lord God Almighty.'"[14]

The ordinary person perceives the sun as a fiery coin-shaped disk, whereas a visionary like Blake sees it as a choir of angels jubilantly singing hymns of praise to the Lord. And a vision such as this, says Blake, has more reality in it than a scientific explanation of the sun. Blake hated Newton, believing that he was capable of no more than a coldly scientific perception of the sun—even believing that Newton, in adding to our scientific knowledge of the sun, was adding to the unreality of the universe and driving away that which was most real: the creations of the visionary imagination, which if authentic are in tune with the productions of God. In his poem "Jerusalem," Blake rails against what he believes to be Newton's desacralization ("de-sacredizing") of the physical universe. He includes John Locke in his scolding.

> *And there behold the Loom of Locke, whose Woof rages dire,*
> *Washed by the Water-wheels of Newton: black the cloth*
> *In heavy wreathes folds over every Nation; cruel Works*
> *Of many Wheels I view, wheel within wheel, with cogs titanic.*[15]

Newton didn't see the sun as a "Heavenly host crying, 'Holy, Holy, Holy.'" But if Blake could somehow have read the Newtonian papers that were stored in two metal boxes not twenty-five miles from where he lived in London, he would have been amazed to discover that Newton saw the sun not only in brilliant scientific detail but also—to name only one scenario that he seemed to be considering—as a divinely ordained accomplice to the Great Comet of 1680 in the upcoming Apocalyptic destruction of our world (see final section of this chapter).

Dobbs writes:

> What Newton needed to lend coherence to his sacramental view of the creation was an agent that operated with majestic regularity and yet was capable of generating the unusual events of natural and sacred history, and he found what he needed in comets....
>
> ... A providential God could use comets to enact upon the earth expressions of his divine will, such as the Flood or the Apocalypse. Comets thus became ... [devices] to explain the divine operation of the universe in 'reasonable' terms."[16]

Of his belief in the God-directed, punitive, aspect of comets, Newton said little in his time, seemingly leaving it to his "surrogates" (such as William Whiston, see below) to float trial balloons of his ideas. But he does put forward in the *Principia,* without mentioning any role that God might play, the notion of the comet as a kind of benevolent cosmic breadbasket, raining essential nutrients down on the Earth whenever the need arises. Newton wrote: "I suspect, moreover, that it is chiefly from the comets that spirits come, which is indeed the smallest but the most subtle and useful part of our air, and so much required to sustain the life of all things with us."[17]

These "spirits," which gave our planet vital nourishment beyond what purely mechanical processes can offer, have many names for Isaac Newton: the "ether"; the "vegetable spirit" (here "vegetable" means "flowering," "flourishing"); "fermental virtue"; the "mercurial spirit" (some of the terms came from alchemy); "light" (which the Stoics, much admired by Newton, associated with God); and Jesus Christ. "It was the Christ," writes Dobbs, "united with God in a 'unity of dominion' though not of substance, that put the ideas [conceived by God] into effect."[18]

As has been discussed, for Newton as the Arian Jesus was subordinate to God. But he was also God's Man in the Universe, a sort of Executive Director of the cosmos. In sculpting the universe as God willed it, he both shaped and animated that which is strictly mechanical, and that which is more than mechanical—that is alive, that is self-actualizing. It was Jesus Christ who drove the comets in their orbits, particularly the Great Comet of 1680. As Dobbs makes clear, "Even though Newton's God is exceedingly transcendent, He never loses touch with His creation, for He always has the Christ transmitting His will into action in the world."[19]

Dobbs continues: "With their periodic returns demonstrated, he [Newton] then used comets as the hidden causes that account in a 'natural' way for the 'miraculous' congruence of natural and sacred histories, especially for the confluence of these histories at the end of time in the final conflagration of a world."[20]

As regards the role of comets as disseminaters of nutrients to Earth, Newton also had this to say: "The diminution caused in the humid parts [of planets] by vegetation and putrefaction . . . by which means the dry parts of the planets must continually increase, and the fluids diminish, may in sufficient length of time be exhausted, if not supplied by some such means."[21]

But God had provided a solution.

> The vapors which arise from the sun, the fixed stars, and the tails of the comets, may meet at last with, and fall into, the atmospheres of the planets by their gravity, and there be condensed and turned into water and humid spirits; and from thence, by a slow heat, pass gradually into the form of salts, and sulphurs, and tinctures, and mud, and clay, and sand, and stones, and coral, and other terrestrial substances.[22]

According to Professor Dobbs, the term "slow heat" points to an alchemical process. Newton regarded the nourishment-recycling comet as an alchemical furnace, with Jesus Christ as the Philosophers' Stone enabling the base metals on the comet to be transmuted into nutrients needed on Earth. Newton believed in the "inertial homogeneity and transformability of matter";[23] that is, he believed all matter was essentially the same, and could be transmuted into any other kind of matter. (This notion of Newton's, one of the basic tenets of alchemy, is elaborated on in chapter 15, "The Secret of Life.") With his assertion that "spirits" as subtle as gossamer wafted down to Earth from comets, help sustain life on Earth, Newton seems to be entering Velikovskian territory.

In 1951, an Austrian psychiatrist named Immanuel Velikovsky published a book that created a worldwide sensation. Its title was *Worlds in Collision,* and in it he argued that a comet while on its way to becoming the planet Venus brushed our world in such a way as to cause all of the biblical disasters. From its gaseous tail it rained down physical phenomena that explain the supernatural events surrounding the exodus of the Jews from Egypt, such as the phenomena of the miraculous "manna from heaven" that fed the Jews as they were crossing the Sinai.*

---

*In Exodus 16:4, after Moses has told God the Israelites are starving, God tells him, "Behold, I will rain bread from heaven for you; and the people shall go out and gather a certain rate everyday, that I may prove them, whether they will walk in my law, or no." And in fact "manna" begins to arrive with the dew every morning, and "it was like coriander seed, white; and the taste of it was like wafer, made of honey" (Exodus 16:31). Velikovsky

Newton never mentions manna, but Velikovsky's explanation fits in well with Newton's assertion that comets serve as God's instruments for the replenishment of depleted planets.

Isaac Newton had no difficulty speaking about the positive benefits of comets. Why, then, if he believed they were also under a divine injunctions to destroy or cleanse worlds, did he say nothing about that? (We will see at this end of this chapter that there was one occasion when he spoke out, and spectacularly, on the subject.)

One reason is that Newton knew these beliefs had a heretical cast. He believed Jesus had a key role . . . as the general manager of the universe; but a Christ so undignified as to lash a comet forward is hardly God's equal. He is an Arian Christ, and for fear of negative consequences Newton had to keep his Arianism to himself.

Newton may also have feared that revealing his thoughts on these matters might make it seem as if he were attributing "occult influences" to the universe. But it was essential to Newton to demonstrate that solely mechanical causes lay behind all the workings of the universe.

Newton got around the problem by finding someone else to tell the world about the dual nature of celestial objects, particularly comets, as physical phenomena and as divine instruments. The man he recruited and trained to do the job (or so it seems to us today) was William Whiston (1667–1752)—the "honest, pious, visionary Whiston,"[24] in Edward Gibbon's words.

▲

(cont. from p. 226) believed fallout from the comet was responsible for the disasters that befell the Egyptians and forced them to release the Jews: rivers turning bloodred; stones falling from the sky; darkness at noon; death of all the Egyptian firstborn; earthquakes; high tides; many others. The Austrian-born catastrophist Velikovsky was convinced that similar calamities had taken place all over the world at that time, including the Ogygian flood. (See chapter 14, "A Glitter of Atlantis.") He found evidence in other contemporary literature of a honey-tasting food falling from heaven, one example being ambrosia, the "food of the gods" of the ancient Greeks. (Velikovsky, *Worlds in Collision*, 145.)

A German colleague, meeting the forty-three-year-old William Whiston in a coffeehouse in 1710, described him as being "of very quick and ardent spirit, tall and spare, with a pointed chin and wears his own hair. . . . He is very fond of speaking and argues with great vehemence . . . [a man who] by his many singular opinions, which he boldly professes, has made himself only too notorious."[25]

This was on the downside of a tremendous divide in William Whiston's life. Up until his thirty-third year, his career rolled forward smoothly and unstoppably. He was born in Leicestershire in 1667 to a blind cleric who homeschooled his brilliant offspring so that young Whiston could help him write his sermons. Pious and precocious, volatile and extremely productive, Whiston entered Cambridge at nineteen. He read voraciously, excelled in mathematics, received his M.A. in 1693, and was ordained a minister that same year.

Newton was still at Cambridge, already famous, almost a national monument, but an almost invisible monument. The great man rarely lectured, and his presence could be felt mainly in the diagrams (drawn with "a bit of a stick") that he left behind in the Trinity gravel beds as he took the occasional walk through the college's gardens. ("The Fellows would cautiously spare [these diagrams] by walking beside them," says a contemporary.[26]) Whiston managed to attend a few of the lectures of the elusive genius, but (or so he tells us) he didn't understand a thing.

The young scholar set himself to mastering the *Principia,* and succeeded to the extent that he was able to become friends with Newton. Whiston later boasted that the friendship had been a close one, but the truth of this is hard to determine. Newton seems to have shared some of his more basic as well as some of his most radical ideas with him, such as, in the latter category, the notion that God, unfettered by necessity, is able "to vary the laws of Nature, and make worlds of several sorts in several parts of the universe."*[27]

---

*This view first became public in Newton's *Optics,* published in 1704.

But these ideas may have been discussed at only a handful of meetings. It's tempting to believe that, from the outset, Newton sized up the brilliant and impressionable Whiston as someone whom he might be able to use as a stalking horse for his more unorthodox mystico-religious ideas, for example, that a comet can be the punitive hand of God in a universe nonetheless strictly controlled by scientific laws.

In 1695, Whiston married and left Cambridge to become a church rector in Lowestoft, Suffolk. The voluble mathematician-cleric was already acquiring the reputation for saintly eccentricity that would prompt Oliver Goldsmith to choose him as the model for Vicar Primrose in Goldsmith's perennially popular novel, *The Vicar of Wakefield* (1766). (Whiston believed clergymen should not get married a second time and had it engraved on his wife's tomb that she was the *only* wife of William Whiston; Goldsmith gives Primrose this peculiarity.[28])

But Whiston had a brilliance in mathematics that far exceeded any quality Goldsmith might want to bestow on his humble characters. At the time of his marriage he was already working furiously on a book that is now regarded as the last important attempt by an equally qualified theologian and mathematician to prove that there is no contradiction between biblical scripture and Newtonian science. This book was titled *A New Theory of the Earth from its Original [Origin] to the Consummation of All Things wherein the Creation of the World in Six Days, the Universal Deluge, and the General Conflagration, Are Shewn to Be Perfectly Agreeable to Reason and Philosophy*. It was published in 1696, and it was immediately acclaimed as a masterpiece.

The book epitomizes an era. As the seventeenth century neared an end, a transformation exciting to some but threatening to many was shaking up the worlds of natural philosophy and theology. The age-old narratives of the Bible were breaking down in the face of the rigorous equations underpinning Newton's radical new interpretation of the universe. The two worldviews could coexist only if it could be proved that they described exactly the same phenomena. It had become necessary to "establish correspondences between the newly understood orbits of celestial bodies and past events and catastrophes known from myths and the bible."[29]

This was the whole thrust of William Whiston's *New Theory of the Earth*. The author never departs from his vision—shared by Newton, but to what extent and with what applications, we cannot know—of one celestial body in particular, the comet, indispensably generating the unusual events of both natural and sacred history. The events Whiston grapples with are: the creation of the Earth (not *by*, but *from*, a comet); the tilting of the Earth on its axis; Noah's Flood; and the apocalyptic destruction of our world, which ends with the Earth's being turned (in Whiston's preferred of two versions) back into a comet it once was. The author strives to demonstrate that all this can be explained in terms of Newtonian physics and also that every bit of it is described in the Book of Genesis.

A New Theory has the flavor of sensationalist catastrophism of many of today's Hollywood disaster movies. The Great Comet of 1680, which Whiston sees as the physical cause of all of these colossal events (though he sometimes brings in Halley's comet), comes across as a sort of Death Star in the *Star Wars* mode. Whiston ascribed a periodicity of 575 years to the 1680 comet after ransacking ancient myth and legend to locate events, spaced an equal number of years apart, suggesting the periodic passing of a comet.*

Whiston's book is no modern-day New Age romp. It is a brilliant, eloquent, and grimly serious attempt to achieve a "union of the scriptural account of God's way of running the world and the physical

---

*Whiston located seven such possible events, spaced 575 years apart. Edward Gibbon lists them in *The Decline and Fall of the Roman Empire* as follows: (1) The statement by Marcus Terentius Varro (116–27 BC) in his *Chronology* that the planet Venus "changed her color, size, and course" in 1767 BC under the reign of Ogyges; (2) the return of this comet in 1193 BC (thought at the time to be the last year of the Trojan War), which was "darkly implied in the fable of Electra, the seventh of the Pleiades, who have been reduced to six since the time of the Trojan war. That nymph, the wife of Dardanus, was unable to support the ruin of her country: she abandoned the dances of her sister orbs, fled from

system presented by the *Principia*."[30] Today only some fundamentalist Christians think this is possible, and they achieve it only by excluding large blocs of religion, or science, or both. And in fact Whiston sometimes has to resort to outlandish and even fantastical hypotheses to make his case. Let's now look briefly at his strange story of the fateful relationship between our planet and a comet.

In 1667, John Milton published his epic poem *Paradise Lost,* which tells the story of how and why Adam and Eve were expelled from the Garden of Eden. The "Fall" of man entailed the "Fall" of the entire planet; Milton describes God's angels tilting the Earth 23½ degrees on its axis and setting it spinning so that what was once a world with a single sunny changeless season becomes one with four seasons. (Roses grew thorns and animals sprouted claws as the whole world slipped into imperfection.) Milton writes:

> ... he [God] bid his Angels turn askance
> The Poles of Earth twice ten degrees and more
> From the Sun's Axle; they with labor push'd day
> Oblique the Centric Globe ... else had the Spring
> Perpetual smil'd on Earth from verdant flowers.[31]

Whiston explains that a comet does this too, in that, "[from] the Impulse of a Comet with little or no Atmosphere ... both the annual Orbit of the Earth ... and a vertiginous Motion about a new and real Axis, would certainly commence."[32]

When Milton describes Adam and Eve's being expelled from the

(cont. from p. 230) the zodiac to the north pole, and obtained, from her disheveled locks, the name of the comet"; (3) the passage of the "tremendous comet of the Sibyl, and perhaps of Pliny," in 618 BC; (4) a "long-haired star" that blazed across the sky in 44 BC, the year of the assassination of Julius Caesar; (5) the return of that comet in AD 531, the fifth year of Justinian's reign; (6) the passage of a comet in AD 1106, "recorded by the chronicles of Europe and China"; and (7) the Great Comet of 1680. (Gibbon, *Decline and Fall,* 2:1426–27.)

Garden of Eden by God and his powerful angels, he brings in the image of a comet (and even a meteor).

> *The Cherubim descended; on the ground*
> *Gliding meteorous.* . . . *High in Front advanc't,*
> *The brandished Sword of God before them blaz'd*
> *Fierce as a Comet*[33]

The impact of Newtonian science was to be so transformative, and even devastating, that, a mere thirty-five years later, Milton's metaphor of God's sword as a blazing comet had become, in Whiston's 1696 *New Theory of the Earth*, a real, live comet streaking past the Earth that tilts it on its axis and triggers off its daily rotation. It was what William Blake, two hundred years later, would abhor: the gobbling up of the poetic, the hyperreal mystical, by cold science. (If Whiston did this, Newton did not; unbeknown to the world, he strove to retain the mystical with the scientific.)

This comet (probably, in Whiston's view, the Great Comet of 1680), which, not accidentally, sped by just as Adam and Eve were expelled from the Garden of Eden, sped by again several orbits later—at the same time, again not coincidentally, that God decided to punish humanity once again for its sinning ways, this time with a Flood that would annihilate all but Noah, his family, and the contents of the ark.

So a comet is the cause of the Flood. How did it do this? Whiston explains that, in that antediluvian era, most of Earth's water was underground. The gravitational pull of the passing comet abruptly yanked these subterranean waters upward so that they struck the underside of the Earth's crust and cracked it open. At the same time, multiple gases in the comet's tail turned into water and fell earthward in a torrent. The rains pounding downward (the "windows," or floodgates, "of heaven were opened," Gen. 7:11–12) and the underground oceans exploding upward through cracks in the Earth's surface (the "breaking up of the fountains of the great deep," Gen. 7:22), together brought to the Earth forty days and forty nights of Flood. Eventually the comet was so far from the Earth that the water drained back into its bowels and the Deluge was over.[34]

This description of the Deluge, some of it based on remarks Edmund Halley made to the Royal Society two years earlier, seems vaguely plausible; perhaps some of this could really have happened. But with Whiston's description of the role the comet plays in the apocalyptic destruction of the Earth we find ourselves entering a never-never land where anything goes and science, fantasy, and desperate contrivance rules. For all we know, Whiston is going to transport us to the planet Mars and show us a row of exhibit cases containing the crystal skulls of ancient Martians.

What does happen is almost as weird. Whiston asks us to imagine that the same comet that caused the Flood (probably, says Whiston, the Great Comet of 1680) is now speeding back from its latest turn around the sun, boiling hot from having just missed falling in. It is heading directly toward Earth, and Whiston has two versions of what happens next. In the first, the gravitational pull of the comet knocks the Earth out of its orbit, sending it hurtling toward the sun, where it is burned to a cinder by the solar orb. Whiston says we can find all this accurately described in Joel 2:30–31, Matthew 24:29, and Luke 21:25–26.

In the second scenario, the one Whiston favors, the comet collides with the Earth. There is an immense conflagration. The incinerated Earth, flickering with scattered fires, rotates more and more slowly and comes to a stop.

Then the Earth takes on—Whiston isn't easy to follow here—the momentum of the comet with which it has collided. It becomes a comet itself, and tears off for the outer reaches of space. Whiston seems to suggest that, amazingly, some humans are still alive on this scorched Earth of a comet-planet. Apparently this is the continuity-ensuring "remnant" of mankind. (See chapter 10, "With Noah on the Mountaintop.")

A long time later—575 years, if we're talking about the orbit of the Great Comet of 1680—the comet-planet returns and settles into the orbit the Earth once occupied. We learn that the creation of the Earth is the destruction of the Earth in reverse. As the comet firms out into a sphere, the husk of the comet it once was detaches itself and hurtles off toward the sun.

What remains is going to be a new planet Earth.

Whiston explains this in terms of Newtonian physics. He also claims it is consistent with the Genesis account of Creation. But how can this be? The Genesis account seems scarcely scientific, given that, to cite just one example, the light is separated from the darkness before the sun and the moon are created.

Whiston insists, with all his colleagues, including Newton, that the "Mosaic" account of Creation is literally true in that it is an account of the creation of the Earth from a comet as you *would have observed it if you'd actually been there.*[35] You had to imagine yourself inside the comet, your consciousness spinning around in its head and then in its tail, not exactly knowing what was happening but observing it all in the same way as is recorded in Genesis. You had to imagine yourself descending to its cooling, coalescing surface of this bewildering ball of matter and then just watching as it metamorphosed into a primordial, pristine, sea-less, planet Earth.

The Mosaic account was simply a birds-eye-view of all this.*

Whiston dedicated *A New Theory* to Newton, insisting that his mentor "well approved of" his book that "depended"[36] on Newton's principles. In fact, says eminent Newton scholar Frank Manuel,

---

*The modern-day catastrophist Immanuel Velikovsky (1895–1979, author of the 1951 worldwide bestseller *Worlds in Collision*) was hugely influenced by William Whiston. His admiration for the brilliant Isaac Newton protégé did not, however, prevent him from including in *Worlds in Collision* a caustic comment about Whiston from the great French naturalist and zoologist Georges Cuvier (1769–1832), who wrote, says Velikovsky, that "Whiston fancied that the earth was created from the atmosphere of one comet, and that it was deluged by the tail of another. The heat which remained from its first origin, in his opinion, excited the whole antediluvian population, men and animals, to sin, for which they were all drowned in the deluge, excepting the fish, whose passions were definitely less violent." (Velikovsky, *Worlds in Collision*, footnote, 57–58.)

"because they are so forthright, William Whiston's works cast important light on the hidden intent and meaning of similar writings by his great contemporary. Where Newton was covert, Whiston shrieked in the marketplace."[37]

If *A New Theory* seems to us today to be a kind of evolutionary dead end in theology and science, it did not seem so to Whiston's contemporaries. The book went into six printings. It was translated into French and German. John Locke wrote: "I have not heard anyone of my acquaintance speak of it, but with great commendations, as I think it deserves." Samuel Johnson's friend Mrs. Thrale enthused that, when she "read Whiston on the expected Comet, how little seem the common objects of our Care!" The philosopher George Berkeley took a copy to the American colonies and donated it to the Yale Library. It became a favorite of Yale president Ezra Stiles, who made sure everyone knew about it. Scholar James Force writes that, in 1849, "Herman Melville, in his novel *Mardi*, was still able to utilize to good effect Whiston's striking vision of the earth after the final judgment, when the earth becomes once again a comet in a radically elliptical orbit on which the damned hurtle between the freezing depths of deep space and the boiling regions near the sun."[38]

In 1701, Newton needed an assistant, and he asked for Whiston. It's not clear whether this was a reward for Whiston's having written *A New Theory*. Whiston took the job, returning to Cambridge with his family. Newton left for London in 1702, leaving Whiston to lecture in his stead and receive Newton's full salary. In 1703, Newton gave up his post as Lucasian professor to become warden (later master) of the London Mint. He had made provision that Whiston should succeed him, and so the young cleric-mathematician did, in that same year.

For several years, William Whiston rode high. He lectured on mathematics and religion; he wrote voluminously; he preached piously; he fathered children (there would be nine in all).

But the eccentricities of this entirely good man were taking their toll. Whiston was a compulsive talker who could keep nothing to himself and could not tell a lie. Despite the warnings of his friends and even of his enemies, he began to speak out about his adherence to the heretical sect of Anglicanism known as Arianism. In 1710, he was summoned before a board at Cambridge, accused of blasphemy, tried, convicted, and fired.

Whiston reacted with his usual fortitudinous buoyancy. He moved his family down to London and plunged into a frenetically productive career of teaching, preaching, writing articles and books, and touring the countryside to give talks. He produced a distinguished translation of *The Genuine Works of Flavius Josephus the Jewish Historian* (1737); his translation is still in print and still widely read.

Newton had quickly broken with Whiston; perhaps he was afraid Whiston would "out" his own, clandestine, Arianism. The great mathematician's antagonism hardened; in 1716, he blocked Whiston from membership in the Royal Society. The relationship deteriorated to the extent that Whiston would ultimately write (although he waited until after his mentor's death) perhaps the most damaging, if accurate, critique of Newton's *The Chronology of Ancient Kingdoms Amended* ever to be published.

The Scientific Revolution was chewing up the scenery, and the Age of Enlightenment was waiting in the wings. Whiston's *A New Theory* was looking more and more like a huge, gorgeous sandcastle that is crumbling on the beach as a whole new tide comes up around it. Whiston became an increasingly isolated figure, a one-man School of Unified Theology and Science heroically striving to hold the sacred and the scientific together. A letter of recommendation written for him in 1714 cautioned prospective employers to "be pleas'd to conjure him Silence upon all Topicks foreign to the Mathematick in his Lectures at the Coffee-house. He has an Itch to be venting his Notions about Baptism & the Arian Doctrine."[39] Believing God would call the Jews back to Jerusalem no later than 1766, Whiston tried to locate the lost tribes of Israel himself. He regularly toured seaside spas with scale models of the Tabernacle of Moses and the Temple of Solomon to back up his talks on the imminent conversion of the Jews.

Whiston's eccentricities were never more strongly in evidence, nor his earliest beliefs ever more stubbornly held, than when he preached to the people of London on the subject of two minor earthquakes that had occurred on February 8 and March 8, 1750. He thundered from coffeehouses that these tremors were God's wrath, evoked by the sin and folly of mankind, and that on account of the "horrid Wickedness of the Present Age" a far more devastating earthquake was in the making for April 8. It could be averted only if mankind instantly changed its ways. Whiston trumpeted to one and all that these earthquakes had been predicted in Isaiah 5:18–20; 2 Esdras 9:3; and Revelation 11:13.

The eccentric, leonine seer, now a very old man, still knew how to roar. His warnings were heeded. On April 7, 100,000 Londoners took refuge in Hyde Park as 730 packed coaches rattled past on the way to safety in the countryside. But nothing happened on April 8. Others, of lesser ability, had predicted the disaster; a wrathful public saw to it that one was incarcerated in an insane asylum. Whiston escaped with only a storm of ridicule and a lampooning in a pamphlet.[40]

It didn't bother him; he was used to it. William Whiston died two years later, aged eighty-four—ridiculed, venerated, scoffed at, and beloved. His books were still selling well. Whiston's superb translation of Josephus's *Antiquities of the Jews* still sells well today; his *New Theory of the Earth* stands out as a little-known monument to the splendor and misery of a humankind that, because of its insecurities, must cling to its old beliefs even as it is brilliantly transcending them.

Edmund Halley, the great English astronomer who is renowned today for having predicted the year of the return of the comet that bears his name, in 1692 tried and failed to obtain the post of Savilian Professor of Astronomy at Oxford University.

It was long thought that Halley's suspected atheism was the reason he didn't get the job; after all, hadn't Halley, in the course of the interviews, made a number of cavalier and even provocative remarks about Christianity, such as that "he believed [in] a God and that was all," and, "I declare myself a Christian and hope to be treated as such"?

It has become clear only in the past few years that Halley failed to get the job not because of his religious beliefs or lack of them, but simply on a technicality; he had received his M.A. from Oxford not because he'd done the traditional course work (the completion of which was essential if you wanted to work there), but because of an achievement of his own.

Halley was a college dropout centuries before there were many college dropouts at all. He left Oxford at the age of twenty to sail to the island of St. Helena in the South Pacific (where Napoleon would later be imprisoned) in time to observe the 1676 transit of Mercury across the sun. During his fifteen-month stay, he set up an observatory and mapped the southern skies; not long after his return to Oxford he published an atlas of 341 stars of the Southern Hemisphere. This remarkable achievement earned him an M.A. from Oxford, but in 1691 he learned that to become a professor at Oxford you needed to have taken the degree in the ordinary way. (However, eleven years later, in 1703, when it had become impossible to ignore Halley's extraordinary accomplishments, he was named Savilian Professor of Geometry at Oxford.)

It has also recently emerged that, contrary to what historians thought for three centuries, Halley didn't believe that the world was eternal, which belief would have put him not on the side of those who fought, like Newton and Whiston, to reconcile Newtonian science and biblical scripture, but in the camp of those who contributed to the destabilization of that relationship. Halley even seemed to believe in the truth of the Mosaic account of Creation, adding, however, an important qualification: the first five days of creation weren't just five years long; they were *extremely* long, providing time for what would come to be called geological epochs. In a paper on the salinity of the oceans, Halley wrote that

> 'tis no where revealed in Scripture how long the *Earth* had existed before this last Creation, nor how long those five Days that pre-

ceded it may be to be accounted; since we are elsewhere told, that in respect to the Almighty a thousand Years is as one Day, being Equally no part of *Eternity;* Nor can it well be conceived how those Days should be to be understood of natural Days, since they are mentioned as Measures of Time before the Creation of the Sun, which was not till the Fourth Day.[41]

So we should not be surprised to learn that, on December 12, 1694—two years before Whiston published his *New Theory of the Earth*—Halley read a paper titled "About the Cause of the Universal Deluge" to a meeting of the Royal Society. In this paper, Halley put forward a theory of periodic catastrophism, specifically suggesting that the Noachic Flood was caused by a comet and trying to explain the mechanisms involved (see below).

The son of a wealthy London soap manufacturer, Edmund Halley (1656–1742) was doing professional work in astronomy by the time he was ten. The young Londoner went up to Oxford—and then, as has been explained, made the kind of zigzag turn often ascribed to genius, leaving Oxford, traveling to the South Atlantic, mapping the stars of the Southern Hemisphere, and, because of that achievement, not long after receiving his M.A. from Oxford "by order of the king."

Halley was diplomatic, friendly, outgoing, and blessed with a swashbuckling zest for life. While on St. Helena he had strapped a barometer on his back and toiled up a mountainside to measure the relationship between altitude and atmospheric pressure. Back in England, he invented a diving bell and stayed in it sixty feet underwater for five hours. At the king's behest, he went pub crawling with Peter the Great when the Russian czar visited London. Halley calculated how high bullets can be shot and how much the wind weighs. He wrote the first annuity tables for life-insurance payments. And he learned Arabic so he could translate into Latin and English an ancient text on mathematics whose Greek original had been lost.

In 1698, captain of the ship this time, he sailed on a scientific expedition to the South Atlantic. En route he had to face down a first mate and quell a mutiny. Halley sailed back to London and brought charges against the officers. In 1699 he once more set sail on the same expedition, voyaging deep into Antarctic waters and mapping both the prevailing ocean winds and variations in the Earth's magnetic field.* Halley was gone for a year on this severely scientific and yet wildly romantic voyage; Voltaire wrote that, compared to this expedition, "the voyage of the Argonauts was but the crossing of a bark from one side of a river to the other."[42]

All of Halley's scientific brilliance, all of his exquisite tact, all of his keenness for living, were to be put to the test, and over a long time, as the wholly unexpected consequence of a single night's conversation in London's Grecian Coffeehouse in January 1684.

A scrounger of garbage behind the coffeehouse the next morning would have been bewildered to find a pile of crumpled napkins on which were sketched out in coffee grounds the curving lines of planetary orbits. These were the worksheets of a heated conversation, whose brilliance must have made the angels listen rapturously, among Edmund Halley, Christopher Wren, and Robert Hooke. Wren (1632–1723) was an astronomer and mathematician as well as the architect who rebuilt Saint Paul's Cathedral after the Great Fire of London. Robert Hooke (1635–1703) was a hyperactive, volatile genius the number of whose mistresses (including his niece, Grace, a ward) was exceeded only by the number of scientific fields—physics, optics, biology, astronomy, and more—in which he had made a significant advance.

---

*New Age enthusiasts will be intrigued to learn that Halley, in a paper read to the Royal Society in 1691, advanced the hypothesis that the Earth is partially hollow. Trying to explain anomalous magnetic compass readings around the world, he speculated that the Earth might be made up of a 500-mile-thick outer shell we walk on, a second concentric shell roughly the diameter of Venus, a third the diameter of Mars, and an innermost core the size of the planet Mercury. He wondered if the inner regions might have atmospheres, or be illuminated, or be inhabited, or if the three spheres rotated on their axes at different speeds.

These three exceptional men accepted, as did their peers, Kepler's description of the orbital motions of the planets. But they didn't know, any more than had Kepler, what kept the planets in those orbits. In trying to understand this, Hooke had actually come up with a kind of proto-theory of gravitation. (This was three years before Newton's *Principia Mathematica*.) All three were playing with the notion that, if gravity were a fact, then the force of a planet's attraction toward the sun was reciprocal to the square of its distance from the sun. The question they were asking themselves that evening was: If all that is so, what is the shape of planet's orbit?

They scribbled down equations but couldn't really decide. Wren asked Halley what Isaac Newton, up in Cambridge, might have to say about this? Newton was already famous for his stunning explanation of the nature of light, and he had acquired a reputation for high genius. Before the evening was out it was decided that Halley, the only one of them who had met Newton even once, would put the question to the already-renowned mathematician the next time he was in Cambridge. On that note, the evening ended.

It was not until August 1684—seven months later—that Halley, on a visit to Cambridge, presented himself at Newton's lodgings and, after the usual exchange of amenities, put the question to him of the shape of a planet's orbit.

Newton answered immediately: "An ellipsis."

"How do you know?" asked a surprised Halley.

"I've written a paper about it," said Newton. He told him he didn't know where the paper was but would send a copy to Halley as soon as he found it.

Six months later, at his home in London, Halley received from Newton a nine-page treatise called *De Motu* ("Of Motion"). The young astronomer read it with growing astonishment. He saw that it was "a step forward in celestial mechanics so immense as to constitute a revolution."[43] Halley rushed back to Cambridge and implored Newton to elaborate on *De Motu*. The Royal Society, he said, would be honored to publish anything Newton might care to send them on the subject, or on any other.

Halley returned to his busy life in London. Newton dropped out of sight for two years. All alone, as he had been when he invented calculus and laid the foundations of gravitation theory while on his mother's farm—all alone, he developed a colossal work that would completely change the way mankind sees the universe.

All alone—except that Edmund Halley watched from the sidelines and intervened whenever he thought Newton needed help or a push. Newton was preparing the work for publication. The Royal Society would publish it. Halley was the editor. With great tact and diplomacy he pushed, prompted, probed, pleaded with, cajoled, and outrageously flattered Newton (though this flattery wasn't really outrageous; it was perfectly justified in light of the magnificent work that Newton was producing). In 1686, the first volume was completed. While it was in the press Newton completed the second and then the final volume. Halley edited every word and supervised the printer (and paid all the expenses). In 1687, the *Principia Mathematica* appeared, delighting, astounding, and confounding (it was very hard to read) the scientific universe.

Would there have been a *Principia* without Edmund Halley? The jury is still out. (See chapter 18, "Son of Archimedes.") Richard Westfall writes: "Halley did not extract the *Principia* from a reluctant Newton. He merely raised a question at a time when Newton was receptive to it. It grasped Newton as nothing had before, and he was powerless in its grip."[44]

One of the last sections Newton completed was on comets and their periodicities. Poring over Newton's groundbreaking text, Halley became fascinated by comets. He had seen the Great Comet of 1680 when, aged twenty-four, he was galloping on horseback to Calais from Paris on a night filled with stars; perhaps, as he looked up at the sky, the still-fresh memories of his amorous conquests on the Continent made the comet seem even brighter than it was. Halley was equally impressed by the mysterious comet that streaked across the sky in 1682—at about the same time, as it happened, that Halley married Mary Tooke, who would bear him three children; here too we can speculate that the gleam in the sky might have been accentuated by the gleam in Halley's eye.

This comet of 1682 traveled in the opposite direction from most other comets. So Halley did not forget it as, working hard on the *Principia* with Newton, his attention was taken up with Newton's work on the Great Comet of 1680. Once the *Principia* was published, Halley began to to ruminate again on the odd, backward-moving comet he'd observed in 1682. He learned of three other backward-traveling comets, those seen by Peter Apion, in Ingoldstadt, Germany, in 1531 and by Johannes Kepler and Danish astronomer Christian Longomontanus in 1607. Halley theorized that the comet of 1682, and these three comets, might be the same. It seemed that the periodicity of this one comet was seventy-six years. Halley set about to demonstrate this mathematically. Newton had demonstrated in the *Principia* that gravity was a property of every celestial body in the universe, huge or small. This meant that the comet of 1682 as it sped along had to thread its way through an invisible jungle of interweaving gravitational fields, not only those of Jupiter and perhaps Saturn but also those of their satellites, of nearby asteroids, and of any celestial object within tens of millions of miles. On top of that, each gravitational field interacted with all the others. Only in 1707 did Halley emerge from beneath his tremendous pile of calculations to publish mathematical proof that the periodicity of the comet of 1682 was indeed seventy-six years. He predicted the comet would return to Earth's skies in 1758.

In the paper on comets and the Flood that he read at the Royal Society in 1694, Halley scarcely mentioned God at all. He was nervous about this, given his reputation for being contemptuous of religion, and asked that the Royal Society not publish his talk; the paper didn't appear in the society's *Philosophical Transactions* until 1723.

Researching years of rainfall records in an English county, he decided it couldn't rain enough in forty days and nights (presumably from a comet's tail) to cover the Earth's surface entirely. Halley thought the gravitational pull of a comet passing nearby could have made the Earth wobble on its axis, so much so that the oceans would slosh up over the continents. This would have constituted a global flood—in fact, a Noah's Deluge. When the oceans flowed back to their original places, they left bodies of water in the larger land cavities, creating, for example, the Caspian Sea.[45]

Halley thus became the first scientist to hypothesize that comets and meteorites (at the time thought to be the same) had impacted the Earth throughout the ages and were responsible for many geographical formations.

On January 14, 1742, Edmund Halley, named Britain's second Astronomer Royal upon the death of John Flamsteed in 1719 and now eighty-five years old, sat behind his desk at Greenwich Observatory, poured himself a tall glass of wine, took a long sip, and expired.

Sixteen years later, in 1758, the comet of 1682 reappeared in the skies. Jubilant Englishmen named it after Edmund Halley, and, ever since, his name has been inseparably linked to the very idea of a comet.

It's likely that Halley died without a prayer on his lips. He had never believed in the sacramental nature of the universe, and he certainly did not feel that it was desperately important, or even important at all, for mankind to hold on to this concept.

Isaac Newton felt very differently.

The Great Comet of 1680 never ceased to haunt Newton's imagination. In the *Principia Mathematica,* he describes it for us.

> The comet which appeared in the year 1680 was, in its perihelion [closest approach to the sun], less distant from the sun than by a sixth part [144,167 miles] of the sun's diameter; and because of its extreme velocity in that proximity to the sun, and some density of the sun's atmosphere, it must have suffered some resistance and retardation; and therefore, being attracted somewhat nearer to the sun in every revolution, will at last fall down upon the body of the sun.[46]

Newton observes that in its perihelion the Great Comet must have been 2,000 times hotter than a red-hot iron. If it was the same size as the Earth, and cooled like other terrestrial bodies, it would take 50,000 years to cool down.[47] If its periodicity were 575 years, as Newton believed, it would have to circle the sun almost 100 times before its temperature returned to normal—except that its temperature would never return to normal, because each time the comet approached the sun it would become hotter than red-hot again.

Newton says nothing in the *Principia* about what grim consequences might ensue if the Great Comet crashed into the sun. He was silent on this subject until a day in March 1724, when he privately made statements about the Great Comet, the sun, and the end of the world that were so uncanny and unsettling that we can be forgiven for asking ourselves what strange alchemical fumes the great man must have been inhaling on that late winter day at the very end of his life.

Perhaps Newton didn't mean to say what he said on March 7, 1724. Perhaps he let these indiscreet words slip merely because he was an old man and subject to such lapses; Newton had turned eighty-one that past Christmas Day. But he was still vigorous: he had all his teeth, his eyesight was undimmed, and later that year he would catch Edmund Halley out in a mathematical error the younger man made while correcting the third edition of the *Principia*. And Newton was still president of the Royal Society and master of the London Mint.

Perhaps he let his guard down simply because of the physical comforts he now enjoyed: his own large luxurious house in an upscale district of London, with its large soft armchairs upholstered in scarlet, the sitting in of which would deprive anyone of all resistance, and a fire blazing in a fireplace whose steady heat could easily lull the most alert into a state of torpor.

Perhaps what disarmed him that day was the loving presence of all that he could call his family in London. That afternoon his stepnephew-in-law, John Conduitt, wealthy, newly elected to Parliament, had come around to see how he was faring in the aftermath of a lengthy attack of the gout. And there may have hovered, behind Newton's armchair, a radiant forty-four-year-old woman to whom all the clichés applied: the beautiful, brilliant, and witty Catherine Barton Conduitt, Newton's niece by his half-sister Hannah Smith and John Conduitt's wife of seven years.

This statuesque, worldly woman was the only one in Newton's family to begin to match him in talent. For twenty years she had lived on and off under her stepuncle's roof. "La Bartica," as the magnificoes called her, had been the toast of London society; great men exchanged witticisms with her, found her irresistible, fell in love with her—were rebuffed by her (though not all of them) with exquisite grace. The French mathematician Rémond de Montmort, visiting London for induction into the Royal Society, met her and afterward wrote, "I am deeply stirred by the honor she does me in remembering me. I have preserved the most magnificent memory of her wit and beauty."[48]

A rumor still circulated in London that Catherine had been secretly married to the powerful Whig politician Charles Montague, First Earl of Halifax, Chancellor of the Exchequer, and patron to Isaac Newton—secretly married and, though it seemed unbelievable, with the passive connivance of the prim and puritanical Newton. The rumor persisted because the public knew that Montague, when he died in 1715, had left her £25,000 ($2 million today), a park, a palace, a mansion, and his jewelry.

That was all in the past now. This afternoon, as Isaac Newton unexpectedly turned the conversation to the subject of the Great Comet of 1680, Catherine Barton Conduitt pricked up her ears, and not only because she was fascinated by astronomy. One of her great friends, when she was the on-again, off-again mistress of a literary salon in her uncle's house and others, had been the great satirist Jonathan Swift. Swift hated science and feared its effects on humankind. Not infrequently he had spoken to her derisively about the obsession of scientists with remote and unknowable objects, like the sun and comets. He'd sworn to her that one day he would steal a paragraph on the Great Comet out of her uncle's *Principia* and lampoon it—but gently, for her sake.

And so Catherine listened intently as Newton remarked soberly to her husband that, "after a certain number of revolutions, a comet, by coming nearer and nearer to the sun, has all its volatile parts condensed and becomes material ready to be drawn into the sun, which has become needy of replenishment because of the constant heat and light that it emits. The comet will be drawn in like this—" He bent forward suddenly and tossed a fagot into the fire. It blazed up with a roar as he continued, "The Great Comet of 1680 has perhaps five or six revolutions left in it. Its size is such, and its temperature will be such, and it will be moving so swiftly, that when it strikes the sun the sun will blaze up just as this fire has done."

He paused, then went on: "And that will so much increase the heat of the sun, that this earth will be burnt, and no animals on it will live. And so sooner or later that will probably be the effect of the Comet of 1680. And perhaps sooner—in 500 years—the next time the comet approaches the sun."[49]

Catherine, beguiled and alarmed and amused, remembered that Jonathan Swift had told her that one day, in a great satiric work, he would pack all the scientists in the world into a floating city where, cut off from normal conversation, they could talk fearfully and exclusively about the sun and sunspots and comets and the upcoming horrible end of the world. Catherine, with her own irrepressible urge to mock, understood why her brilliant friend would want to do this, and she understood all the more when her uncle now observed, with an unnerving sobriety, that there were already "marks of ruin" on the Earth's surface that could not be explained by Noah's Flood and that this implied that the Earth's surface had been incinerated once before.

Catherine knew that, perhaps at that very moment, a thousand miles away in Dublin, Jonathan Swift was putting the finishing touches to book three of his great lampoon *Gulliver's Travels,* in which—as he had told her only a month before, in the correspondence they maintained— he had finally created that floating city, and called it Laputa, and filled it with all those scientists in the world who were so blinded by the sun and other celestial bodies that when they met an acquaintance in the morning the first question they asked was about, not the friend's, but the sun's health, whether it looked well at its rising and setting, and what were their chances of avoiding the stroke of an approaching comet.[50]

Now her husband was calmly asking her uncle how, given that such a catastrophe would not possibly leave survivors—how could the Earth be repeopled? Isaac Newton seemed not to hear. Then he replied brusquely that that would "require the power of a creator." He added: "I have always maintained that all of the revolutions of the heavenly bodies are by the direction of the Supreme Being."

John Conduitt leaned forward. "But exactly how might the Supreme Being use His powers, given the urgency of a case like this?"

Isaac Newton replied: "All this might be superintended by superior 'intelligent beings' under God's direction. It may be that God keeps in reserve superior beings, say on a moon of Jupiter, to step in and do what is required in a case like this."[51]

John Conduitt was silent. And Catherine Barton Conduitt, hearing this, found it easy to repress her amazement and her laughter and even to expel from her mind the mocking image of Jonathan Swift. For she had learned not to dismiss a single word her illustrious uncle uttered, even when it was a matter of such extravagant words as she had just now heard.

So she stored away, under advisement, what Isaac Newton had said about superior beings on a moon of Jupiter. And very quickly the subject changed though the conversation went on, and that evening she and her husband left her uncle's house with the usual fond good-byes.

We don't know what conclusions Catherine Barton Conduitt eventually reached regarding aliens on a satellite of Jupiter prepared to help the world at the time of the ultimate crisis. But we can speculate about the careful thinking that lay behind everything Isaac Newton said, and about how it might have applied in this case: that, given Jupiter's immense distance from the sun (467 million miles, five times farther from Earth), and its huge size (a diameter of 86,000 miles, four times greater than that of Earth); given that any one of its moons, provided it happened to be on the far side of Jupiter from the sun, would be amply protected from harm in the event of a solar catastrophe—given these circumstances, a Jovian satellite might be spared so that its inhabitants might succor our world. Newton must have concluded that this was the nearest point from which help could come to help the planet Earth in the throes of incineration.

We might expect God to have positioned Jupiter at its closest approach to Earth at the time of the Great Comet of 1680-triggered eruption of the sun. And, in fact, twenty-six years after Newton's eerie conversation with John Conduitt, in 1750, when the earthquake predicted by Whiston and others failed to eventuate, there were, writes James Force, "even reports circulating that Newton had himself scientifically predicted the earthquakes on the basis of the close approach of Jupiter."[52]

But, in a letter to London's *Daily Advertiser* of March 14, 1750, William Whiston wrote irately that he did not "in the least believe that Sir Isaac Newton foretold any Earthquake; and is sure that *Jupiter*, at the Beginning of of this Year 1750, was, and is 400,000,000 Miles off the Earth, and so could not possibly have any Influence on Earthquakes here below."[53]

So Newton's strange comments about Jupiter and the destruction of the world seem to have acquired a life of their own and continued to vibrate up through the temporal ether. Though Whiston got it backward: Newton did not suggest that the planet Jupiter would cause destruction; he essayed, rather, that strange beings from one of its satellites would come to our ailing planet and somehow secure it.*

---

*In his conversation with Conduitt, Newton revealed "that he thought the inner planets of the solar system would be devastated by a comet that would crash into the Sun and cause it to dramatically expand. Newton thought that there was life on other planets, and he told Conduitt that the 1572 and 1604 supernovas were the result of the same process happening in other star systems." Late in the seventeenth century Newton had also "told David Gregory the satellites of Jupiter were proto-Earths that were 'held in reserve' by God to repopulate the solar system after the cometary cataclysm." (The Newton Project, www.newtonproject.ox.ac.uk/tour/newton-on-science-and-religion.)

▲

Now that we've seen the importance of comets in the scientific and the religious life of the natural philosophers of the seventeenth century, it's time to return to Mount Ararat and watch Noah and his family as they make their way down the slopes of the sacred mountain and begin to repopulate the Earth. Before that, though, we'll take a look at how Newton fought to revise the chronology of the ancient world so that he could better understand just how it was that the post-Flood remnant of mankind grew into the mighty nations of the world.

CHAPTER TWELVE

DECONSTRUCTING TIME

In the early 1970s, a group of Soviet mathematicians at Moscow State University made an astonishing announcement: All the dates in history are wrong. The recorded history of mankind started no earlier than AD 900, and most historical events took place after AD 1300.

The leader of this group of time iconoclasts was a professor named A. T. Fomenko. He and his fellow researchers were building on the work of the Russian mathematician and topologist Nicolai Aleksandrovich Morozov (1854–1946), who in 1924 challenged traditional chronology with his book *Christ: A History of Human Culture from the Standpoint of the Natural Sciences*. Claiming to utilize cutting-edge techniques in mathematics, astronomy, linguistics, philology, and geology, Morozov had reworked the chronologies of ancient Greece, Rome, Egypt, and China and come up with the conclusion that no recorded events took place in the ancient world before the third century AD.[1]

Fomenko spells out his group's assertions in a four-volume work, *History: Fiction or Science?* published between 2005 and 2008. To most of us, these claims seem preposterous, and almost no modern historian takes them seriously (although former world chess champion and Russian opposition leader Gary Kasparov has shown a keen interest). How could all the dates in all the history books be wrong?

Fomenko says the answer lies in rapacious Western capitalism. He asserts that in the West the facts of history have long been up for sale to the highest bidder. In Europe, "the court historians knew only too well how to please their masters," who were, to cite one example, "*condottieri* [mercenary soldier-captain] upstarts who were seeking legitimacy in days of yore in order to become popes, cardinals or to found regal dynasties such as the Medici. They paid exceedingly well for a glorious but fictitious past."[2]

Lambasting European capitalism with a vehemence that threatens to heat up the cold war, Fomenko fulminates that

> The corporations of Petrarch and Dante, Bracciolini and Machiavelli, Giotto, Bernini, Da Vinci and Michelangelo not only created immortal masterpieces [that were] exceedingly well-paid for by Roman popes *et al* and Medici princes of Florence, but they also mass-produced "ancient" manuscripts, frescoes, statues [that were] very much in demand by wealthy customers from England, France, Germany and Russia.... Oxbridge scholars earned their daily bread and butter by cooking very ancient Greek and Roman Empire history, mostly from Italian ingredients.[3]

Fomenko's assault on history may seem to be politically motivated and to derive mainly from conspiracy theory. But it serves to remind us that the dates we learned in school ("1066 and all that") and thought were set in stone are far more the creation of human desire, error, and malfeasance than we would like to think.

It was Isaac Newton who by and large devised the tools the Russian mathematicians used to draconically shorten the chronology of the world. When it came to stripping centuries from history, Sir Isaac literally wrote the book: in his landmark work *The Chronology of Ancient Kingdoms Amended,* published posthumously in 1728, he hacked five hundred years out of the chronology of the ancient world.

Newton begins his book in about 1100 BC with the "first memory of things in Europe" and goes on to minutely trace the chronologies of the Greeks, Egyptians, Assyrians, Babylonians, and Medes and Persians up to Alexander the Great's conquest of Persia in 334 to 331 BC. There is no chapter on the Israelites, because Newton believed the chronology of the Jews in the Old Testament was without error; in place of a chapter on ancient Israel and Judaea he substitutes a five-page diagram of the Temple of Solomon.

The Chronology of Ancient Kingdoms Amended is one of the strangest books ever written. Newton toiled over it all through the second half of his life and was still working on it when he died; clearly, it was of the greatest importance to him. Newton scholar Richard Westfall, who wrote the definitive biography of Sir Isaac, thought it was one of the most excruciatingly boring books ever written: "A work of colossal tedium, it excited for a brief time the interest and opposition of the handful able to get excited over the date of the Argonauts before it sank into oblivion. It is read today only by the tiniest remnant who for their sins must pass through its purgatory."[4]

The chronologists of France, who happened to see a pirated abridgment of the book twelve years before it was actually published, found it infuriating and pigheaded, because Newton had drastically abbreviated some of the time periods most cherished by French historians. They hurled insults across the Channel at the English mathematical giant; Newton howled back, and this seemingly harmless little volume, which made Richard Westfall numb with boredom, provoked a literary scandal ten years before it was published.*

Three centuries later, another Frenchman, Bernard Chazelle,† declared that the book must have made many Europeans of a romantic turn of mind happy when it was published, because, in cutting 500 years out of history, Newton brought the mythological lovers Dido, Queen of Carthage, and Aeneas, future founder of Rome, within

---

*This scandal was one reason (the main one being that it had been written by Isaac Newton) there was a bidding war for this obscure chronology/history book when Newton's heirs offered it for publication. The winning team of two publishers bought the rights for £350, a huge sum at the time.

†Chazelle is Eugene Higgins Professor of Computer Science at Princeton University.

thirty-two years of each another. In the *Aeneid,* Virgil tells the story of their passionate love affair; but traditional chronology had put three hundred years between the two. Chazelle believed that Newton, though no romantic, must have delighted European lovers of romance by (quite unintentionally) proving that Dido and Aeneas lived at virtually the same time—close enough, at any rate, to have had a love affair.[5] Possibly; but only a few hundred scholars and clerics read *The Chronology of Ancient Kingdoms Amended* when it was first published.

In the 1980s, maverick Newton scholar David Castillejo (who later became an actor and drama historian) declared that Newton's *Chronology* must be a whole new art form. He explained that on the surface the *Chronology* is "so thick and boring as to be almost impenetrable" but that, since Newton was a towering genius and "working at the height of his cunning and subtlety" when he began the *Chronology,* it surely must contain hidden depths of esoteric meaning.[6]

Castillejo pointed out mysterious number patterns in the book and linked them to similar patterns in Newton's writings on alchemy and his interpretation of Saint John's Revelation. The scholar-actor speculated that the *Chronology* was a skein of hieroglyphs encoding original information about gravity and an unknown force opposing gravity. The trouble was that Newton's book was such an "incredibly difficult work to decipher that it will probably have to be passed one day through a computer" to yield up its hidden treasure.

Castillejo begins his discussion of the *Chronology* by stating that it consists of "lists of names, names, almost nothing but names. Names of princes, names of heroes, names of kings. Names following names, multiplying names, transformed into other names—all intertwined together in a tight mesh."[7]

This gives us fair warning of the thick jungles of weird nomenclature we are about to enter. Newton's book is like a gigantic telephone directory; it brims over with names of kings, queens, generals, villains, gods, heroes, explorers, and inventors, not connecting them to telephone numbers but locking their lives with specific events in the Bible. Newton's sentences are as terse and spare as quadratic equations. They avoid adjectives or adverbs at any cost. Sometimes a paragraph springs into being that looks like a simple statement of historical fact but then blossoms out into a description of a crucial turning point in cultural history, such as the invention of music;* almost always, Newton cites a breakthrough in technology as the cause of a particular cultural leap forward.

The reaction of modern-day readers to *The Chronology of Ancient Kingdoms Amended* seems to be a function of how seriously they take it. James Joyce said he wanted the ideal reader with ideal insomnia to devote his life to reading *Finnegan's Wake*;[8] some believe the *Chronology* requires the same degree of commitment. Others, though, seem able to immerse themselves in this incredibly rich farrago of ancient history with a certain zestfulness. L. Cress writes:

> As I read, I would wonder why we were touring a small isle in the Mediterranean, debating the heritage of a prince—and suddenly, aha! Newton would say, "therefore," and prove that the four generations of history we had endured proved that someone was the same age as someone else, and that his great grandson therefore could not have been as ancient as some believe. Many events are connected by degrees of association to the Trojan War, to the reign of King Solomon, or to Nebuchadnezzar's destruction of Jerusalem. From thence an ancient history spiders off into tales of a king whose fifty daughters married his brother's fifty sons, then at their father's command all murdered their husbands so as to defend against their uncle's betrayal; to the way Philistia was overrun by exiles from Egypt, making the Philistines more powerful and land-desperate when they fought the

---

*"Where they [a branch of the Phoenicians called the Curetes] settled [in Greece] they wrought first in copper, 'till iron was invented, and then in iron; and when they had made themselves armor, they danced in it at the sacrifices with tumult and clamor, and bells, and pipes, and drums, and swords, with which they struck upon one another's armor, in musical times, appearing seized with a divine fury; and this is reckoned the original [origin] of music in Greece." (Newton, *Chronology of Ancient Kingdoms Amended*, 146–47.)

Judges, Saul, and David. Apparently all the ancients ran around conquering each other, erecting pillars, kidnapping princesses, and stealing them back. A great king of one country would be worshiped by his colonies in other countries until no one remembered he was a king and everyone thought he'd always been a god.[9]

The Chronology of Ancient Kingdoms Amended is organized around Newton's core conviction that Jewish civilization was the first in the world and nourished all other civilizations. As we've seen, ancient semitic languages scholar Abraham Yahuda characterized Newton as a "Judaic monotheist of the school of Maimonides," regarding him as "more a monotheist than a trinitarian" for whom "Jehovah is the unique God."[10] In the *Chronology,* as in many other of his works, Newton's "secret faith"—his fierce Judaic monotheism—shines through like a beacon.

Newton seems to have begun the *Chronology* in 1685 with the aim of proving that the philosophers of the ancient world knew the Earth went around the sun and that everything was made of atoms. He seems to have ended the *Chronology* by veering into an amazingly prescient attempt to understand, in the vein of philosophers of history like Giambattista Vico and Arnold Toynbee, the origins and rise of civilizations. His overarching concern seems always to have been to prove to the world that "the Israelites at about the time of David gave writing, art, science and commerce to countries like Egypt, Assyria, Babylonia and Phoenicia, and that the transmission of culture had not been the other way around"[11] (although, as we will see, Castillejo had suggested that Newton's "historical" text was a cover masking something far different).

Newton had his work cut out for him. He first of all had to deal with the enormous time spans of ancient nations that had been handed down by the world's first historians: for Chaldea-Babylonia, 473,000 years (Berossus, third century BC); for Egypt, 36,525 years or thirty dynasties or 131 generations (Manetho, third century BC); and for Greece, 18,000 years (as inscribed on the Parian marble, a 2,000-year-old pillar transported from Athens to London in 1627).

As a Christian with certain fundamentalist views, Newton couldn't countenance these figures, because according to the Anglican bishop James Ussher, the world had been created only in 4004 BC (the Flood occurring 1,656 years later, in 2340 BC).

So how did the great chroniclers of antiquity get their numbers so wrong?

Modern-day chronologer Larry Pierce, editor of the updated version of Newton's *Chronology of Ancient Kingdoms Amended,* explains that

> in the centuries before Christ, a war broke out to see which nation had the oldest pedigree, whether real or invented. Just as an arms race raged between the super-powers in the 1960s, so an age race raged among the ancient civilizations in the centuries before Christ's birth. Each claimed to have the oldest history. While some writers seemed interested in the truth, others were playing a game to see who could spin the biggest and most convincing yarn about the antiquity of their nation.[12]

Frank Manuel adds that "great age was a chronological sign that one's kings and one's gods were closer to the fount of pure knowledge."[13] And Newton declared, at the beginning of the *Chronology,* that "all nations, when they begin to keep exact accounts of Time, have been prone to raise their Antiquities; and this humor has been promoted, by the Contentions between Nations about their Originals [origins]."[14]

Even non-patriotic historians faced multiple difficulties in getting their chronologies right. War, famine, treachery, rapine, revolt, the murder of monarchs, dynastic marriages, and the fury of rebellion kept the Mediterranean world in a state of semi-chaos. Historical records were easily lost, destroyed, or corrupted. In 525–523 BC, the Persians under Cambyses II conquered Egypt and destroyed or carried off the nation's entire store of historical archives—except for that part of it which had already been carried off, a century before, when the Assyrians conquered Egypt. Never-ending war kept the priests scrambling to reconstruct a part or all of their historical heritage. It's understandable that they made mistakes, and even that they seized opportunities to fabricate. Moreover, "except for the Bible itself, the other histories of early nations were not recorded until well after the events had passed," writes Manuel.[15] The first historian to write about ancient Egypt, for example, was Herodotus (ca. 484–425 BC). Newton believed that no oral tradition older than a century should be trusted.

Sir Isaac had some worthy predecessors in the business of creating chronologies. Theophilus of Antioch (115–180) produced a brief catalog of biblical events based on the "divine authority" of Christianity. Later on, a pagan convert to Christianity, Sextus Julius Africanus (ca. 160–240), matched up Greek and Latin history with biblical history, labeling every event from the Creation in 5499 BC to year three of the emperor Eliogabalus's reign (AD 221) as "in the year of Abraham."

Eusebius of Caesarea (260/265–339/340), whom we have already met as a judicial presence at the Council of Tyre, wrote a "global" chronology but did little to shorten the astronomical genealogies of Chaldea, Egypt, and Greece. More than 1,200 years later, the formidable polymath Joseph Justice Scaliger (1540–1609) ransacked the oldest libraries in Europe and came away with fifty ancient calendars that he synchronized to create the most comprehensive chronology yet. But even he did not seriously grapple with the vain and bloated time lines of the ancient world. That task would be left to Sir Isaac Newton.

Newton tackled the overgrown chronologies of the first empires on Earth by inventing two new laws. The first, we might call the law of parsimony. This law was to be used with breathtaking effectiveness by A. T. Fomenko and his peers in their breakout chronological multivolume *History: Fiction or Science?* The law of parsimony is based on the principle of Occam's razor: that entities should not be multiplied beyond necessity; or, to put it more simply, that, all things being equal, the simplest answer is the right answer.

Deploying this law was one of the things Isaac Newton was born to do. "It was Newton's peculiar art to reduce unbelievable complexity to almost unbelievable simplicity," writes David Castillejo.[16] The best illustration is Newton's universal law of gravitation, the elements of which began to form in Newton's mind when it occurred to him that the force that makes an apple fall might be the same force that prevents the moon from flying off into outer space. John Conduitt wrote in his *Memoir* of Newton:

> In the year 1666 he retired again from Cambridge ... to his mother in Lincolnshire & whilst he was musing in a garden it came into his thought that the power of gravity (which brought an apple from the tree to the ground) was not limited to a certain distance from the earth but that this power must extend much farther than was usually thought. Why not as high as the moon said he to himself & if so that must influence her motion & perhaps retain her in her orbit.[17]

Eventually, Newton would prove that gravity is a function of every discrete body in the universe.

The great mathematician would display his skill in "conflation"—compressing into one entity (or into a very few) a multiplicity of entities all bearing a resemblance to one another—in his interpretation of Saint John's Revelation. Here, he usually calls conflating "synchronizing"; he demonstrates it when, for example, he declares that four prophetic hieroglyphs—the woman in the wilderness, the seven-headed beast, the downtrodden court, and the witnesses in sackcloth—are one and the same event because they each have the exact same duration in time.

Never did Newton apply the principle of Occam's razor more vigorously than when he lit into the tangled and monstrously overblown time lines of the ancient world. Here he wields this razor like a machete. The champion of the Jews hacks away superfluous underbrush, tears out whole vineyards, and sometimes fells entire orchards—and finally history becomes rational and manageable. Newton excoriates the priests and ancient historians for inserting names of kings in lists with no mention of their accomplishments, or extending the reigns of kings into hundreds of years, or for insisting that one king is two. Reading that both Apappus, twentieth king of Thebes, and Phiops, twentieth king of Memphis, reigned for nearly a hundred years and were followed by a king named Nitocris who reigned for a year, he declares that Apappus and Phiops are the same king and Thebes and Memphis are the same kingdom. He does this—as we'll see—and then he uses his second new law to mightily compress the time period during which this one king reigned. He wrote: "It is further to be observed that the kings are often set down in the wrong order & their names corrupted and repeated again & again & intertwined with the names of other great men & women who were only the relations of kings or their viceroys or secretaries of state."[18]

Newton is diabolically wily when it comes to identifying other, perhaps unintentional, ways in which the record has been distorted: Some kings are called one name at home, another in each conquered country, and another as a royal title or one earned through military prowess. Some legendary figures are pseudonyms for the same kings. Through prodigies of etymology and phonetics, Newton demonstrates how the names of ancient kings have been distorted and multiplied over time "producing an apparent succession of kings where there was often originally only one."[19]

In Newton's sparse and rather forbidding sentences, mythological figures jostle cheek to jowl with historical personages. "Sesac has in his army Ethiopians commanded by Pan," writes Newton, or, "Laomedan King of Troy is slain by Hercules."[20] Newton is ever the dedicated euhemerist; for him, mythology is history in disguise. One of his lasting achievements in the *Chronology* is to show that the Egyptian pharaoh Sesostris is one and the same as the biblical king Sesac. He does this by providing numerous variations on the name "Sesac," many of which he is able to link etymologically with the name "Sesostris"; he also establishes that the actions and conquests of Sesostris and Sesac are one and the the same. This correction of history by Newton—which has been accepted by historians—is one of the ways in which the mathematician was, however indirectly, able to postpone the Trojan War until eighty-five or ninety years after the death of Solomon. As we'll see, Newton seeks to show that the kingdom of Egypt wasn't of long duration but consisted of a number of kingdoms of short duration unfolding side by side through history (see below). In the *Chronology,* and in conversation with his colleagues, he insisted that the diffusion route of temple architecture ran from Solomon through Sesostris to the Greeks. He told William Stukeley that all Egyptian temples were modeled on the Temple of Solomon: "'These heathens first imitated the Temple of Solomon in their own temples, then plundered it, and finally destroyed it.' And in the *Chronology* he mentions that the chief temple at Babylon was modeled on the pyramids."[21]

The other new law of chronology Newton invented was that of the length of reigns. He had noted that the pagan writers of antiquity tended to reckon the length of ruling dynasties in terms of three successive kings to a century; Newton believed they were mixing up the length of a monarch's reign with the length of a generation of mankind, which was usually about thirty-three years. It seemed to him that early death, the succession of brothers, and the overall wear and tear of running a kingdom in ancient times, must surely make for an average reign of only eighteen to twenty years. (The writers of the Old Testament, on the other hand, had almost always gotten the length of dynasties right.)

Newton culled all of history, both ancient and modern, to see how long on average that rulers ruled. He came up with the figure of eighteen to twenty years. Voltaire agreed with him, writing:

> It is very evident that mankind in general live longer than kings are found to reign, so that an author who should write a history in which there were no dates fixed, and should know that nine kings had reigned over a nation—such a historian would commit a great error should he allow three hundred years to these nine monarchs.
>
> Every generation takes about thirty-six years; every reign is, one with the other, about twenty. Thirty kings of England have swayed the scepter from William the Conqueror to George I, the years of whose reigns added together amount to six hundred and forty-eight years;

which, being divided equally among the thirty kings, give to every one a reign of twenty-one years and a half very near. Sixty-three kings of France have sat upon the throne; these have, one with another, reigned about twenty years each. This is the usual course of Nature. The ancients, therefore, were mistaken when they supposed the durations in general of reigns to equal that of generations.[22]

This may not be of great interest to us today; in our age, royalty has no political clout and suffers mainly from the wear and tear of media exposure. But Newton's insight was a brilliant one, and, applying this law of regnal length, he was able to knock a considerable number of years off the astronomically long time spans with which the first historians had credited the first nations.*

The Moscow chronology dissidents, led by *Chronology: Fiction or Science?* author Fomenko, have taken Newton's Occam's razor technique of conflation to whole new new heights: they've rammed together entire ruling dynasties! The mathematicians assert that because there were roughly the same number of popes from 911 to 1376 as there were kings of Judah from 931 to 586 BC, then these two groups must be the same group. They've found similarities between the popes of 141–314 and those of 314–532 and decided these represent a single line of popes. The Russian time revisionists assert that the chronology of the Carolingian kings and that of a string of Roman emperors refer to the same rulers. In this way, the Soviets have been able to hack roughly one thousand years out of the orthodox history of the world, backing up their assertions with, among other things, what they claim to be corrected scientific data about eclipses.

---

*It's much easier to be a king or queen today; does Newton's law of regnal length still apply? If we add up the lengths of reign of the last six rulers of Great Britain—Victoria (64 years) + Edward VII (9) + George V (26) + Edward VIII (1) + George VI (16) + Elizabeth II (64) = 179—and divide by 6, we get 29.833... (almost 30) years. Victoria ruled for 64 years, and Elizabeth II is in her 66th year as queen, but Edward VIII's mere one year helps bring the average down to only ten more than Newton calculated.

▲

Newton was so infinitely resourceful in coming up with time lines that he seems able to reach any conclusion he wants, observes Castillejo. "His etymology and his identifications and ramifications of mythologies, between them, allow him an almost complete fluidity and freedom to rearrange the facts and events of history in any way he wishes." Newton's inclusion of a five-page diagrammatic description of Solomon's Temple in the middle of the book prompted the actor-scholar to speculate that the *Chronology* is really about the "heathen kingdoms and empires that surrounded the Jewish people and Temple on all sides." By placing the Temple of Solomon in the center, Newton "crystallizes the whole of Middle Eastern history round the history of the Jews."[23] By shrinking the size of the ancient nations on either side of the temple, Newton greatly increases the predominance of the Temple of Solomon, again giving weight to his assertion that it was the Israelites who had introduced civilization to the world. As Manuel puts it, "Newton has animated the political world with the equivalent of a principle of universal gravitation.... The sun of the human world is the monarchy of Israel. This movement toward coalescence is a real law and admits of no exceptions."[24]

The Chronology of Ancient Kingdoms Amended shades into a number of other thematically related treatises by Newton, all of them written over decades, and all of them surviving in multiple manuscripts. These are "Original [Origin] of Religions," "Original [Origin] of Monarchies," and "Theologiae Gentiles Origines Philosophicae" ("Philosophical Origins of Pagan Theology"); we have dealt with this last at some length in chapter 10, "With Noah on the Mountaintop."

Often these works continue Newton's process of "deconstructing time," sometimes by filling in the period between Noah and "the first memory of things in Europe in 1100 BC," other times by leaping ahead. The "Original of Monarchies" begins with a description of how Noah and his family repopulated the world. At the end of chapter 10 we left Noah, his wife, his three sons Ham (founder of the "Hamitic" race), Shem (founder of the "Semitic" race), and Japheth (founder of northeastern regions) and his three daughters-in-law atop Mount Ararat and preparing to embark on their spectacular adventure of repopulating the world. When Newton picks up the story, the "remnant" family, having finally descended the holy mountain, has spent a hundred years in Mesopotamia. (We'll recall that human life spans were still extremely long; Noah would live to more than five hundred).

After the conflicts around the Tower of Babel had shattered the original Noachic language into a multitude of languages, Noah kept his eldest son, Shem, with him in Shinar and sent Ham to Egypt and Japheth to Asia Minor. When it came to be Ham's turn to divide his dominion among his four sons, he kept his eldest, Chus, with him on the Arabian side of the Nile and sent Mizraim to Thebes, where he was expected to expand southward. Then he sent Phut to the western side of the Nile; this son was expected to spread out into Northern Africa. Ham sent his son Canaan to lower Egypt, from where he was expected to expand eastward toward Syria.[25]

From here on in, the permutations and combinations of the sons and all their progeny become almost impossible to follow. Edward Gibbon gives us a summary, with introduction, of the nation-building activities of Noah's son Japheth and his immediate successors.

> Among the nations who have adopted the Mosaic history of the world, the Ark of Noah has been of the same use as was formerly to the Greeks and Romans the siege of Troy. On a narrow basis of acknowledged truth, an immense but rude superstructure of fable has been erected; and the wild Irishman, as well as the wild Tartar, could point out the individual son of Japheth from whose loins his ancestors were lineally descended. The last century [the eighteenth] abounded with antiquarians of profound learning and easy faith, who, by the dim light of legends and traditions, of conjectures and

etymologies, conducted the great-grandchildren of Noah from the Tower of Babel to the extremities of the globe.[26]

Gibbon says that Askenaz, son of Gomer, grandson of Japheth, and great-grandson of Noah, founded Germany, while, some centuries later,

> the giant Partholanus, who was the son of Seara, the son of Esra, the son of Sru, the son of Framant, the son of Fathaclan, the son of Magog, the son of Japheth, the son of Noah, landed on the coast of Munster [Ireland], in the 14th day of May, in the year of the world one thousand nine hundred and seventy-eight. [Subtracting 1,978 years from 2400 BC, the year of the Flood, we arrive at AD 422 as the year Noah's descendants founded Ireland.] Though he [Partholanus] succeeded in his great enterprise, the loose behavior of his wife rendered his domestic life very unhappy, and provoked him to such a degree that he killed—her favorite greyhound. This . . . was the first instance of female falsehood and infidelity ever known in Ireland.[27]

Today, we can't imagine how the progeny of a single family could manage the feat of populating an entire planet. Newton's age had no conception of the vast stretches of geological time over which *Homo sapiens* developed nor of the laws of evolution. They believed man was created pretty much as he is today, except that he was much smarter. God had told the story quite clearly in the Bible, with words that were meant to be taken literally. Nevertheless, the scholars of the day, while believing in the truth of Noah and his family, applied themselves diligently to working out the practical details of just how this amazing repeopling of a planet could have taken place.

Almost all pondered the command God gave Noah, not once, but twice, in Genesis 9:1 and 9:7, to "be fruitful and multiply, [and] replenish the earth." Did this command license man to be licentious? The Calvinist clergyman Andrew Willet interpreted it to mean God was giving permission to all post-Flood men and women, married or not, to practice polygamy and the "obscene fecundity" of "unlawful copulations."[28] The Protestant jurist Jean du Temps interpreted God's command to mean that once the floodwaters had receded Noah's three sons set about producing twins annually, a male and a female; these offspring in their turn procreated in the same way when they turned twenty, and so forth. This extraordinary method of birth control, which would have required an amazing knowledge of DNA and genetics, jumped the world's population to 1,554,420 people—518,140 descendants per son—101 years after the Flood. (Athanasius Kircher, accepting du Temps's bizarre scenario, somehow came up with a figure of more than 23 billion people on Earth at the end of 101 years—perhaps indicating not even a meager knowledge of biology on the part of the renowned Jesuit polymath.)[29]

The Jesuit theologian Petavius believed God's wish was that two of Noah's sons should sire only males while the third son should father "an ample number of females" to be impregnated by the male offspring. This would put 623, 612,358, 728 males on the face of Earth just 285 years after the Flood. Voltaire, ridiculing all of the above, jested that the priests produced these figures because they didn't know how babies were made.[30]

Though Newton first of all thought that mankind had recovered quite quickly from the Flood, perhaps in a hundred years or so, he became more and more pessimistic about the ability of our species to recoup its forces. He thought the world must have been bleak and harsh for the first thousand years: the first Greeks had been cannibals who "lived in caves and dens of the earth like wild beasts," while the early Egyptians "lived in mountainous syringes or subterranean vaults."[31] The latter at least provided the Egyptians with a cultural breakthrough: because they lived inside mountains, they discovered metals and invented metallurgy.

Newton's changed time frame was, perhaps, partly a strategy for stretching out the development time of ancient nations so that they could not attain to civilization before the ancient Jews did. In the earliest drafts of "The Original of Monarchies," he was inclined to fix the origins of all kingdoms at shortly after the deluge. Then he became convinced that elaborate kingdoms and city life had emerged only toward the beginning of the first millennium BC—that neither Egyptian nor Greek civilization could have existed much before the time of Solomon's reign.

Those of us who've seen Cecil B. DeMille's movie *The Ten Commandments* and watched the enslaved Israelites toiling in the shadow of the towering palaces of ancient Egypt find it hard to believe that the land of the pharaohs was not fully a civilization while the Jews were still twelve wandering tribes. But, as has been recounted, Newton believed there was no single monolithic kingdom of Egypt but rather a collection of many tiny Egyptian city-states that did not follow one another chronologically but existed side by side; Newton turned an extension in time on the part of the ancient Egyptians into an extension through space.

This was only one of Newton's many strategies to slow down the evolution of the Egyptian, Greek, and Mesopotamian empires so as to make Jewish civilization preeminent. Thus the first great achievements of mankind came not from the pagan classical world but from the Jews from whom Christianity had sprung.

This belief in the primacy of Jewish civilization wasn't confined to Newton's time. It went back at least two thousand years. Clement of Alexandria (ca. AD 150–215) insisted in his *Stromata* ("Lists") that the Greeks had stolen their culture from the Egyptians, who had stolen it from the Jews. Eusebius of Caesarea (AD 260/265–339/340) made the same claim a century later, declaring that Solon and Pythagoras were disciples of Egyptian prophets. Eusebius stated that those Syrians who had reputedly invented the alphabet were actually a tribe of Hebrews who were then living in the part of Phoenicia later called Judaea.

Moses had possessed a knowledge that would ferment among the Jews for generations, then blossom forth at the time of David into a civilization. Newton believed the great patriarch possessed knowledge of the *prisca sapientia* delivered up by Noah from the antediluvian world. In the second century BC the Greco-Jewish historian Eupolemus emphatically declared that: "Moses was the first wise man, and the first that imparted grammar to the Jews."[32] The Jews passed the knowledge of grammar along to the Phoenicians, who passed it on to the Greeks.

"Homer had read over all the books of Moses, as by places stolen thence almost word for word may appear [as evidenced by passages stolen almost word for word)," wrote the great Elizabethan adventurer-scholar Sir Walter Raleigh. Because Homer had copied his epic poems from the Torah, which many (though not Newton) thought was written by Moses, the *Iliad* and the *Odyssey* were merely warmed-over Moses. Scholar John Wallis pronounced contemptuously that Plato was no more than "Moses disguised in Greek dress," the "Attic Moses, [who] stole everything he taught about God and the world from the books of Moses."[33] Christian scholars went so far as to bring Moses's wife Zipporah into the act, venerating her not as a fountainhead of knowledge but as the template for all the pagan goddesses of the ancient world.[34]

All this seems very strange to us today. Though ancient Greek literature and ancient Hebrew literature developed at about the same, these two bodies of literature were very different, and often even in conflict. Paul Johnson writes that the Jewish revolt against Rome, culminating with the war of AD 66–70, was

> at bottom a clash between Jewish and Greek culture. Moreover, the clash arose from books. There were only two great literatures, the Greek and the Jewish, for Latin texts, modeled on the Greek, were only just beginning to constitute a corpus. More and more people were literate, especially Greeks and Jews, who had elementary schools. Writers were emerging as personalities: we know the names of as many as 1,000 Hellenistic authors, and Jewish writers too were just beginning to identify themselves.... In many respects Hebrew literature was far more dynamic than Greek. Greek texts, from Homer onwards, were guides to virtue, decorum and modes of thought; but the Hebrew texts had a marked tendency to become

plans for action. Moreover, this dynamic element was becoming more important.[35]

But there was no curbing Isaac Newton's enthusiasm to prove the primacy of the civilization that produced Abraham, Moses, David, Solomon, and Jesus Christ. He pursued this "vendetta" until his very last days.

Historians point out that there are no human beings in the histories of the rise of civilizations as set forth by Isaac Newton. Manuel writes:

> In his law of the growth of the great kingdoms Newton was performing for political history a function similar, *mutatis mutandis*, to his discovery of the laws of motion (it was universal and it was simple), though he considered that prophets like Daniel had anticipated him by depicting the same history of the "four kingdoms" in hieroglyphic language. Newton never wrote a history of men—they do not seem to count as individuals in his narrative—but of bodies politic as he had written a history of bodies physical. These agglomerations did not spring into being suddenly: like the physical planets they too had an "original," a history of creation, and extension in space which could be marked chronologically, as they too would have an end. Newton's chronological writings might be called the mathematical principles of the consolidation of empires because they dealt primarily with quantities of geographic space in a temporal sequence; the individuals mentioned in his histories . . . were merely signposts . . . they have no distinctive human qualities. . . . His kings are automaton-like agents in the acquisition of power and the extensions of dominance.[36]

While admiring Newton's prodigious efforts, Manuel and others point out that Newton, in treating history as pure mathematics, was incapable of ascribing to the shakers and movers of civilization any more than the simplest "stock" emotions, for example, "vanity," and "lust"; he scarcely noticed the passions of the men and women in history but was concerned only with mechanics, with the passionless dynamics of the physical growth of kingdoms.

"Everything human is alien to him," concludes Manuel, "at least insofar as he expressed himself on mankind. His history hardly ever records a feeling, an emotion. . . . Newton's passion for factual detail shriveled the past to a chronological table and a list of place names."[37] "In what appears to be an early exercise in social science history, Newton had incinerated the fiber and substance of human life in a numerical furnace," add Professors Buchwald and Feingold.[38]

Still, however dispassionate Newton may have been in his deliberations, it's now become clear that he was far ahead of his time in the insights he gained into the causes of civilization. A man of his brilliance could not acutely observe and catalogue the rise and fall of ancient Egypt, of Assyria, of Babylonia, and of Persia without seeing with uncommon penetration into the beating heart of mankind's profound impulse to gather and grow within a collectivity. Critic Jonathan Reé writes that Newton's *Chronology* "may be the source of one of the formative ideas of our time: that every society must pass through the same stages—savagery, pastoralism and agriculture—on the way to civilized maturity."[39] Contemporary historians call this the "stadial view" of societal progression; its basic principles were taken up in full by Enlightenment historians such as Adam Smith.

The insightful David Castillejo may be right in crediting *The Chronology of Ancient Kingdoms Amended* with unknown metaphysical depths. But until such time as (as Castillejo has suggested) Newton's great history book is "passed through a computer," we may have to stay with the far less romantic but every bit as impressive judgment that Isaac Newton was the greatest sociologist of history of his time—and is still, today, a figure to contend with in that field.

CHAPTER THIRTEEN

CHIRON AND THE STAR GLOBE

Chiron the centaur cut a tremendous figure in the classical age. And this because of, or perhaps despite, the fact that he was half man, half horse, his human body ending at the waist and becoming a horse's body. But, unlike most other centaurs, who were lawless and warlike, he was gracious and regal, a being of wisdom and justice. It was Chiron who, according to that second-century AD gossipmonger of genius Saint Clement of Alexandria, "first led mortals to righteousness."[1]

Chiron (or Cheiron) was a historical figure. He was the sacred king of the centaurs, a race that worshipped the horse and were part of the mixed population of Pelasgians ("seafarers") that first streamed across mainland Greece in about 1600 BC. It was on two very human feet that Chiron led his people into warfare with firmness and dignity, and then only if absolutely necessary. The word *cheiron* is linked with the Greek word *cheir*, a "hand," and *centaur* with *centron*, a "goat."[2] "Chiron" is usually translated as "the Handy One."

Chiron was very rapidly deified, becoming the son of Chronos and the ocean nymph Philyra. Apollo and Diana doted on him and taught him everything they knew about botany, music, astronomy, divination, medicine, hunting, and gymnastics. The chroniclers don't know where history ends and myth begins when they say Chiron directed a famous school on the slopes of Mount Pelion whose pupils included Hercules, Achilles, and Jason, the leader of the Argonautic expedition that circumnavigated the ancient world in search of the Golden Fleece.

Isaac Newton was greatly interested in Chiron. He believed it was this brilliant centaur who had first traced out the constellations in the sky and given them names. He believed Chiron had fashioned the first star globe and, delineating those constellations on it, had given it specifically to his old pupil Jason because Jason's unusually large longboat would be sailing out of sight of land for long periods of time and would need the guidance of the stars at night.

There was more. Chiron and his constellations were to become indispensable elements in a fantastically original mathematical-astronomical-chronological stratagem that Newton would employ to pin down the dates of King Solomon's death and the fall of Troy and establish once and for all an accurate chronology of the ancient world.

In the same way that the stained-glass windows of some of the great cathedrals of Europe sometimes tell, pane by pane, the story of the life of Christ, Isaac Newton belived groupings of constellations in the night sky sometimes recount the adventures of the heroes of ancient Greece. In his *Chronology of Ancient Kingdoms Amended,* Newton lists sixteen constellations that refer to the adventures of Jason and the Argonauts.* These include:

---

*"There's the golden *RAM*, the ensign of the Vessel in which *Phryxus* fled to *Colchis;* the *BULL* with brazen hoofs tamed by *Jason;* and the *TWINS, CASTOR* and *POLLUX,* two of the *Argonauts,* with the *SWAN* of *Leda* their mother. There's the Ship *ARGO,* and *HYDRUS* the watchful Dragon; with *Medea's CUP,* and a *RAVEN* upon its Carcass, the Symbol of Death. There's *CHIRON* the master of *Jason,* with his *ALTAR* and *SACRIFICE.* There's the *Argonaut HERCULES* with his *DART* and *VULTURE* falling down; and the *DRAGON, CRAB* and *LION,* whom he slew; and the *HARP* of the *Argonaut Orpheus.* All these relate to the *Argonauts.* There's *ORION* the son of *Neptune,* or as some say, the grandson of *Minos,* with his *DOGS,* and *HARE,* and *RIVER,* and *SCORPION.* There's the story of *Perseus* in the Constellations of *PERSEUS, ANDROMEDA, CEPHEUS, CASSIOPEA* and *CETUS:* That of *Callisto,* and her son *Arcas,* in *URSA MAJOR* and *ARCTOPHYLAX:* That of *Icareus* and his daughter *Erigone* in *BOOTES, PLAUSTRUM* and *VIRGO. URSA MINOR* relates to one of the Nurses of

Centaurus for Chiron; Argo for the fifty-oared longboat *Argo;* Taurus the Bull for the sacred bulls Jason kills to claim possession of the fleece; and the Cup for the cup that the priestess Medea, madly in love with Jason, fills with a potion Jason needs to defeat the bulls. That these constellations were in the night sky at all was a matter of great significance to Newton, for a reason intimately bound up with his great mathematical-astronomical adventure that we will describe in this chapter.

We know today that there never was a voyage of the Argonauts in search of a legendary and hallowed golden fleece. (Even the ancients didn't quite swallow the legend; Strabo, for one, "mentions the gold of the Caucasus as a reasonable motive for Jason's expedition, pointing out that the natives collected gold dust from the river on fleecy skins."[3]) And so we can't help but feel, without knowing anything about it, that Newton had no chance of success in the experiment he was planning to carry out.

But Newton himself didn't believe entirely literally in the legend of the Golden Fleece,[†] since for him all gods and goddesses had once been, euphemistically speaking, mortal men and women. Certainly Newton's notion of what the Argonautic expedition was all about shimmers and shifts as he writes about it. Sometimes he sees it as heraldic, a symbol of an important breakthrough in Greek technological history; namely, the invention of the fifty-oared iron boat that enabled the ancient Greeks to travel out of sight of land for long periods of time. (Today's historians date the invention of the fifty-oared iron boat at about 700 BC.) Newton

---

(cont. from p. 272) Jupiter, *AURIGA* to *Erechthonius*, *OPHIUCHUS* to *Phorbas*, *SAGITTARIUS* to *Crolus* the son of the Nurse of the Muses, *CAPRICORN* to *Pan*, and *AQUARIUS* to *Ganimede*. There's *Ariadne's CROWN*, *Bellerophon's HORSE*, *Neptune's DOLPHIN*, *Ganimede's EAGLE*, *Jupiter's GOAT* with her *KIDS*, *Bacchus's ASSES*, and the *FISHES* of *Venus* and *Cupid*, and their Parent the *SOUTH FISH*. These with *DELTOTON*, are the old Constellations mentioned by *Aratus:* and they all relate to the *Argonauts* and their Contemporaries, and to Persons one or two Generations older." (Newton, *Chronology*, chap. 1, "Of the Chronology of the First Ages of the Greeks," www.newtonproject.sussex.ac.uk/view/texts/normalized/THEM00186.)

†The Golden Fleece was another name for the Philosophers' Stone, and so the argonautic mission must also have had resonances for Newton in terms of his attempts, through years of alchemical practices, to transmute base metals into gold.

wrote: "The Ship *Argo* was the first long ship built by the *Greeks*. Hitherto they had used round vessels of burden, and kept within sight of the shore; and now, . . . by the dictates of the Oracle, and consent of the Princes of *Greece*, the Flower of *Greece* were to sail with Expedition through the deep, in a long Ship with Sails, and guide their Ship by the Stars."[4]

Other times, Newton sees the voyage of the *Argo* as a specific historical event, telling us that *"Egypt* was in its greatest distraction: and then it was, as I conceive, that the *Greeks,* hearing thereof, contrived the *Argonautic* Expedition, and sent the flower of *Greece* in the Ship *Argo* to persuade the Nations upon the Sea Coasts of the *Euxine* and *Mediterranean Seas* to revolt from *Egypt,* and set up for themselves, as the *Libyans, Ethiopians* and *Jews* had done before."[5]

But however much Newton's vision of the voyage of the Argonauts seems to vary, there is one constant he always comes back to, and that is Chiron's gift of the delineated star globe to Jason. This was important to Newton because the star globe showed those positions as they were in the actual year, or so Newton believed, that the Argonautic expedition was launched. And that meant that if you could just have a look at this globe, you could figure out the actual date of the voyage of the Argonauts.

But, what star globe? There was no record that any such star globe had ever existed. What was Newton thinking of?

To begin to get an answer to this question, we need to pick up the thread of the history of chronology that we left dangling in the middle of the previous chapter.

Though he was hailed by his peers as "the light of the world" and "the sea of sciences," the outrageously arrogant Dutch French scholar Joseph Justice Scaliger (1540–1609) thought these epithets fell far short of truly describing his genius. His colleagues found this infuriating, but it's not hard to see why Scaliger thought so well of himself. He seems to have read virtually all classical literature, and he spoke thirteen languages. He wrote landmark works on drama, music, linguistics, and philosophy (most have not been translated from Latin, since they are a little too encyclopedic and exacting for modern tastes).

Scaliger ransacked the great libraries of Europe and the Middle East and managed to come up with fifty ancient calendars. He harmonized them to create a single, global chronology putting the Greeks, the Romans, the Jews, the Persians, the Babylonians, the Assyrians, the Egyptians, the Ethiopians, and forty-two other cultures, all together on the same timescale. (The previous record was a matchup of five cultures.)

When Scaliger was still a schoolboy, Copernicus's precise calculations of the movements of the planets and their moons, upon which the astronomer based his paradigm-smashing demonstration that the solar system was heliocentric, were finally published. Astronomers could now calculate with great accuracy the position in the sky of the moons and planets the year before, a century before—millennia before—and in so doing pin down many a date in ancient history.

Scaliger led this field. Early on, he worked out an accurate date for the Battle of Marathon, a pivotal event in the first Greek-Persian War, changing that date from 776 to 491 BC. Today, historians put it in September 490 BC.

In his *History of the Peloponnesian War,* the great Greek historian Thucydides (460–396 BC) links three eclipses of the moon to three specific events in the twenty-seven-year-long conflict pitting Athens and its allies against Sparta and its allies. Scaliger used Copernicus's tables to determine that the eclipses took place in 431 BC, 424 BC, and 413 BC. Modern-day historians date the Peloponnesian War at 431–404 BC.*

Scaliger blazed a trail that more and more European natural philosophers followed. Some of their projects seem whimsical to us today. The Polish astronomer Hevelius (1611–87) calculated just where the sun stood in the sky above the Garden of Eden on the first day of Creation.

---

*Our chronologist friends from Soviet Russia (see chapter 12) will have nothing to do with these figures, which they consider to be hopelessly distorted by politics. In *Christ: A History of Human Culture from the Standpoint of the Natural Sciences,* the Russian mathematician-topologist Nicolai Aleksandrovich Morozov calculated the dates of the three eclipses of the Peloponnesian War at AD 1133, 1140, and 1151!

Edmund Halley (1665–1742) used his consummate mathematical and astronomical skills to determine exactly in what place and on what day Julius Caesar first invaded in Britain.

Isaac Newton quickly added this brilliant new tool to the repertory of techniques he was using to date the ancient world. He located a second-century AD document quoting an earlier source to the effect that, in the fourth year of the 202nd Greek Olympiad, in Bythnia (now northwest Asian Turkey), an earthquake and a solar eclipse had occurred at exactly the same time. Newton decided these concurrent happenings were the literally earthshaking events accompanying Christ's crucifixion and death on Golgotha, when "the veil of the temple was rent in the awful darkness that followed; when the rock of Golgotha was split apart by an earthquake; when the artillery of heaven thundered, and in the baleful glare of the lightnings the shrouded dead flitted about the streets of Jerusalem."[6]

Newton determined that this fourth year of the 202nd Olympiad was 34 BC and declared that this was the year of Christ's death. The mathematician met with some objections from his colleagues, who reminded him that a solar eclipse lasts no more than six minutes while the darkness over Golgotha lasted nearly three hours. Superbly confident as always, Newton, felt no need to defend his conclusions.

But even while he was coming up with this audacious calculation concerning the birth of Christ, Newton's mind was already roaming restlessly backward in time to the expedition of the Argonauts. As has been said, he had concluded that Chiron had given Jason a newly minted star globe, and he was beginning to ponder the possibility of using a particular astronomical phenomenon to read the delineations on that star globe and figure out when Jason and his crew of fifty heroes had set sail.

This particular astronomical phenomenon had been known to the ancient Chaldeans. It had been rediscovered by the greatest of ancient Greek astronomers, Hipparchus, in the second century BC. Newton himself had plotted its final decimal point, giving us the figure that we use today.

It was called the precession of the equinoxes. Here is Isaac Asimov's description of that phenomenon:

> About 150 BC, the Greek astronomer Hipparchus noticed an odd thing in connection with the sun's motion against the stars. Twice a year, it seems, the sun crosses the celestial equator and shines directly above the earth's equator. These moments are called the equinoxes. Hipparchus compared astronomical records made over hundreds of years, first by Babylonian astronomers, then by Greek astronomers. In doing so he found that the sun slowly changed position against the stars at the time of the equinoxes. Each year the sun crossed the celestial equator at a point a little farther to the east than it had the year before. The east is the direction of morning, so that the time of equinox, as measured by the stars, is a little earlier each time. Each equinox precedes ("goes before") the one before. . . . For this reason, the shift is called the precession of the equinoxes. . . . Hipparchus decided that this shift could be most easily explained by supposing that the position of the north celestial pole was changing. If that were so, the whole sky would seem to shift in a single piece, carrying the stars (but not the sun, moon or planets) with it.
>
> As the earth turned, its axis moved in a slow circle. As it moved, it kept pointing to different parts of the sky, so that the north celestial pole moved as well.[7]

In other words, the plane of the terrestrial equator is inclined to the Earth's orbit at approximately 23½ degrees, and it is this inclination that accounts for the existence of the equinoxes and solstices. As the Earth moves in its orbit, and rotates to give what we call day and night, its axis also performs a slow gyrating motion with a period of 25,800 years. This means that over a period of 25,800 years the location of the equinoctial and solstitial lines, or colures, changes in relation to the fixed stars. It is a very slow change. The colures inch their way across the sky, returning to their starting point 25,800 years later.

This tiny, continuous, observed change of place of the constellations is distinct enough that we can say each night sky has its own fingerprint. Since it takes the equinoxes 25,800 years to process through a full cycle, the sun passes through a different sign of the zodiac every 2,156 years. In the time of Hipparchus (ca. 190–120 BC), the sun rose and set in the sign of the Ram; over the past two thousand years it has been passing through the sign of Pisces; and, as all New Agers know, it has passed into the sign of Aquarius.

The change in the sky's fingerprint is virtually imperceptible from one night to the next. But it is visible over the seventy-two-year life span of a man or woman; in this time the constellations move en bloc one degree through the sky. (Imagine yourself standing in a field at night and turning in a full circle until you've scanned the entire horizon. You've moved through three hundred and sixty degrees.)

But you had to have an actual star globe, or a star map, delineated not only with the stars of the night sky but also with the colures, or equinoctial and solstitial lines, to identify a particular year using the phenomenon of the precession of the equinoxes. And if Isaac Newton actually had such a star globe, however did he obtain it?

To get an answer to that question, we first have to go to the palace of King Antigonus Gonatas of Macedonia, in Pella, on a day in the summer of 276 BC.

Antigonus Gonatas, having summoned the renowned poet Aratus into his royal presence, was finding it difficult to speak to this physician-poet who seemed to hold him in contempt. Though fearless on the battlefield, Antigonus Gonatas was tongue-tied in the face of genius. Suddenly Aratus bowed politely in his direction. Antigonus Gonatas, finding his voice at last, told the poet he wanted him to write a Homeric poem that combined Zeus, the stars, and the weather.

This would be an epic poem intended to protect the people of Macedonia from disaster. For two years, Antigonus Gonatas's subjects had been suffering from storms, drought, crop failure, shipwreck, and starvation. The king wasn't without humane instincts, and certainly not without the desire to remain in power. But how could he retain the support of his people in the face of these calamities?

Antigonus Gonatas, and Aratus—and indeed most of the people living in Macedonia and Greece in the third century BC—believed that Zeus had created the night sky for one purpose only: to provide mankind with a gigantic clock and weather channel enabling us to tell the time, predict the weather, and know when to reap and sow.

The god had supplemented his meteorological night sky with quirks of animal behavior to further signal weather changes, as in, "And it is no sign of good weather when goats are busy about the prickly evergreen oak, or sows go frantic with the blown straw."[8]

In the ninth century BC, the poet Hesiod in his *Works and Days* had given the ancient world a quasi almanac deciphering the seasonal changes that Zeus set forth in the sky with the rising and setting of stars and the passage of mists that meaningfully riffled our view of the constellations.

Other poem-almanacs appeared, such as the recent one by the poet Theophrastus. But, as mankind's needs grew, more celestial devices were needed to successfully interpret those heavens. This was a problem that Gonatas hoped to address, by having the very renowned poet—and his subject to boot—Aratus update with an almanac written with a certain Homeric verve.

Something happened that must have been coincidence but was surely a gift from the gods. It had emboldened the king to summon Aratus to help address the project. What happened was that, the week before, Antigonus Gonatas's old and wizened royal astrologer, with whom the king had been discussing his project, had slipped into his hand a yellowed book at least as old and wizened as the royal astrologer himself.

"This book," the astrologer had intoned, "is a hundred years old! It was written by the great astronomer Eudoxus," and the astrologer showed the king some of the pages. "It is a most exact and mathematical description of the fifty-five star families of the night sky. It is appropriate for your new project."

Antigonus Gonatas was about to take this under advisement when the old astrologer gripped him by the arm and, leaning forward, whispered urgently: "There is something divine in this book! It is said that Eudoxus copied these star families from a most ancient primitive sphere, a star globe so ancient as to have been furnished by Zeus himself. A star globe that embodies the graciousness of the stars."

The king took the book—it was titled *Phaenomena* (*Phenomena*, or *Visible Signs*)—and strode away. Antigonus Gonatas considered himself an expert on astronomy and, more importantly, a pious man. He knew about this Eudoxus. He knew he had been the greatest mathematician of his time after Archimedes. He also knew that Eudoxus had preached that pleasure was the highest good, that he had written a piece of mockery, *Dialogue of Dogs*, and that he loved to make fun of the philosopher Plato, sometimes even to his face. Could reworking the work of such a man possibly be a good thing? The king was inclined to dismiss this *Phaenomena*.

But, over the next week, as Antigonus was daily shown evidence of the mounting plight of his subjects, he decided there was little time to lose and that he must use the book of that impious astronomer. So he summoned Aratus (who was living in his palace) and made the request to him that has been noted above.

Aratus hesitated—which was, of course, futile, because Antigonus was the king. But just as Aratus was hesitating, the Stoic philosopher Persaeus, who was a permanent resident at Antigonus Gonatus's court, suddenly appeared at the king's elbow and admonished Aratus (apparently the philosopher was also in on this project too) to be sure and present Zeus in his new almanac as, not the stingy god of the proto-Stoicism of Hesiod, a god not able to offer mankind a way out of his ceaseless toil, but the full-blown Stoic god of Theophrastus, a kinder, gentler Zeus.

So Aratus—who would stick mainly with Hesiod's *Works and Days* as a structural model but would adopt the Theophrastian spirit—went off and wrote his poem.

▲

We know very little about Aratus (ca. 315–245 BC). It's generally agreed that he was born in Soli, Cicilia, and went to Athens to study with the Stoic philosophers, particularly Zeno. He was probably sixty, and a poet of renown, when Antigonus Gonatus summoned him to his palace in Pella.

It took Aratus two years to write his book, which he called the *Phaenomena* after Eudoxus's learned tome. It was published in Athens. We don't know what success Aratus's poem had among the farmers, fishermen, and sailors of Macedonia, or even if it helped at all. We do know that it became an instant hit among the upper classes of Athens. If there had been a New York Times bestseller list in that great Greek city-state in 276 BC, this new *Visible Signs* would have gone straight to the top.

Aratus's *Phaenomena* is a combination *Astronomy for Idiots,* pint-size Homeric epic, and *Farmers' Almanac* (two thousand years before Ben Franklin's *Poor Richard's Almanac*). Perhaps part of its charm came from the fact that it made it easy for the elite aristocracy of Athens to learn about the stars. And it contained a powerful dose of the Stoicism then popular (or feigned thus) among the elite of ancient Greece and Macedonia. Most of all, perhaps, it caught the powerful if subtle cadence of Homer in a manner that was universally appealing. Soon blue and golden paintings of the forty-seven stellar families of Aratus's *Phaenomena* (he had left out eight of Eudoxus's) were emblazoning the concave ceilings of the mansions of the rich and famous in Athens; the domiciles of the upper classes took on the aspect of what we today call planetariums.

Today we may not appreciate lines like

> *A murky manger [Cancer] with both stars*
> *Shining unaltered is a sign of rain.*
> *If while the northern Ass [Pegasus] is dimmed*
> *By vaporous shroud, he of the south gleam radiant,*
> *Expect a south wind; the vaporous shroud and radiance*
> *Exchanging stars harbinger Boreas [the Wind].*[9]

But in ancient Greece and Rome these lines were repeated by many lips over the centuries. Aratus's *Phaenomena* was the third most widely read book of the classical era after the *Iliad* and the *Odyssey*. In 89 BC, the Roman philosopher-orator Cicero translated it into Latin; a century later, the Roman poet Ovid followed with another translation. In AD 15, the popular and learned Roman general Julius Caesar Germanicus, eventually the adopted son of Tiberius, retranslated and updated the *Phaenomena* from his armed camp between military engagements. The book caught the fancy of the upper classes of Rome just as it had that of the aristocrats of Athens. It soon became, writes one scholar, "the polite amusement of the Roman ladies to work the celestial forms in gold and silver on[to] the most costly hangings."[10] Late in the first century AD, the apostle Paul, preaching to a crowd from the steps of the Acropolis in Athens, turned Aratus's Hymn to Zeus that begins the poem into a hymn to the Christian God and paraphrased it thus: "In Him we live, and move, and have our being; as certain also of your own poets have said, for we are also His offspring." This text appears in Acts 17:28.

There were more Latin translations. Fifty original manuscripts of *Phaenomena* are extant in museums around the world today, and there are numerous modern editions.*

The treatise on which Aratus's *Phaenomena* was based—the

---

*That *Phaenomena* survived as well as it did is extraordinary. Eighty percent of the literary works of high antiquity have not survived. Like the archives, annals, and historical records of the ancient world mentioned in the previous chapter, they were destroyed in the blistering wars and social convulsions of ancient times.

Of the Greek tragedians, Aeschylus (ca. 525–456 BC) wrote 70 to 90 plays; 7 have survived. Sophocles (497–406 BC) wrote 123 plays; 7 have survived. Euripides wrote 92 plays; 18 have survived. The great Greek comedian Aristophanes (447–386 BC) wrote 44 plays; 11 have survived. Even Homer's works did not escape the juggernaut of war; there is a legend that the author of the *Iliad* and the *Odyssey* wrote a third, now lost, comedic epic called *Margites*. Every one of the works of Greek New Comedy, a genre that flourished from 323 to 260 BC, was lost until 1952, when a papyrus copy of a play by the genre's master, Menander (342–291 BC)—*Dyscolos* (*The Curmudgeon*)—turned up in Egypt.

Phaenomena of the astronomer Eudoxus of Cnidus, written in 366 BC—did poorly compared to its successsor. The text has long been lost; we know something of its contents because Hipparchus (ca. 190–120 BC) summarized it and Aratus's poem in a *Commentary* in which he gave both works a lambasting. The Alexandrian astronomer Ptolemy (ca. AD 90–168) swept much of Hipparchus's *Commentary* up into his magisterial compendium of astronomical lore called the *Almagest*. *Almagest* means "Greatest Book" and is the title the Arab translators of the book bestowed on Ptolemy's work. This "greatest book," famous for its elegant and definitive description of the geocentric or "Ptolemaic" universe, was the astronomy textbook of record for the Middle East, Asia, and Europe up until the sixteenth century. In Newton's day it was still required reading.

That is how knowledge of the Eudoxean-Aratean constellation catalogue may have made its way into the English village of Grantham where Isaac Newton went to school—if it did make its way into the school. We can't be sure it did. We do know that the Grantham school was a very well-equipped school by any standard, and if a copy of Ptolemy's *Almagest* lay on one of its bookshelves, Newton likely had a look at it. And if the book wasn't there, Newton would have eagerly perused a copy early on in his career at Cambridge, which would have led him to Aratus's *Phaenomena*.

Let's recall the star globe newly delineated with constellations that, Isaac Newton was certain, Chiron the centaur had given Jason at the time of the launching of the expedition of the Argonauts in search of the Golden Fleece.

Let's also recall the "primitive sphere" on which, legend has it, Eudoxus based his description of the stars in the *Phaenomena*.

Isaac Newton decided they were one and the same sphere, which meant that the constellations as described by Eudoxus and by Aratus were in the same positions in the night sky as those on the star globe fashioned by Chiron the centaur. And, since Chiron had, or so Newton believed, traced out the constellations and marked them on his star globe in the same year that the *Argo* sailed, then the constellations as Eudoxus and Aratus described them were a picture of the night sky in the same year that the *Argo* set sail.

Because this globe indicated, along with the location of the stars, the position of the equinoctial and solstitial lines relative to the rising and setting of the sun, Newton could use those phenomena to determine the actual, historical year when the expedition of the Argonauts sailed.

Newton believed Chiron the centaur had created the first star globe. And Newton believed that the "primitive sphere" pointed toward a "first" star globe. Moreover, Newton came up with proof—or so he thought—that Chiron, along with his many other accomplishments, was an astronomer.

Sometimes we get the impression that there were certain brilliant men of high antiquity who got themselves born solely to supply Isaac Newton, a couple of millennia later, with the one, tiny, obscure, but indispensable fact he needed to confirm one or another of his more daring hypotheses.

Such a man was Saint Clement of Alexandria (AD 150–213), founder of the famous Empathetical School at Alexandria. Clement's *Stromata* ("Patchworks") was the definitive text for those who believed all Greek science and philosophy originated with the Jews. Along with much that was conventional, it was as crowded with esoteric, eccentric, and anonymous facts as an ancient map of the globe is packed with griffins, mermaids, dragons, unicorns, and headless cannibals with mouths in their chests.

It is to Clement of Alexandria that we owe the quirky information that the god who strummed Apollo's lyre to tune up the planets and get them playing the music of the spheres was Jesus Christ; that Pythagoras studied with the Druids in England and with the Brahmins in India; that the Brahmins got their name from the patriarch Abraham; and that when you were dancing in the celebration of the Greek mysteries and had reached the final stage of ecstasy, then (in Isaac Newton's words), *"one no longer has to learn, one can see and grasp with one's mind the nature of things."*[11]

It was Clement who originally supplied Newton with the information that, in Newton's words, "the constellations were invented by Chiron at the time of the Argonautic expedition and inscribed on a celestial sphere for the use of Jason and his men as they navigated uncharted waters."[12]

Saint Clement also intimated that, while it was Chiron who had invented the globe, it was the Argonaut Palamedes, or perhaps the Argonaut Musaeus, who, in the words of Newton:

> made a sphere for his fellow sailors and is reputed the first among the Greeks who made one, that is he made a celestial sphere upon which he delineated the asterisms of Chiron . . . Chiron invented them and Musaeus drew them upon a globe while the ship Argo was building: not sooner because that ship was one of the Asterisms; not later because Chiron was at that time very ancient.[13]

Now, Newton found in the *Stromata* what was a critically important (according to Newton) fact: that an ancient epic called the *Gigantomachia* ("War of the Giants"), more than 95 percent of which is lost, and whose authorship is unknown, had clearly stated that Chiron the centaur was a "practical astronomer."

Or so Isaac Newton insisted on believing. His colleagues were quick to point out that the words "practical astronomer" as used in the *Gigantomachia* were a loose translation from the Greek of words that actually meant "stars and constellations." These words would allow him, they said, to translate the words as "astrologer," but Newton's colleagues would permit him to go no further.

Their opinions, however, had no effect on Newton: "practical astronomer" it would be.

But, even if Chiron really had traced out the constellations in the night sky, what proof did Newton have that he had done it in the very year the *Argo* sailed?

To answer this question, Newton wants us to visit a series of others:

Is there a constellation Helen?
Is there a constellation Achilles?
Is there a constellation Hector?
Is there a constellation Andromache, or Achilles, or Patroclus? And so on.

No, there are no constellations named for these heroes and heroines of the Trojan War, which was thought to have taken place only thirty years after the voyage of the Argonauts. And this is proof, says Newton, that Chiron did the naming; he would have been dead by the time of the Trojan War.

Newton believed the fact that the names of sixteen constellations are related to the Argonautic expedition was proof that Chiron did the naming at the time of the voyage; he did not know about the expedition earlier, and therefore would not have named the stars in that way earlier.

But Chiron seems to know too much about the voyage of the Argonauts! How could he, before it has begun, name a constellation for the sacred bulls that Jason would have to kill, or for the chalice full of poison that Medea would use to help Jason secure the Golden Fleece, an event that took place halfway through the voyage?

Newton's colleagues raised these points against him. But Newton saw no problem.

In our discussion we have avoided the issue of just how incredibly difficult it must have been for Newton to date a millennia-old "primitive sphere" using rather primitive equinoctial and solstitial markings and positioning of constellations. For most Newton scholars, this is the real issue. Professors Jed Z. Buchwald and Mordechai Feingold comment:

Using the description of the night sky in Aratos' *Phaenomena* and Eudoxus' description of the night sky as summarized in Hipparchus and reproduced in the *Almagest,* Newton came up with a tight range of dates in the area of 933 BC. This was incredibly difficult thing to do, since the nomenclature of star maps and celestial globes was very different in ancient times and not very well known in Newton's time (and not completely understood in our own). But Newton did succeed.[14]

The figure Newton came up with was 933 BC. It was generally assumed in Newton's time that the Trojan War took place thirty years after Jason's voyage, and that King Solomon died forty-two years before the voyage; thus, in nailing down 933 BC as the date of the first year of the Trojan War and 975 BC as the date of Solomon's death, he had added five hundred years to the chronology of the world and he had—he thought—fixed all the ancient times of the Earth once and for all.

But he met with much opposition (especially when historians decided that there had never been a voyage of the Argonauts and that the Trojan War had taken place not in the tenth century BC but in the eleventh century BC), and now this great experience of Newton's in time travel is taken seriously only because it shows, once again, the sheer originality of his imagination and the immense audacity of his genius.

CHAPTER FOURTEEN

A Glitter of Atlantis

The island of Gozo, part of the archipelago of Malta in the Mediterranean Sea, is one of the sacred places of the Earth.

The size of Manhattan, called Ogygia ("first memory") by the ancient world, it is the site of the oldest freestanding structures in the Mediterranean: three Neolithic limestone temples of "extreme architectural sophistication and complexity" dating back to 3700 BC. These "magic and potent island-sanctuaries"[1] predate the Great Pyramids of Egypt by a thousand years.

Gozo boasts lofty cliffs on which nature has carved out vaguely human forms, deep caves, and ancient grottoes giving on to the sea. The seventeenth-century Jesuit polymath Athanasius Kircher climbed to the cave high above Ramla Bay, where, according to legend, Odysseus spent seven years as the love slave of the sea nymph Calypso.[2] Homer says Calypso, the daughter of Atlas and sister of Prometheus, lived alone on the island with her three handmaidens until the Greek hero, returning from the Trojan War, was shipwrecked on her shore.

Homer testifies that this lustrously beautiful granddaughter of Saturn knew "the depths of the whole sea."[3] The ancient Greeks called the island of Ogygia Omphalos Thalasses, the Navel of the Sea. They revered it as Saturn's island, where the god slept in a mountain cave made of gold. The Chaldeans believed Ogygia was the home of the divine barmaid Siduri, who "dwells by the deep sea"[4] and entertained and consoled the epic hero Gilgamesh. Gilgamesh is associated with Utnapishtim's flood, on which the story of Noah's Flood is based. Many in the ancient world believed Ogygia was the last remaining mountaintop of the sunken continent of Atlantis. The fourth-century BC historian Eumalos of Cyrene wrote:

> The summit of Mount Atlas, which was situated in the middle of the island Atlantika, was not submerged. This summit of Mount Atlas has preserved the name of Ogyge from that of its last king, and it is in fact because of this circumstance that we still know as Ogygia that island which still exists between Libya and Sicily; it is nothing more than the summit of the Mount of Atlantika.[5]

Today, there has been an enthusiastic revival of interest in the theory that not only Gozo but other islands of the ten-island Maltese archipelago are the last remaining bits of the legendary lost continent of Atlantis. For brief moments in his *Chronology of Ancient Kingdoms Amended,* Isaac Newton shared that belief.

It's surprising to find Newton mentioning Atlantis, since he is loath to admit the existence of any great civilization prior to that of the Jews. But he does indeed mention the legendary lost continent, and more than once, albeit briefly, though he seems to regard it as only a tiny city-state. Newton seems to locate Atlantis in or around Ogygia/Gozo, and he connects it with the legendary Ogygian flood.

But, wait a moment! Was there ever really an Atlantis at all? Millions of bottles of ink have been spilled on the subject of where it was and why it sank, but has anyone ever really come up with concrete proof that it existed?

The Egerton Sykes Collection, housed in the Association of Religious Enlightenment library in Virginia Beach, Virginia, is the largest collection of Atlantis-based materials in the world. It contains more than six thousand books, magazines, pamphlets, photos, slides, tapes, personal letters, unpublished manuscripts, and newspapers on the subject of the lost continent.

The basic story in every one of these documents can be traced to a single source: the Greek philosopher Plato's description of Atlantis in his dialogues *Critias* and *Timaeus*.

Newton had a comprehensive knowledge of the works of Plato (427–347 BC), but he might never have lingered over the story of Atlantis if it had not been for the towering stature of the man who, according to Plato, first introduced the story of the vanished continent to ancient Greece and Rome. This was Solon, the paradigm-bending Athenian lawmaker who lived from about 638 to 558 BC.

Plutarch tells us that Solon was "entirely original, and followed no man's example, and, without the aid of any ally, achieved his most important measures by his own conduct."[6] In about 594 BC, Athens found itself mired in its own unequal-distribution-of-wealth predicament, one quite a bit more serious than the "1 percent/99 percent" inequality-of-income problem that led to mass demonstrations in the United States in 2012. Plutarch writes that, in the Athens of Solon's time, "All the people were indebted to the rich . . . [many were] sent into slavery at home, or sold to strangers; some (for no law forbade it) were forced to sell their children, or fly the country to avoid the cruelty of their creditors."

The Athenians decided they needed a despot—a single ruler holding all the power of the state—to tell them how to get out of this predicament. They called on Solon "to take the government into his own hands, and when he was once settled, manage the business freely and according to his pleasure." Solon gave the Athenians a whole new set of laws, many of them entirely original. He used these laws to engineer the cancellation of almost every debt in Athens. The city's creditors complied with this because Solon's "own private worth and reputation" outweighed "all the ordinary ill-repute and discredit of the change."[7]

His laws having been passed and implemented, Solon left Athens, "under the pretense," writes the Greek historian Herodotus, "of wishing to see the world, but really to avoid being forced to repeal any of the laws which, at the request of the Athenians, he had made for them. Without his sanction the Athenians could not repeal them, as they had bound themselves under a heavy curse to be governed for ten years by the laws which should be imposed on them by Solon."[8]

This extraordinary lawgiver arrived in Egypt in about 590 BC. A tireless gatherer of knowledge, he sought out the priests of the temple of Neith in Sais, Egypt's administrative capital. Solon's great fame easily gained him entry to the temple. In a dimly lit chamber, surrounded by stelae so ancient the hieroglyphs had faded from some, he opened the discussion by recounting to the two priests assigned to him the history of Athens.

Solon seems to have been largely an egoless man, but even he must have been startled when the older of the two priests replied tartly: "O Solon, Solon, you Hellenes are never anything but children, and there is not an old man among you." The younger priest chimed in, "Those genealogies of yours which you just now recounted to us, Solon, they are no better than the tales of children!"[9]

Solon inquired as to what they meant. And it is at this point that the story of Athens became one with the history of the world. The priests told Solon the story of a devastating war they claimed had been fought nine thousand years before. Athens had been a combatant in this conflict—not, the priests hastened to add, the Athens Solon knew, but a proto-Athens, an earlier version of the present-day city-state. This proto-Athens had gone to war with a huge country named Atlantis that lay beyond the Straits of Gibraltar. The warriors of Atlantis conquered the southern shore of Africa as far east as Egypt and the northern shore of Europe up to Italy. The proto-Athenians raced to the rescue of the subjugated nations on the other side. A spectacular battle was fought between Athens and Atlantis. Athens won, and all of Atlantis's conquered nation-states around the Mediterranean were liberated.

There was no time for the victors to rejoice or for the vanquished to weep. Almost immediately a great flood overwhelmed that part of the world, sweeping Atlantis beneath the ocean. Only a tiny portion of the proto-Athenians survived, to live in the forests in a primitive state for centuries until finally a new Athens began to emerge.

This is all Plato has to say, in *Critias,* about what the priests told Solon of the last days of Atlantis. (Plato's *Timaeus* contains a lengthy if incomplete history and description of Atlantis, apparently from the same Egyptian priests.) Solon left Egypt not long afterward, continued on his travels, and eventually arrived back in Athens.

The great lawgiver had intended to turn his notes on Atlantis into an epic poem. But age and new responsibilities intervened. Plato tells us the lawgiver passed his notes on to his nephew, Dropides, who bequeathed them to his son, Critias. At age ninety Critias gave the Atlantis material to his ten-year-old grandson, also named Critias. This was the Athenian citizen for whom Plato's dialogue *Critias* is named, and the Critias who, in that dialogue, tells Socrates (Plato's literary persona) the story of Atlantis.

Despite the thousands of books that have been written on the subject, this, and the pages from Plato's *Timaeus* describing Atlantis, are the only sources we have for the story of Atlantis. (Solon died about 560 BC; the younger Critias was born about 460 BC and died in 403 BC, when Plato was twenty-four.)

Some twenty-two hundred years later, Isaac Newton, reading about Atlantis in *Timaeus* and *Critias* in his rooms in Cambridge, smelled something fishy in the grandiosity of the Egyptian priests and their mocking rejoinders to Solon. Let's recall the words of modern-day chronologer Larry Pierce (see chapter 11, "Deconstructing Time"): "In the centuries before Christ, a war broke out to see which nation had the oldest pedigree, whether real or invented. . . . Each claimed to have the oldest history. . . . While some writers seemed interested in the truth, others were playing a game to see who could spin the biggest and most convincing yarn about the antiquity of their nation."[10]

No one understood this better than Newton, who, as has been discussed, began the *Chronology of Ancient Kingdoms Amended* with these words: "All nations, when they begin to keep exact accounts of Time, have been prone to raise their Antiquities; and this humor has been promoted, by the Contentions between Nations about their Originals [Origins]."[11]

This "vanity-of-nations" syndrome is mainly why Newton asserts that Atlantis wasn't nearly as old as the priests claimed, and very likely not as big. He claims that the Egyptian priests "magnified the stories and antiquity of their Gods so exceedingly as to make them nine thousand years older than Solon, and the island Atlantis bigger than all Africa and Asia together, and full of people."[12]

The Egyptian priests had to make Atlantis disappear, because that would make it impossible for Solon, or anyone else, to prove or disprove their assertions about the "lost" continent. Newton writes: "And because in the days of Solon this great island did not appear, they [the priests] pretended that it was sunk into the sea with all its people: Thus great was the vanity of the Priests of Egypt in magnifying their antiquities."[13]

According to Plato, Solon recorded that the priests told him Atlantis was located in the Atlantic. If this is so, why did Newton place it in the area of Ogygia in the Mediterranean? He may have believed the priests had really told Solon that it was in the Mediterranean, but that Plato himself changed the ocean to the Atlantic ("beyond the Straits of Gibraltar").

Such is the theory of Professor J. V. Luce, who argues in *The End of Atlantis: New Light on an Old Legend* (1975), that the story of the destruction of Atlantis is the story of a violent volcanic eruption that took place on the island of Thera (Santorini), seventy-five miles north of Minoan Crete, in the Mediterranean Sea about 1500 BC.

Examining numerous ancient Egyptian and Greek documents, Luce decided that the Egyptians of Solon's time "knew little and cared less about foreign countries. They were not great travelers or seafarers, and their geographical horizons were quite restricted." Their knowledge of the Mediterranean extended no farther than the "isles which are in the midst of the Great Green [Sea]"; that is, Crete, and they had no reason to assume that the Mediterranean wasn't big enough to swallow an entire continent.[14]

Two hundred years later, the Greeks knew enough about the size and depth of the Mediterranean to decide that actually it wasn't big enough to contain an entire sunken continent. So—perhaps to make the story sound more authentic—Plato crossed out "Mediterranean" (or whatever name the Egyptians and Solon used for this body of water) and substituted "beyond the Straits of Gibraltar."

Luce also offers a more complex motive: Plato "was influenced by an imagined parallel between the Persian invasion of Greece from the east and the antediluvian aggression by Atlantis from the west. In the interests of poetic symmetry the vast land empire had to be balanced by the vast sea empire."[15]

Newton places the destruction of Atlantis in the same time period. What prompts him to override the statement in Plato that Atlantis is in the Atlantic is the legendary "Ogygian flood," which Newton considers actually took place, and which he connects with the destruction of Atlantis.

Newton explains that for the ancient Greeks the word *ogygian* had the connotation of "as old as the first memory of things." (Modern-day dictionaries list *ogyges* and *ogygian* as synonyms for "primeval," "primal," and "earliest dawn.") He suggests that the word originated with the (certainly) legendary King Ogyges of Thebes. The Ogygian flood has its name because it took place during the reign of Ogyges. Newton says the Ogygian flood was "1,020 years older than the first Olympiad." In the second-century AD *Chronicon I,* Eusebius states that the first Olympiad took place in 776 BC. So Newton is able to place the Ogygian flood at 1796 BC.

The Ogygian flood "covered the whole world and was so devastating that Attica remained without kings until the reign of Cecrops," he says. This was the flood that "consumed Atlantis and much of Greece."[16] Thus Newton sets the date of the destruction of Atlantis at 1796 BC.

▲

Clearly, the Ogygian flood is connected with the island of Ogygia. And so the island of Ogygia is associated with the sinking of Atlantis. In the *Chronicle,* Newton explains this, if a little bit obscurely.

> Homer writes that Ulysses [Odysseus] found the Island Ogygia covered with wood, and uninhabited, except by Calypso and her maids. . . . In that island [Gades] Homer places Calypso, the daughter of Atlas, presently after the Trojan War when Ulysses' being shipwrecked, escaped thither. Homer calls it the Ogygian Island and places it 18 or 20 days' sail westward from Phoenicia or Corcyra. . . . This island is by Homer described a small one, destitute of shipping and cities and inhabited only by Calypso and her women who dwelt in a cave in the midst of a wood, there being no men in the island to assist Ulysses in building a new ship or to accompany him thence to Corcyra: which description of the island agrees to Gades.[17]

Here we run into a contradiction. As we'll see farther on in this chapter, Newton seems to have believed that Calypso was the last surviving inhabitant of Atlantis. But throughout the *Chronology* he has been at pains to establish the true date of the fall of Troy as 904 BC. He tells us Odysseus was shipwrecked on Ogygia eight years later. That put the year of Odysseus's arrival at Ogygia at 896 BC—and makes Capypso, if she is the last surviving member of Atlantean society, at least nine hundred years old!

Perhaps we can get to the bottom of this by looking carefully at the account Newton gives us of the history of Atlantis, which he takes almost word for word from Plato. Newon writes: "The Gods, having finished their conquests, divided the whole earth amongst themselves, partly into larger, partly into smaller portions, and instituted Temples and Sacred Rites to themselves; and that the Island Atlantis fell to the lot of Neptune, who made his eldest Son Atlas King of the whole Island, a part of which was called Gadir."[18]

Plato (and Newton) depart slightly from traditional chronology by making Neptune, not Saturn, Atlas's father; Saturn becomes the father of Neptune and grandfather of Atlas. But in general Newton is here operating in full euhemeristic mode. All of these figures—Saturn, Neptune, Atlas—we regard today as mythical. (Atlas, for example, we think of as the mythical Greek god of astronomy who bore a globe of the world on his shoulders.) But Newton thought of them all as entirely mortal members of an Atlantean dynasty of rulers.

For Isaac Newton, Atlantis wasn't a vast continent supporting a highly advanced civilization. It was a tiny city-state that had made territorial gains that aroused the ire of the Athenians. Neptune, then Atlas, were kings of this feisty rogue state of Atlantis—and that seems to make Calypso, who was Atlas's daughter, and who survived on the mountaintop of Ogygia/Gozo, not only the last surviving Atlantean but also the putative queen of Atlantis. (She is all by herself, a sort of Newtonian "remnant," and it's almost as if her adoring subjugation of Odysseus is her attempt to play Eve to the Greek hero Adam and revive the lost civilization of Atlantis—an attempt that's thwarted when the ancient gods force her to set Odysseus free.)

Calypso is, of course, a sea nymph, an immortal goddess—and that could explain why she survived for nine hundred years. But Newton didn't believe she was a goddess; he didn't believe there were gods or goddesses. There seems to have been some sort of error in his connecting the destruction of Atlantis with Calypso and the arrival of Odysseus on Calypso's island; that is, on Atlantis/Ogygia/Gozo. Over the past two decades, there has been a huge upsurge of interest on the part of the people of Malta itself in the possibility that the islands of the Maltese archipelago, and Gozo in particular, might be remnants of Atlantis. Let's see if Malta's own experts can help us with this problem.

Running from 60 miles south of Sicily to 100 miles north of Libya, ten rocky islands stick up from the sparkling Mediterranean like the fingertips of a drowning giant yearning to be free. Or, at least, such is the image that presents itself to those (and they are not a few) who believe these ten islands are the topmost parts of a long-submerged Atlantis.

The seven northernmost islands in this straggling chain constitute the Republic of Malta. Four of them, mere foam-encircled specks in the glittering ocean, are uninhabited—though on the four farthest north temples of extreme antiquity rear up to the sky. The other three islands are Malta, a bit smaller than Martha's Vineyard at 122 square miles; Comino, barely 1 square mile; and Gozo, 8 by 4 miles and home to 31,000 people (one-tenth of the population of the republic).

Almost six thousand years ago, a race of peoples appeared on these islands, spread across the entire archipelago, built limestone temples, raised crops, hewed ports out of the rocky coast—and then, about 3700 BC, melted away to nothing. The traces of these enigmatic peoples have to a large extent been erased in the successive waves of conquest and assimilation that have swept across the archipelago: the Phoenicians, the Carthaginians, the Romans, the Byzantines, the Fatimids, the Normans, the Sicilians, and the Aragonese. After the Knights of Saint John were ousted from the island of Rhodes in 1522, Charles V, the Holy Roman Emperor and king of Spain, ceded the property to the order for an annual symbolic rent payment of one live Maltese falcon. Thereafter the history of Malta was to some degree stabilized, although Napoleon rudely conquered the archipelago on his way to Egypt in 1795, and the Nazis pulverized it with bombs in World War II.

In the early nineteenth century, the islands began to renew their acquaintance with the distant past. A local architect, Giorgio Grongnet, examined the temples at Gozo minutely and decided that this tiny isle, with its lush green terraced fields, was the last surviving mountaintop of Atlantis.

You might have expected Grongnet to know what he was talking about; he was a religious architecture expert, who had designed the rotunda of the Church of the Assumption of Our Lady, at Mosta, Malta, the ninth largest unsupported church dome in the world. But it was discovered that he had forged a key inscription in his report, and his work was largely discredited.

In 1995, a Maltese pediatrician named Anton Mifsud went to Gozo to look around and complete the job of discrediting Grongnet. But, instead, he was converted to the idea that Gozo in particular, and also, perhaps, the six other islands in the archipelago, were the remains of Atlantis.

Mifsud assembled a great deal of material and, along with three colleagues, in 2000 published *Malta: Echoes of Plato's Island*. It attracted so much attention that, in 2005, documentaries on Malta as Atlantis are regularly aired by the U.S. National Geographic Channel and Japan's national channel TBS-1.

Mifsud maintains that the Neolithic civilization of Malta is the only one in the West old enough to have existed at the same time as the Atlantis described by Plato. He proudly points out a number of similarities between the two.

Gozo, Malta, and Comino are crisscrossed with hundreds of deep and ancient cart ruts cut in the stone. These often extend for long distances out into the sea; sometimes a cart rut ends at a cliff on one islet and recommences at a cliff on another.

Mifsud contends these ruts, and the very many (usually submarine) canals on the archipelago, are identical with the large grid of irrigation canals Plato claims once crisscrossed Atlantis. He says the continuity of some ruts and canals from one island to another proves the separate islands of the archipelago were once much closer together—that, in fact, they once constituted a single landmass that was torn apart in an aboriginal past by a catastrophe that also sank parts of the landmass.

As in Plato's Atlantis, says Mifsud, the sacrifice of bulls was a feature of Neolithic religious life in Malta. Bull horns of high antiquity, along with flint knives used in the sacrificial slaughter of bulls and ancient images of the taurines, have been found beneath the floor of the famous temple at Tarxien and at other locations as well.

There are fifty-three temples of high antiquity on the Maltese archipelago today, many of them well preserved, a number of them partially or wholly underwater. A nineteenth-century researcher wrote ecstatically that "these anthropomorphic structures indicate the islands [of Malta] in pre-history as probably the Holy Shrine of the Middle Sea." During the nineteenth century it was still possible to see that the temples had been constructed from rectangular blocks of "unbelievable sizes . . . stones five to six feet long, and laid without mortar."[19] The American journalist Mark Adams, visiting the Neolithic structures of Mnajdra and Hagar Qim in 2014, described them as "stunning, clusters of oval rooms built from giant slabs of cut yellow limestone and set on a desolate bluff overlooking the water."[20]

Mifsud asserts that these temples, in their great antiquity, in their deployment and orientation, and in the motifs and ornamentation that adorn them, are almost identical with Atlantean temples described by Plato in *Timaeus*. Mifsud doesn't mention that some of the Neolithic temples of Malta, especially on Gozo, resemble, by virtue of their central hearth and their circular shape, *prytanea*—the ancient temples whose origins Isaac Newton believed could to traced to the time before the Flood. Did Newton know of these formidably ancient temples on Malta? Certainly he never visited that archipelago country; this great genius, who discovered what makes the tides rise and fall, probably never saw the sea and lived all his life inside a tiny triangle of English land whose three points were his birthplace—Woolsthorpe, the town of Cambridge, and the city of London. But Nostradamus knew of the temples of Malta and mentions them in century 1, quatrain 9, of his *Prophecies,* and it's almost certain that Newton had read, however skeptically, Nostradamus's famous volume of prognostications. So we can probably assume that the great man knew about Malta's beguilingly venerable temples.*

So nothing new about Calypso—who, as we'll recall, does have a "cave" on Gozo—emerges from the minute research of Anton Mifsud and his

---

*Century 1, quatrain 9, of the *Prophecies* reads: "From the East will come a treacherous act / To vex Hadrie and the heirs of Romulus [Italy] / Accompanied by the Libyan fleet, / The temples of the Maltese [trembling] and its islands emptied." (Hogue, *Nostradamus,* 73.)

colleagues. Calypso is only one brilliantly colored thread in the bright and vivid tapestry Newton weaves of the "embroilments" (his word) of a multitude of squabbling city-states in the ancient world surrounding the Mediterranean Sea. Let us let go of the Calypso thread and turn over some of the other threads that Newton introduces into his brief story of Atlantis. We may find (especially on account of Newton's habit of turning mythical heroes into humans and conflating them with each other) that the scientist has some tantalizing hints to offer, especially for Atlantis buffs, on the true history of the lost continent of Atlantis.

Newton sets the stage for something like this when he tells us that Neptune, Atlas's father and Atlantis's second king, invented ships with tall sails and originated the concept of seaborne fleets. We'll recall that, often when he makes such statements, Newton is simultaneously using the name of, for him a historical person, for us a mythical figure, as a symbol or emblem for a technological innovation he considers has been important for the advancement of mankind. Here, though, Newton is perhaps giving is some idea of the very seaworthy nature of the Atlanteans (whom, after all, we associate with oceans).

Newton says about Neptune:

> The Cretans affirmed that Neptune was the first man who set out a fleet, having obtained this Prefecture of his father Saturn; whence posterity reckoned things done in the sea to be under his government, and mariners honored him with sacrifices: the invention of tall Ships with sails is also ascribed to him. He was first worshiped in Africa, as Herodotus affirms, and therefore Reigned over that province.[21]

Newton goes on to describe a war fought between Neptune's son, Atlas, and the Greek warrior hero Hercules. Before doing so, he conflates Atlas with Antaeus.

Antaeus is the son of Neptune and of the Earth goddess Gaia. Therefore, he is Atlas's half brother. He is a mythical (for us, though not for Newton) giant who was in the habit of engaging everyone he met in a fight to the death, and who invariably won. To maintain his great strength, Antaeus was dependent on regular physical contact with his mother, the Earth. Hercules (also for Newton a human and not a god) defeated Antaeus by holding him aloft, and so unable to touch the earth, for a long enough period that the giant's strength drained away and Hercules, still holding him aloft, was able to strangle him to death.

Newton justified his conflation of Atlas and Antaeus in this way:

> Antæus and Atlas were both of them sons of Neptune. Both of them Reigned over all Libya and Afric, between Mount Atlas and the Mediterranean to the very Ocean; both of them invaded Egypt, and contended with Hercules in the wars of the Gods, and therefore they are but two names of one and the same man; and even the name Atlas in the oblique cases seems to have been compounded of the name Antaeus, and some other word, perhaps the word Atal, "cursed," put before it.[22]

But we shouldn't trust Newton entirely on this. He keeps changing his mind. Elsewhere he argues that both Antaeus and Atlas are actually Phut, the evil brother of the Egyptian ruler Sesostris; somewhere else he asserts that Atlas is the son of Antaeus and the nephew of Sesostris. And then he says that Atlas and Sesostris are brothers! Finally, he declares disingenuously: "But it is difficult to state these things exactly. So it's of small consequence."[23]

If Atlas and Antaeus are really the same person, Newton has given Atlantis buffs something to chew on. Might it be that when he writes, "In these wars Hercules took the Libyan world from Atlas, and made Atlas pay tribute out of his golden orchard, the Kingdom of Afric,"[24] he means by Hercules, proto-Athens, and by Atlas, Atlantis—and that he's talking about the last, great battle between these two nation-states? It may be that in Newton's conflation of Atlas with Antaeus—which brings into the story the image of Hercules holding the son of Gaia aloft in order to kill him—we may have a kind of veiled allegory of a critical feature of the final battle between the Atlanteans and the proto-Athenians. (Note that Newton says mysteriously, "Hercules overthrew him several times, and every time he grew stronger by recruits from Libya, his mother earth."[25])

Not even all of Plato's admiring contemporaries believed his Atlantis story was true. His great pupil and successor Aristotle believed Atlantis wasn't real and said famously that "the man who dreamed it [Atlantis] up, made it vanish."[26] (On the other hand, Crantor [ca. 300 BC], the first editor of *Timaeus,* thought every point in the narrative was literally true, and he is even supposed to have sent emissaries to Egypt who were able to verify Solon's story by examining the appropriate stelae in the temple of Neith.[27])

The classical scholars of the nineteenth century, schooled in the skeptical-rationalist mode of thinking of the Age of Enlightenment, didn't believe Atlantis had ever existed. Benjamin Jowett, perhaps Plato's greatest translator, observed caustically: "The world, like a child, has readily, and for the most part unhesitatingly, accepted the tale of the Island of Atlantis." Jowett noted that the story rested on the authority of the Egyptian priests and that historically the Egyptian priests took pleasure in deceiving the Greeks. What we never seem able to to realize, he tells us, is that "there is a greater deceiver or magician than the Egyptian priests, that is to say, Plato himself."[28]

Should we regard the story of Atlantis as simply a great Platonic poem, a multilayered philosophical truth that nonetheless is not anchored in concrete reality? Perhaps; but for those of us less rational than Benjamin Jowett, the findings of the Maltese researchers and the brief observations of the "universally learned" and preternaturally intuitive Newton perhaps deserve some careful thought.

CHAPTER FIFTEEN

THE SECRET OF LIFE

At the beginning of August 1669, shortly after being appointed Lucasian Professor of Mathematics at Cambridge, Isaac Newton, aged twenty-six, disappeared into London.

This was only his second trip to the city. We can be sure this haughty Puritan didn't visit any of London's notorious fleshpots. He probably didn't do much sightseeing at all. Isaac Newton, already the greatest scientist of the age (though nobody knew it except his mentor, Isaac Barrow), was in London on a mission. This reclusive genius, who had barely survived his own birth, was about to take his first steps toward discovering how anything got born at all.

We know little of what Newton did during that month of August he spent in London. But we can imagine him hurrying through the streets, glancing swiftly from side to side, and fastening his attention on whatever raucous and mud-splashed event reflected his intellectual passion of the moment, which, just then, was how life could ever have emerged from nonlife. London was shimmering in a heat wave; Nefertem, the Egyptian god of perfume, ruled the streets and alleys that morning, but Nefertem was in a bad state of mind; the stench of rotting excrement and decomposing carrion rose up like a thick and rancid mist.

Newton stopped at the mouth of an alley and watched intently an army of maggots swarming out of the belly of a decomposing dog. What, Newton asked himself, made this phenomenon of "spontaneous generation" happen? Hamlet had demanded: Does the sun breed maggots out of a dead dog? (2.2.180) and the answer was yes. Robert Boyle had written that decomposing lion corpses gave birth to bees; decaying twigs stirred, wriggled, and became worms; and horsehairs stood up, weaved back and forth, and metamorphosed into snakes. The slow putrefaction of dead plants and animal corpses, brought to a fine simmer by the summer heat, birthed buzzing, squirming creatures that were completely different from their sires.

How could this happen? The youthful Newton was coming to believe in a vegetative spirit—a subtle and tiny spark of the divine that, when diffused through brute mechanical matter, animated it with nonmechanical life. But how did God introduce this vegetative spirit, this divine spark, into dead and putrefying bodies and rouse them to put forth other forms of life?

Young Newton strode forward quickly and, suddenly crowded against the curving belly of a young woman as pretty and bright as a new farthing—and knowing from the way she lowered her eyes that she was not stout but pregnant—asked himself, as he darted along the sidewalk milling with Londoners, Why is it that conception doesn't really take at conception, but some time later, when the male and female semen, having immediately mingled, have undergone putrefaction to the point of becoming a neutralized sexless mass[1]—after which the vegetative spirit slips into this undifferentiated mass and is somehow transformed into a foetus? How had the vegetative spirit been able to cross the chasm that separates the world of the wholly divine from the world of the entirely physical?

A closed sedan chair swayed by in the street, conveyed by four sweating liveried servants. He peered in the window and saw what looked shockingly for a moment like a heap of bones wrapped in a velvet shroud. The sedan chair lurched forward: it was a man as old as Methuselah, wearing a frock coat as young as the morning. The chair suddenly sped up, as if trying to deny that ultimately it was bearing its passenger to his death; and Newton, falling back, wondered: Why does this man have to die at all, be he ever so old? Why is it that the vegetative spirit, having performed that tremendous miracle of leaping the gap between the eternal and the temporal, must ever have to leave the body? Roger Bacon said that it might be possible for the human body to live forever since the soul is immortal. Hadn't Methuselah lived to 969, and Noah to 370? A mixture of the steady heat of a hot sun, abstemiousness, and perpetual piety had enabled the desert monk Saint Anthony and the Stoic leader Cleanthes to live to 105, and this despite the fact that the perpetual piety of Anthony had been directed at the perpetually evil doctrine of the Trinity.

With these captivating and concerning questions on his mind, young Newton arrived at the borough of Little Britain, which was his destination—and immediately stumbled against a two-foot high rock that stood sentinel-like on one side of the city gate. He immediately flopped down on the rock, and, rubbing his shin, asked himself thoughtfully: How does the vegetative spirit enter soulless ore?

If the rock he was sitting on was filled with ores, why did these metals die when the rock was taken from the earth? For it was well known that metals flourished alive in the Earth's womb and that they grew like plants and animals. They grew until, after a thousand years, or two thousand years, or twenty thousand—nobody knew—they were transmuted into gold. What made this happen? How had the vegetative spirit been introduced into the dead matter of the incipient ores? Was it possible to grow metals in a laboratory furnace and, controlling their growth, observing them minutely—manipulating them—discover how it was that the spirit of life animated that which had once been so brutally mechanical?

Newton got up suddenly and, navigating energetically the first narrow, twisting streets of the borough, found himself before the door of the Pelican, which was his destination and whose proprietor was William Cooper, a man (to quote a later source) "of impeccable reputation but also a dealer in illegal and highly-sought after manuscripts."[2]

Newton rang. The door opened. He was smoothly and hurriedly

pulled inside by a hand stained black with tobacco. Inside, there was the glitter of large lamps (the windows were shuttered), smoke everywhere, dimly lit bookcases, and several men gesticulating, sitting and standing in the light that seemed like a twilight. The tobacco-stained hand belonged to a bewhiskered man who peered at him carefully and then said: "Yes! The Cambridge don. Precisely on time! Of course. I'm Cooper! Come, let us enter the inner sanctum."

He was ushered into a small room at the back, occupied by a young man of vigilant foxlike appearance with a pistol stuck in his belt, seated on a small, narrow chair, and reading a ragged book. There was a small table in the room, and another chair; Cooper sat Newton down and then withdrew from a large locked wardrobe, one after another, seven leather-bound volumes that he placed on the table in front of Newton.

"My assistant," said Cooper, nodding toward the other chair, "will help, if wanted. Just call." The proprietor of the Pelican slipped out of the room. Newton bent forward and opened the first volume.

If the vigilant foxlike man—who was there to make sure nothing happened to the books—had gotten up at that moment, crossed the brief space of floor, and looked over Newton's shoulder, he would have been instantly convinced that this young Cambridge don (for so Mr. Cooper had identified him) was a vile and thuggish creature who was hopelessly mired in the Bogs and Swamps of Everlasting Sin. Graphically visible on the opened page were the words, "THE MENSTRUAL BLOOD OF THE SORDID WHORE," and below was a dark Gothic woodcut of a naked crone holding up a dripping goblet. If the young man hadn't turned his eyes away quickly he would have glimpsed other bizarre phrases, not as dark but bewildering, like "Green Lion," "Doves of Diana," and "Jove's Eagle"—and many more.

But the young man didn't get up, for he was engrossed in a book more salacious than anything in front of Newton, and Isaac Newton spent an hour looking carefully through the volumes, called for William Cooper, and paid him a quantity of bright new guineas for the books and to have them sent by stagecoach to Cambridge in September.

Those six volumes were the *Theatrum Chemicum* of Elias Ashmole, published in 1652. They contained poems about alchemy by many famous English poets, including Geoffrey Chaucer, John Gower, and John Tyler, and many other phrases, such as "Green Lion," which were a part of the enigmatic "riddling" language making up an alchemical recipe.

Isaac Newton was becoming an alchemist. He would undertake the care of his own furnaces and try to discover how the vegetative spirit entered nonliving matter and fired it with life. This would be the first focus of his daunting alchemical project. There would soon be other foci; Newton would seek to penetrate to the primal elements of the universe and understand their true relationship to the framing, forming activities of God. Some of this would happen under the rubric of alchemy. But the first question Newton would put to the muse of alchemy was, What is the active unifying celestial principle necessary for life?[3]

He spent the entire month of August in London. The only thing we know for sure, apart from his visit to the Pelican, is that he purchased a seven-shilling tin furnace, an eight-shilling iron furnace, a variety of glass equipment (such as beakers and crucibles), and a multiplicity of chemicals (such as antimony and sulphur) for a total of £10.[4] All this material, Newton took back with him when he returned to Trinity College at the end of August. The tin and iron furnaces would be replaced in two years by two brick furnaces that Newton "made and altered himself, without troubling a bricklayer,"[5] placing them against the wall of his garden that abutted Trinity Chapel.

There is no doubt that in that month of August 1669 young Isaac Newton became a member of a clandestine network of alchemists that had been extending its tentacles all across Europe for many centuries now.[6] This network had to be clandestine, because practicing alchemy was against the law in England. In 1404, Britain's Parliament, fearing that alchemically produced gold would flood the market, debase the currency, and probably sink the economy,* had passed the Act Against Multipliers ("multipliers" of gold), which made the practice of alchemy illegal. The law was still in force in Newton's time. It would be repealed only in 1696 due to the forceful lobbying of Robert Boyle, who was not only the founder of modern chemistry but an enthusiastic alchemist.

In 1317, Pope John XII—an entirely immoral pope—issued a decree condemning to death all alchemists who practiced the crime of "falsification." Alchemists caught passing off "false or adulterate" or "alchemic" metal as real gold or silver would have to pay the public treasury an equivalent amount in gold or silver. If they could not, they were prosecuted as criminals. For some time to come, John XII's edict provided great lords with an excuse for hanging alchemists who had promised them the Philosophers' Stone but hadn't produced it. Sometimes the executions took place on gold-painted gibbets; other times (to make the public spectacle even grander), the miscreants were hanged wearing tinsel suits so that they glittered brightly while swaying in the wind.[7]

So Isaac Newton as an alchemist had to conduct himself with as much caution as he deployed when writing or talking about Arianism. Apart from the alchemical experiments he began to carry out in his rooms, the only change in his lifestyle was that he increasingly received in the mail mysterious letters and packages signed off by individuals with odd-sounding names like "W.S.," or "Mr. F.," or "Fran," or "Meheux," or "hee." These letters and packages contained alchemical recipes and books and powders; the names were the pseudonyms of those involved. Newton sent materials in exchange, following the age-old alchemical practice; one of his alchemist's pseudonyms was Jeova Sanctus Unus, or "One Holy God," which was based on an anagram of the Latinized version of his name, Isaacus Neuutonius. From time to time Newton also had brief visits, often at night with strangers whose

---

*In 1965, the French secret service attempted to topple the president of Guinea, Ahmed Sekou-Touré, from power by flooding the country with counterfeit French francs. Guinea's economy was ruined. Sekou-Touré remained in power.

faces he couldn't always see and who stopped by solely to hand him the bulkier alchemy-related materials he had requested.

There were, however, distinguished men and women with whom he could discuss alchemy. In the early part of the seventeenth century, that clandestine circle of alchemists who rendezvoused at William Cooper's Pelican came to be associated with another, larger, circle known as the Invisible College. This latter, developing slowly over the years, was the brainchild of Samuel Hartlib (ca. 1600–1662), a half-English, half-Polish scholar who came to England in 1630 to acquire all knowledge and to dispense it, as equitably as possible, to everyone.

Versed in science, medicine, alchemy, agriculture, politics, and education, Hartlib was called the Great Intelligencer. Many learned men and women gravitated toward him. There weren't really any meetings; everyone knew Samuel Hartlib, but they didn't necessarily know each other. Small "cells," like that comprising John Milton, Robert Boyle, John Locke, and Samuel Pepys, discoursed freely among themselves but knew little of the other brilliant groups who existed throughout Britain (and on the Continent).

Sometimes several of these "cells" came together casually, almost inadvertently, at the country mansions of the great political and cultural elites of seventeenth-century England. Over the thirty years Newton practiced alchemy, he was often absent from Cambridge; we know little or nothing about where he went or why. But he may have spent some time at the loosely organized gatherings that took place under the unstated aegis of the Invisible College (also called the "Hartlib Circle").

At these gatherings, discussing alchemy was not only condoned, but it was almost de rigueur. Henry More, the saintly Cambridge Platonist scholar with whom Newton shared impassioned conversations about the Book of Revelation (see chapter 2, "The Newton Code"), was a member of the Invisible College/Hartlib Circle and often spent time at "Ragley"; this was the Warwickshire estate of his student and great friend Anne Finch, the Viscountess Conway. Ragley became a focal point for the growing body of intellectuals interested in alchemy. It seems likely that Newton attended from time to time.* So Isaac Newton, when he practiced alchemy, was not entirely alone; he was a member of a sympathetic network of aristocrats who, if they never breathed the name of "alchemy" in public, seized upon the subject when alone among their kind.

But what exactly is alchemy?

The French alchemy scholar Serge Lequeuvre asserts that this science/art "began in the smelting pots of China in the fourth century BC."[8]

As far back as then, the Chinese believed that metals were alive; that they were formed by the copulation of male and female metals; and that they were composed of some combination of mercury, which was female, and sulphur, which was male. The offspring of mercury and sulphur was usually a bloodred sulphide, cinnabar, which the ancient Chinese associated with royalty.†

Alchemy quickly went in three separate ways in ancient China. Using metals, fire, soupçons of other rarer minerals and even admixtures of exotic plants, the Chinese developed (1) *aurifaction,* transmuting

---

*Anne Finch, Viscountess of Conway (aka Anne Conway)(1631–1679), whom Newton almost certainly knew, is another one of those unsung female geniuses who only with great difficulty were able to express themselves before the nineteenth century. Anne's father, who died a week before she was born, was speaker of the House of Commons. Growing up in a privileged milieu, often in the company of scholars, she learned French, Latin, Hebrew, Greek, mathematics, and philosophy. Between 1671 and 1674 (or 1677 and 1679), she wrote an anti-Cartesian work *The Principles of the Most Ancient and Modern Philosophy;* Anne died without having a chance to revise it or correct it. Her famous friend and colleague Francis Mercury von Helmont translated *Principles* into Latin, published it, and showed it to his friend the philosopher/mathematician Gottfried Leibniz. Scholars consider Anne Finch to have been in many ways a perceptive forerunner of Leibniz. Von Helmont's translation was published anonymously, which is one reason why it was virtually unknown until the past century. It is available at http://digital.library.upenn.edu/women/conway/principles/principles.html.

†Mercury is a clear, yielding, metallic liquid, still called "quicksilver" ("quick" means living); sulphur is "solid, yellow like the sun, arousing thought associations of raucous, stifling heat." It is not yielding, but aggressive. (Needham, *Science and Civilization in China,* 4:455–56.)

metals into gold; (2) *aurifiction,* as the name implies, changing living metals into an artifact as beautiful as gold; and (3) the pursuit of the "drug of deathlessness," or *elixir of immortality.*

Aurifaction ("making gold") and aurifiction ("faking gold") never ceased up through the thirteenth century to be popular in China, but it was the pursuit of alchemically enhanced elixirs of deathlessness from the third century BC to the third century AD that held emperor and serf alike in thrall.

The ancient Chinese believed they were surrounded by invisible sages who had liberated themselves from death and wandered freely, not in heaven or hell or purgatory (there is no afterworld in Chinese thought) but on this Earth, in these skies, in our cosmos. These legions of the weirdly undead, traveling in attenuated wispy versions of their earthly bodies, had achieved through diet, meditation, and often the ingestion of metals a peculiarly Chinese state called *hsien,* or material immortality. Joseph Needham tells us that in this state "the body was still needed, preserved in however etherealised or 'lightened' a form, whether the deathless being remained among the scenic beauties of earth or ascended as a perfected immortal to the ranks of the Administration on high—in either case within the natural world suffused by the Tao of all things."[9]

Chinese alchemy was under continuous pressure from its rulers, who sought immortality to develop new drugs of longevity and deathlessness. These were subtle mixtures of cinnabar, sulphur, mercury, gold, even arsenic, and others, that had to be ingested in measured quantities on a regular basis. The resources of the entire nation were ruthlessly mobilized so that a ruler could realize his dream of cheating death. The best minds of the day—scholars, naturalists, alchemists, magicians—were summoned to court (bringing in their wake armies of con men, hangers-on, and lunatics) to join in the vast research project into the nature of *hsien.*

The "cult of *hsien*" could be lethal; there may have been emperors who died of mercury poisoning; but the cult became so popular that a mythos emerged. In the earthly paradise of *kun-lun* ("for holy immortals"), peaches and gems of immortality grew on trees. On sacred islands lost in distant oceans, white and gold were the only colors, and elixirs of immortality sprouted beneath mulberry bushes. In all of these alchemy-oriented places, there were underground substances so refined that only a material immortal could mine them.

Apart from its popularity in China, immortality alchemy did not—or at least not for a long time—travel well outside the country. But Chinese mercury-sulphur theory spread by trade routes to Persia and Arabia. Its precepts mingled with those of Egyptian alchemy that had developed from the need to perfect mummification techniques. These gave rise to proto-scientific arts embraced by the magi of Hellenistic Greece and Rome and particularly Alexandria. Out of this emerged the notion of the "Philosophers' Stone," an agent of universal transmutation—a catalytic substance that would drive impurities from metals and return them to their primal state as gold (and also, hoped many, cure people of illnesses). This tangled skein of art and science, making its way from the Middle East across the Roman Empire, became the heart of alchemy as it flourished in Europe in the Middle Ages.

The greater part of Isaac Newton's alchemy manuscripts, acquired at the Sotheby's auction of 1936 by John Maynard Keynes and bequeathed to Cambridge when Keynes died in 1946, are housed in the King's College Library of the university. There are more than one hundred auctioned lots, of all sizes, from thick treatises to sheaves of one or two pages. Altogether, Newton wrote a million words on this divine art. As of 2015, probably no more than 20 percent of his alchemical documents have been read. Newton was, apparently, not at all original in his alchemical researches; rather, he was the great cataloger and organizer of all the alchemical literature that came before him. He transcribed, he compared, he commented on, and he quoted, other men's alchemical work.

Newton organized this literature in an ambitious "alphabetical compendium of a hundred and thirteen pages covering the usage of alchemical terms in scores of authors [he] had consulted."[10] This compendium he called the "Index Chemicus"; it contains 879 separate entries and cites 150 different works. There is a list of 47 alchemy axioms, drawn up in a manner similar to the geometry axioms of Euclid. But Newton published only one paper on an alchemy-related topic, "On Acids"; his best-known writings on alchemy are, perhaps, his translations with commentary of the *Hermetica* and the Emerald Tablet, allegedly by the Egyptian god Hermes. Appendix F of this book consists of Newton's translation of the Emerald Tablet.

Newton probably conducted thousands of alchemical experiments. These were, and certainly Newton's were, far more complex and far different from anything we can even imagine today. In the twenty-first century, when we think "alchemist," we think the three witches in Shakespeare's *Macbeth,* gamboling around a bubbling cauldron and tossing into it an amazingly unseemly bunch of condiments such as baboon's blood, the finger of a child strangled at birth, a dragon's scale, a Turk's nose, the liver of a blaspheming Jew, a Tartar's lips, a witch's mummy—and so on and so forth—in order to conjure up "a charm of powerful trouble" that "like a Hell broth [will] boil and bubble" (4.1.5–37).

Alchemy was actually a subtle, complex, and delicate process reminiscent more suggestive of the rearranging of DNA and gene structures by geneticists than the machinations of bloodthirsty witches. Genetics concerns itself with living tissue; so did alchemy, since metals were thought to be alive. Just as fetuses grow in wombs, minerals grew in the womb of the Earth and then in the furnaces of alchemists, in their own inscrutable way, which it was the alchemist's passion to understand, to manipulate, and to accelerate. The basic ores of alchemy had to be "putrified"—artificially made to rot, decomposed, melted down slowly in the alchemist's furnace until they had become the primal matter that underlay all matter; then they were "regenerated," or recomposed to attain the desired effect. This could take months, even years; there were innumerable stages of distillation; exquisite care had to be taken in adding the precise chemicals and minerals, weighed out in minute amounts, to the fermenting mass.

All this would come to be seared into Isaac Newton's soul. Humphrey Newton (no relation), his lab assistant from 1685 to 1690, gives us a glimpse of the mathematician's work schedule.

> He rarely went to bed before 2 or 3 of the clock, sometimes not till 5 or 6 . . . especially at spring and fall of the leaf, at which time he used to employ about six weeks in his laboratory, the fire scarce going out either night or day, he sitting up one night and I another, till he had finished his chemical experiments, in the performance of which he was most accurate, strict, exact. What his aim might be I was not able to penetrate into, but his pains, his diligence at those set times, made me think he aimed at something beyond the reach of human art and industry. . . . He would sometimes, tho' but very seldom, look into an old moldy book which lay in his laboratory, I think it was titled *Agricola de Metallic*, the transmuting of metals being his chief design, for which purpose antimony was a great ingredient.[11]

Was there a general drift, a guiding theme, to Newton's alchemical experiments? David Castillejo tells us that "Newton's work was centered around the marriage of sulphur and mercury," with Newton adding no terminology of his own but juggling the images of different authors. Sir Isaac identified "two simple forces or sperms in nature which are fundamental to the alchemists' work," one being "sulfurous, solar, masculine and non-volatile," the other being "mercurial, lunar, feminine and volatile."[12] These are the same two forces that were identified by the ancient Chinese. Joseph Needham, in his magisterial *Science and Civilization in Ancient China*, expresses delight at "the *shui yin* [mercury] and the *liu huang* [sulphur] of ancient Chinese alchemy coming through to the threshold of modern science" as exemplified in Newton's works; he tells us, almost in awed tones, that, "with Newton, mercury and sulphur found one of their ultimate incarnations in what he called particles of 'earth' and 'acid,' but so penetrating was his insight that these sound almost like protons and electrons, the sub-atomic particles out of which all sorts of matter—'one catholick matter,' as Boyle put it—would be built by variants of their stable configurations."[13]

When we try to follow Newton's experiments, we immediately encounter the bewildering, "riddling," hieroglyphic, beguilingly quasi-sexual language that the vigilant foxlike young man at the Pelican bookstore would have seen if he had looked over Newton's shoulder. We struggle to pay attention as Newton tells us he uses "the regimen or phase of Mercury" to show how "the caduceus of Mercury" is formed and then uses "the regimen of Saturn" to show how this caduceus "is fermented"—and digested with the two serpents until they die and grow black. We're relieved to read finally that the caduceus will open other substances by "striking on them with love."[14] And so on, and so forth.

But there are beguiling glimmers, sparks from the divine, and vast tracts of ancient history open out to us in these strange depths inhabited by creatures that seem half god and half metal.

In the 1990s and early 2000s two American historians of science, William Newman of Indiana University and Lawrence Principe of Johns Hopkins University (who is also a chemist), made replicas of the alchemical glassware and furnaces used in Newton's time and used this replicated equipment to duplicate a number of Newton's alchemical experiments. In one, they "cooked" copper and iron together in a slow fire until they obtained a purple alloy with a striated netlike surface. They called this experiment, as had Isaac Newton, the "Net"; the purple alloy was thought to be one step toward the Philosophers' Stone.

This experiment is intriguing because, surprisingly, it relates to a myth found in both Ovid's *Metamorphosis* and Homer's *Odyssey*. This is the story of how from afar the sun god Apollo showed Vulcan, the blacksmith of the gods, his wife Venus in bed with Mars. To punish the lovers, Vulcan made a net whose meshes were as fine as cobwebs but as strong as iron. He caught Mars and Venus in this nearly invisible net in the act of lovemaking and hung this net from the ceiling for all the Olympians to see.*[15]

How exactly does this mythological story connect with the alchemical experiment called the Net? The "William Newman Project" website explains:

> In alchemy, "Venus" usually means "copper," Mars means "iron," and "Vulcan" means "fire." Hence "Venus" referred to the copper in the alloy [in the Net experiment], and "Vulcan" to the intense heat used in making it. Since the antimony regulus that is added to the copper is itself reduced from stibnite (antimony sulphide) by the addition of iron, "Mars" (iron) was thought to be present in "the Net" as well. *Voilà*—the whole myth becomes a recipe for "the Net."[16]

Isaac Newton believed that the alchemical recipe for the Net came first and that the myth was a corrupted form of the recipe. The great Roman poet Lucian of Samosata (AD 125–180), who wrote the first trip-to-the-moon sci-fi story, adds yet another wrinkle, remarking that this story of the Vulcan-Venus-Mars triangle, while ludicrous, wasn't an idle fantasy, "since it must have referred to a conjunction of Mars and Venus, and it is fair to add, a conjunction in the Pleiades."[17]

This is enigmatic indeed! It begins to take on meaning when we learn that Isaac Newton believed that the myth of the Net was not only a corruption of the alchemical recipe of the Net but that the alchemical formula itself was a corruption of some far more ancient aspect of the *prisca sapientia*—the original, primal religion of mankind or, as

---

*Homer tells the same story in the *Odyssey*, using the Greek names Hephaistos (Vulcan), Aphrodite (Venus), and Ares (Mars): "Now to his harp the blinded minstrel sang / of Ares' dalliance with Aphrodite: / how hidden in Hephaistos' house they played / at love together, and the gifts of Ares, / dishonoring Hephaistos' bed—and how / the word that wounds the heart came to the master / from Helios [Apollo], who had seen the two embrace; / and when he learned it, Lord Hephaistos went / with baleful calculation to his forge. . . ." (Homer, *Odyssey*, 8.280–88.)

Newton sometimes calls it, "astronomical theology." (See chapter 18, "Son of Archimedes.")

The "Gaia" theoretician James Lovelock has presented us with the beginnings of an explanation of this enigmatic triple presence of the Net story as myth, alchemical recipe, and astronomical event—one that Isaac Newton might have found captivating. Lovelock makes the beguiling suggestion that one of the reasons man emerged as a sentient species was so our planet could remember itself.[18] That primordial Earth that seeks to remember itself was aware of the great planetary movements that were taking place in the sky, and perhaps somehow felt their presence or was somehow associated in that these celestial movements caused eructations and disturbances in the Earth itself; the alchemical recipes replicate deep dynamic processes in the history of the Earth. Alchemy was the memory of of God's most ancient activities in the world; as such, it had to be not only respected but feared. William Irwin Thompson observes in *Imaginary Landscapes* that, "for the mythopoeic imagination of the ancients, knowledge, and complex astronomical knowledge, was stored in images and hieroglyphs."[19] This is something that we do understand only poorly today; it seems that Isaac Newton understood it well.

Professors Principe and Newman, who have examined scores of Newton's alchemical notebooks, believe he was trying to produce the Philosophers' Stone because he thought that "alchemy promised tremendous control over the natural world."[20]

That power could be used to help the world or it could be used for personal aggrandizement. Newton transcribed an ancient treatise, *The Epitome of the Treatise of Health,* that described an elixir of good health and immortality and also a "highest" Philosophers' Stone, the "Angels' Stone," which opened out in the practitioner the power of remote viewing, of making others bow to one's will, of foretelling the future—of being surrounded by beautiful smells!—of not having to eat, and of living as long as the patriarchs, or perhaps forever. Manuel believes this elixir held some interest for Newton: "That Newton might discover an elixir of life which would bestow immortality upon him—if not upon his fellowmen, for whom he had less concern—may be a remote motivation for his search, but one not to be wholly excluded in the light of his hypochondria and his omnipotence fantasies."[21]

Newton may have known of the Taoist state of *hsien,* or material immortality, if not through the writings of Roger Bacon then through the learned French Jesuits who, recently established in China, were regularly communicating information about ancient Chinese philosophy to the West. Newton's hair was turning gray when he was thirty, and when his roommate, John Wickins, remarked that this must be "the effect of his deep attention of mind," Newton jocularly replied that it must be due to "the experiments he made so often with quicksilver [mercury], as if from hence he took so soon that color."[22] Newton ingested mercury daily, tasting it to check its composition for the chemical experiments he was performing. As we saw earlier, this was the same practice that great emperors of China (and many another lowly official) followed; they ingested mercury daily as an essential step to achieving material immortality. Newton gave up the practice when he went down from Cambridge to London, but he still managed to live to the age of eighty-four.

In 1678, a blazing fire in Newton's study destroyed many of his papers on calculus, optics, and alchemy; he had left a candle burning on his desk when he went out intending to return immediately. John Conduitt wrote that, after that, "he would never undertake that work [alchemy] again, a loss much to be regretted."[23]

Did Newton interpret this fire as a divine rebuke, sent to him because he was beginning to regard alchemy as a potential fount of personal power? The fact is that Newton had been coming more and more to believe that a certain godliness was necessary for the practice of alchemy. One of his favorite alchemists, the American George Starkey (aka Eirenius Philalethes), had published in 1658 *Pyrotechny Asserted and Illustrated,* in which he asserted that the "son of Pyrotechnic" [the alchemist] ... must resolve to give himself up wholly unto it, and the prosecution of the same, next unto the service of God," a quote that Newton excerpted.[24] In old age Newton told John Conduitt: "They who search after the Philosophers' Stone [are] by their own rules obliged to a strict and religious life. That study [is in that way] fruitful of experiments."[25] Years before, in papers affixed to an alchemy manuscript, Newton wrote:

> For alchemy does not trade with metals as ignorant vulgars think, which error has made them distress [twist and distort] that noble science; but ... God created handmaidens to conceive & bring forth its creatures [angels attended its practice]. ... This philosophy is not of that kind which tends to vanity & deceit but rather to profit & to edification, inducing first the knowledge of God & secondly the way to find out true medicines in the creatures [substances created by the alchemical process]. ... The scope is to glorify God in his wonderful works, to teach a man how to live well.[26]

You had to practice the ancient art of alchemy, then, as if you were worshipping God. This was no different from Newton's attitude toward every one of his pursuits: the scientist, in unlocking the secrets of the universe, was worshipping God, and the sole purpose of the scientist's activities was to demonstrate God's own activities in the world, and therefore enhance man's belief in a ruling and benevolent deity.

Robert Boyle (1627–1691), a wealthy Anglo-Irish natural philosopher, contributed to the foundations of modern chemistry even while he practiced alchemy passionately. Frank Manuel writes that Boyle "had a passionate belief in the transmutability of differing forms of matter by the rearrangement of their particles through the agency of fire."[27] Newton shared Boyle's belief in the universal transmutability of matter, writing in the *Principia*: "Any body can be transformed into another, of whatever kind, and all the intermediate degrees of qualities can be induced in it."[28]

Newton and Boyle also shared the conviction that there were concepts of a higher and nobler nature that should not be disclosed to the world. In 1676, Boyle published in the Royal Society's house organ *Philosophical Transactions* a description of an alchemical experiment in which he did not reveal the composition of the key substance, a certain "Philosophical Mercury," which he used. In a letter to the editor of the *Transactions,* Newton commended Boyle for his discretion, declaring that the great alchemists of the past (the "Hermetick writers") had also not revealed the nature of Philosophical Mercury, and their example should be followed. Newton seems to be hinting at the possibility of unleashing dangerous forces when he asserts that alchemy "may possibly be an inlet to something more noble, not to be communicated without immense damage to the world if there should be any verity in the Hermetick writers."[29]

Boyle seems to have agreed. In 1678, he published anonymously in London a pamphlet, *Of a Degradation of Gold Made by an Anti-Elixir: A Strange Chymical Narrative,* in which he tells a "backwards" alchemy success story: gold is transmuted into an inferior substance. "Pyrophilus" ("Lover of Fire") tells the narrator the story of a nameless traveler from an unnamed land who appeared, almost out of nowhere, just long enough to give Pyrophilus an almost microscopically small portion of a red powder and explain to him in the barest of detail how to use it in an alchemical experiment. The mysterious stranger's visit is so brief that he doesn't have time to tell Pyrophilus what the experiment is all about.

The red powder is the "anti-elixir" of the pamphlet's title; there is so little of it—a mere speck, one-tenth of a grain or one-sixty-fourth of an ounce—that Pyrophilus is afraid to verify its weight for fear of losing it in the scales. The experiment is non-repeatable; there's only enough powder for one. Joseph Needham describes the experiment:

> The anti-elixir, the nature of which was never disclosed, was a dark red powder which could convert pure gold into a brittle silvery mass

and baser materials such as yellowish-brown powder partly vitrified. The de-aurification ["de-golding"] purported to have been proved by the touchstone, cupellation and the hydrostatic balance [material produced had a lesser specific gravity than gold].[30]

Pyrophilus argues forcefully that transmuting a "noble" metal into a "less noble" metal is as great a proof of the truth of alchemy as the reverse. He ends *Of a Degradation of Gold Made by an Anti-Elixir: A Strange Chymical Narrative* on a provocative, if disturbing, note: "I will allow the Company to believe that, as extraordinary as I perceive most of you think the Phenomena of the lately recited Experiment; yet I have not (because I must not do it) as yet acquainted you with the strangest Effect of our admirable Powder."[31]

What is this "strangest Effect"? It must have manifested in the course of the experiment, for there was no red powder left for a second experiment. Why can't it be revealed to the reader? Apparently because no reader can be trusted to use this knowledge wisely. There is something slightly menacing about this Pyrophilan story: The power of alchemy can corrupt as much as it ennobles. And it can effect things so problematical that their nature must remain a secret.

Boyle's story, told anonymously, is emblematic of an uneasiness about alchemy that made its great practitioners in the seventeenth century tread warily. We'll recall that, in chapter 2 ("The Newton Code"), we watched as Isaac Newton, Henry More, and Richard Bentley circled warily around each other as they individually vied for the hand of the muse of prophecy; each tried to win her, overpower her—penetrate to her deepest secrets! No one was afraid of her.

But now we see the great triumvirate of Isaac Newton, Robert Boyle, and John Locke (the English philosopher of "innate ideas," who was also an earnest alchemist) not circling the muse of alchemy, or perhaps the goddess of the red earth, so much as retreating from her even as they approach her. They seem paralyzed in her presence; they are anxious to find out all they can, and then suddenly, on the point of possessing her—they stop and retire. Ambivalence controls these three great men who are capable of powerful intellectual decisiveness.

This frustrating drama plays itself out no more vividly than around the deathbed of Robert Boyle in 1692. "Red earth was thought to be about as close to the Philosophers' Stone as you could get," explains Principe. "It was assumed that if you could create red earth, it would be relatively simple to get to the Philosophers' Stone from there."[32] Newton and Locke knew Boyle had possessed some of this priceless red powder and had known how to use it. Locke was the executor of Boyle's will. Historian David Brewster writes:

> Boyle had, before his death, communicated this process [of using the red powder, aka "red earth," as a sort of mini Philosophers' Stone] both to Locke and Newton, and procured some of the red earth for his friends. Having received some of this earth from Locke, Newton tells him, that, though he has "no inclination to prosecute the process," yet, as he had "a mind to prosecute it," he would "be glad to assist him," though "he feared he had lost the first and third of the process out of his pocket." He goes on to thank Locke for "what he communicated to him out of his own notes about it," and adds, in a postscript, that "when the hot weather is over, he intends to try the beginning (that is the first of the three parts of the recipe), though the success seems improbable."[33]

Locke sends Newton transcripts of two of Boyle's papers, which apparently deal with how to "multiply gold" using the red earth. But Newton in his reply "dissuades Locke against incurring any expense by a too hasty trial of the recipe." He declares that several chemists were engaged in trying the process and that Mr. Boyle, in communicating it to himself, "had reserved a part of it from my knowledge, though I knew more of it than he has told me." David Brewster continues: "In 'offering his secret' to Newton and Locke, he [Boyle] imposed conditions upon them, while, in the case of Newton at least, he did not perform his own part in the arrangement. On another occasion, when he communicated two experiments in return for one, 'he cumbered them,' says Newton, 'with such circumstances as startled me, and made me afraid of any more.'"[34]

Locke's letter also suggests that at this time a company had been established in London to multiply gold with this recipe. Brewster resumes:

> Although Boyle possessed the golden recipe for twenty years, yet Newton could not find that he had "either tried it himself, or got it tried successfully by any body else; for," he says, "when I spoke doubtingly about it, he confessed that he had not seen it tried, but added, that a certain gentleman was now about it, and it succeeded very well so far as he had gone, and that all the signs appeared, so that I needed not doubt of it."[35]

The reader may by now have had his or her fill of the extraordinarily neurotic behavior of these three natural philosophers who were otherwise capable of extraordinary intellectual achievements. Did they simply not want to admit to each other or to themselves that they were beginning to feel that alchemy might not be a substantive pursuit? Such a conclusion seems to be belied by the mysterious success of the Pyrophilan experiment as described anonymously by Boyle and Newton's conviction that alchemy contained enormous potential power, the dangers surrounding which sometimes made the pursuit of the secrets of alchemy almost not worth the effort.

In the unread 80 percent of Newton's one million words about alchemy housed in the university library at Cambridge, it is likely there are answers to these questions.

Newton's numerous nonscientific, even "occult," endeavors may have helped open his mind to some of his most original ideas.[36]

"I think that his early optical theories owed a debt to alchemy," writes William Newman. Newman believes Newton's alchemical experiments in which he deconstructed and reconstructed compounds suggested to him the idea of using prisms to separate sunlight into a rainbow spectrum, then recombine those colors to form white light again.[37]

In *Isaac Newton: The Last Sorcerer,* Michael White argues that Newton's lifelong efforts to purify the metal antimony with iron to produce the "Star Regulus of Antimony" may have helped him conceive the idea of a gravitational force. White says that this specific regulus "does look like a star, and its radiating shard-like crystals may be imagined as lines of light radiating from a starlike center. But the crystal may just as easily be visualized as representing shards or lines of light pointing inwards—a star at the center with lines of light, *or force,* traveling *towards* its center."[38]

It may be that Newton looked beyond light and gravity to try to open up the possibility of total human transformation with his alchemical experiments. The question with which he began, that August day in London in 1669 when he first entered William Cooper's Pelican, was, How could the vegetative spirit (which he later came to call the active alchemical principle) enter mechanical matter in such a way as to imbue it with the divine spark of life? Dobbs summarizes the question he ultimately sought to answer, even in the last days of his life: "Will the rediscovery of the pure, potent fire that is the ultimate secret of the active alchemical principle lead to the restoration of true religion and the ushering in of the millennium?"[39]

CHAPTER SIXTEEN

MASTERS OF THE PRISCA SAPIENTIA, PART 1

Aristarchus, Anaxagoras, Numa Pompilius

In 1922, in *Fantasia of the Unconscious*, the British novelist D. H. Lawrence made an astonishing assertion about the prehistory of the world. He wrote:

> I honestly think that the great pagan world of which Egypt and Greece were the last living terms, the great pagan world which preceded our era, once had a vast and perhaps perfect science of its own, a science in terms of life. In our era this science crumbled into magic and charlatanry. But even wisdom crumbles.
>
> I believe that this great science previous to ours and quite different in constitution and nature from our science once was universal, established all over the existing globe. I believe it was esoteric, invested in a large priesthood. Just as mathematics and mechanics and physics are defined and expounded in the same way in the universities of China or Bolivia or London or Moscow today, so, it seems to me, in the great world previous to ours a great science and cosmology were taught esoterically in all countries of the globe, Asia, Polynesia, Atlantis, and Europe.

> ... In that world men lived and taught and knew, and were in one complete correspondence over all the earth. Men wandered back and forth from Atlantis to the Polynesian Continent as men now sail from Europe to America. The interchange was complete, and knowledge, science was universal over the earth, cosmopolitan as it is today.
>
> Then came the melting of the glaciers, and the world flood. The refugees from the drowned continent fled to the high places of America, Europe, Asia, and the Pacific Isles. And some degenerated naturally into cave men, neolithic and paleolithic creatures, and some retained their marvelous innate beauty and self-perfection, as the South Sea Islanders, and some wandered savage in Africa, and some, like Druids or Etruscans or Chaldeans or Amerindians or Chinese, refused to forget, but taught the old wisdom, only in its half-forgotten symbolic forms. More or less forgotten, as knowledge: remembered as ritual, gesture, and myth-story.[1]

It would be a vast exaggeration to say that the natural philosophers of the seventeenth century subscribed to D. H. Lawrence's description. But certainly a good many of them believed that the first peoples of the world possessed a perfected body of knowledge that the men of the Renaissance called the *prisca sapientia,* or "pristine wisdom."* They believed this body of knowledge had become corrupted as the centuries progressed but that Noah had been able to bring bits and pieces of it through the flood to the new world.

No comprehensive picture of this "land of the *prisca sapientia*" has come down to us. But, here and there in classical literature, we sometimes get tantalizing hints of a reality—or perhaps they are only ancient myths, figments of the archaic mind—aglow with more knowledge and power than our own. The church father Origen quotes the Jewish historian Josephus as asserting that "the constellations were known long

---

*The term *prisca theologia,* or "pristine theology," is also used. The two terms are almost interchangeable. Throughout this book, we use the term *prisca sapientia.* It should be thought of as including the term *prisca theologia.*

before the days of the patriarchs by Noah, Enoch, Seth, and Adam—indeed, were mentioned in the *Book of Enoch* as 'already named and divided.'" This knowledge was acquired not with telescopes, but through longevity, which, Josephus declares, "was a blessing specially bestowed to give opportunity for a long-continued period of observation and comparison of the heavenly bodies."[2]

The thirteenth-century English Franciscan philosopher and protoexperimental scientist Roger Bacon believed the first men and women in the world acquired longevity through alchemical practices carried through to perfection. He quoted ancient sources of alchemical lore to the effect that there are four levels of the Philosophers' Stone and that those who attained to the highest level—the "Angels' Stone"—were constantly surrounded by beautiful odors, didn't have to eat, could foretell the future, and could live to a very great age. Did the first men and women in the world function at the fourth level of the Philosophers' Stone?

For most of us today, all this is mere fancifulness; we don't believe there ever was a "land of the *prisca sapienta*." Ages not much earlier than our own believed differently. In a lecture he gave shortly after Isaac Newton's death, Lord Francis Atterbury, dean of Westminster, declared that the great mathematician believed the ancients were "men of genius and superior intelligence who had carried their discoveries in every field much further than we today suspect, judging from what remains of their writings. More ancient writings have been lost than have been preserved, and perhaps our new discoveries are of less value than those that we have lost."[3]

Isaac Newton pushed a belief in the superiority of the thinkers of high antiquity to the furthest extreme: he claimed they had known all about the atomic structure of the universe and that the Greek seer Pythagoras had anticipated his own law of universal gravitation, veiling it in the legend of the music of the spheres.

During the final decade of the seventeenth century, Sir Isaac began to gingerly feel out his colleagues as to what they thought about these extraordinary claims. He was likely the power behind the letter his brilliant protégé Nicolas Fatio de Duillier sent on February 5, 1692, to Europe's greatest scientist after Newton, the Dutch natural philosopher and mathematician Christiaan Huygens (1629–1695).

The gregarious and voluble Fatio had multiple contacts. He flitted like a brilliantly colored hummingbird from drawing room to drawing room of the scientific elite of Europe, as learned and gifted as any of them, but with an emotional vulnerability that eventually proved his undoing. Fifteen years later, the French Prophets, a Protestant sect whose members channeled the Holy Ghost while trembling and fainting, won his heart and soul; Fatio never broke free of the attachment and ended up a broken old man dabbling in new kinds of watches and a history of apple growing. In 1692, though, Fatio was Newton's presumed successor and his most trusted friend, and he wrote in his letter to Huygens:

> M. Newton believes he has discovered quite clearly that the Ancients, such as Pythagoras, Plato, etc., had provided all the demonstrations he provides of the true System of the World, which is based on the principle of a Gravity whose force diminishes in inverse proportion to the square of the increasing distance. The ancients made this knowledge into a great mystery. But fragments of it remain, and it appears, he [Newton] says, that if these fragments are put together then in effect they express the same ideas as he sets forth in the *Principia Mathematica*.[4]

Huygens was a late Renaissance genius who discovered the rings of Saturn, invented the pendulum clock, and composed a revolutionary treatise on optics. In 1692 he wrote a proto-science-fiction novel, *Cosmotheoros* (*The Celestial Worlds Discover'd*), in which the narrator tours the solar system and discusses the music of the celestial worlds with extraterrestrials. Few minds were as brilliant and open as Huygens's, but his reply to Fatio was dismissive: the ancients had knowledge of some general principles, he wrote, but they were not capable of grasping the details of the new science that Isaac Newton had invented.

The amiable and able mathematician-astronomer David Gregory (1659–1708), Newton's Scots disciple, was often unctuous and ingratiating in Newton's presence but fully appreciated the achievements of the world's greatest mathematician. In May 1694, Gregory traveled to Cambridge from Edinburgh to burn some incense at the great man's altar and pry as much information out of him as he could. Gregory was there for a month. Newton was unusually forthcoming; the Scotsman filled several notebooks, recording in one that Newton intended to

> spread [stretch] himself in exhibiting the agreement of his own findings with that of the ancients and principally with that of Thales, the legendary founder of Greek philosophy. He would demonstrate that what Epicurus and Lucretius had affirmed [about the atomic structure of matter] was true and valid, and that the charge of atheism laid on them was unjust . . . [that] Thoth, the Egyptian Hermes or Mercury, had been "a believer in the Copernican system," while Pythagoras and Plato had "observed the gravitation of all bodies towards all."[5]

Gregory accepted Newton's assertions with few reservations. In 1702, he incorporated most of the mathematician's classical references into the preface of his leading-edge Newtonian textbook *Elementa astronomiae physicae et Geometriae* (*Elements of Astronomy, Physics and Astronomy*).[6]

Newton slipped some of his early observations on the subject into the "classical" draft scholia he had intended to add to the second edition of the *Principia*, then thought better of it and withdrew them. These remarks eventually found their way into his posthumously published *De Mundi Systemate* (*The System of the World*) of 1728 (see appendix E).

It seems strange that the proud and pricklish Isaac Newton, who harassed Robert Hooke and Gottfried Leibniz to their graves for daring to suggest that they had anticipated him in the discovery of the law of gravitation and of calculus, respectively, would be willing to suggest that anyone at all, even the supreme geniuses of the ancient world, had anticipated his discoveries. Edward Dolnick remarks:

The notion is both surprising and poignant. Isaac Newton was not only the supreme genius of modern times but also a man so jealous and bad-tempered that he exploded in fury at anyone who dared question him. He refused to speak to his rivals; he deleted all references to them from his published works; he hurled abuse at them even after their deaths. . . . But here was Newton arguing vehemently that his boldest insights had all been known thousands of years before his birth.[7]

Scholars J. E. McGuire and P. M. Rattansi suggest that Newton's textual analyses of ancient natural philosophy may have served to "provide a pedigree for his own doctrines, to legitimate them for an audience which still widely accepted the idea of a *prisca sapientia*. He could use them as a direct defense for his own doctrines, as he does in the *Opticks,* and, on one occasion, during the controversy with Leibniz [over who first invented calculus]."[8]

Newton found it particularly difficult to defend his idea that gravity could act over a distance because of the presence of God; but he could point to many instances where the seers of high antiquity, discussing in veiled, symbolical form what Newton was sure was gravity, credited its existence entirely to God.

One of the first "masters of the *prisca sapientia,*" so to speak, that Newton engaged with was the Greek astronomer Aristarchus of Samos (ca. 319–230 BC), who, seventeen hundred years before Copernicus, published a brief work demonstrating mathematically that the Earth went around the sun.

Newton may also have wanted to remind his colleagues that the scientists of antiquity had suffered as much from the censure of religious authorities as had, in his own time, Giordano Bruno and Galileo (and potentially Isaac Newton himself). If Newton was in any kind of conflict with the church, it simply came with the territory of introducing revolutionary ideas into the world.

Though Newton could not possibly have put it in these terms—

modern psychology was only invented, in his own time, by his friend John Locke—we sometimes wonder if, in alluding to the realm of the *prisca sapientia,* he is not alluding to an expanded state of consciousness. William Blake saw man as devolving down through successive states of consciousness. Did Newton, on an intuitive level, see mankind in somewhat the same way?

For Blake, we moderns are merely shrunken versions of our true selves; but, by using our imaginations, which are part of the mind of God, we can return to a higher state; we can expand our beings infinitely. The fall of man was the partial fall of God along with the fall of man, a fall from a state of perfection down through seven stages during which we contracted and fragmented and shriveled. In the earliest ages, the consciousnesses of men and women were much more fully expanded. In talking about this descent through seven stages, couldn't you just as well be talking about the progressive corruption of the *prisca sapientia*?

Was it the expanded consciousness of higher beings that Newton intuited when he connected the great sages of antiquity with the idea of a *prisca sapientia*?

Let's take a close look at three eminent figures of ancient times whom Newton held in high esteem because of the revelatory nature of their achievements—revelatory because they seemed to reveal a portion of the wisdom of the *prisca sapientia,* the eternal wisdom that Newton himself thought he was restoring.

Let's start by spending a night with Aristarchus of Samos, mentioned above. And let's assume that the shade of Newton, having torn himself away from the ravishing sight of the planet Saturn, has come along for the ride.

When Nicholas Copernicus's (1473–1543) *On the Revolutions of the Celestial Spheres,* which demonstrated the heliocentric nature of the universe, was first published, just a few months before he died, it failed to include a manuscript remark that would have turned the Inquisition violently against the brilliant Polish astronomer. This was Copernicus's observation that "the mobility of the earth [its motion, specifically around the sun] . . . was the opinion of Aristarchus of Samos." This was Copernicus's only reference to the man about whom historian of science J. L. E. Dreyer writes: "When we consider that seventeen hundred years were to elapse before the orbital motion of the earth was again taught by anybody, we cannot help wondering how Aristarchus can have been led to so daring and lofty and idea."[9]

Newton thought he knew how. He believed there lingered in the great Greek astronomer some of the power of the *prisca sapientia*.

Let's descend upon a high sloping hill on the outskirts of Athens some time in 240 BC. It's midnight. The sky is clear. The stars gleam like diamonds; the moon is full. All across the hill, groves of fig trees alternate with luxuriant vineyards.

A gray-bearded old man with a lantern is hurrying between the groves and the vineyards. This is Aristarchus of Samos. He owns this hill, not in the sense that he has purchased it and is its proprietor, but creatively, intellectually, in the sense that he has invested huge quantities of his intellectual and creative capital in it. Aristarchus has increased the yield of the fig trees by placing male branches on female branches to make it easier for the gall wasps to do their work of pollenization. He has facilitated the growth of the grapes by introducing wild asses ito gnaw excess twigs off the vines.[10] In a cleared space between a vineyard and a fig orchard, he has placed a tall, hemispherical, stone sundial, the *scaphe,* which he invented himself and which is the most advanced of the day.

Indeed, Aristarchus of Samos is the Leonardo da Vinci of his day, and the first-century BC Roman military engineer and architect Vitruvius will include him among "the few great men who possessed an equally profound knowledge of all branches of science, geometry, astronomy, music, &c."[11] He does not own the sky glittering with stars above the mountaintop, but he has pushed its boundaries farther away: in a book called *On the Sizes and Distances of the Sun and Moon,* Aristarchus has used an arcane form of geometry to determine that the sun is 18 to 20 times farther away from the Earth than the moon and that its radius is 18 to 20 times greater than that of the moon. He was wrong (the sun is 400 times farther away than the moon and 109 times bigger than the Earth), but Aristarchus's figures doubled those of his contemporaries. They seemed unnatural to many of his contemporaries, and they drew away from Aristarchus.

These new distances are far from Aristarchus's greatest achievement. Five years before he has used an even stranger breed of geometry to conclude, in the words of Archimedes in his *Sand-Reckoner*, that "the fixed stars and the sun remain unmoved, that the earth revolves around the sun in the circumference of a circle, the sun lying in the middle of the orbit."[12]

(We can assume that the shade of Newton, flitting in and out of the decades, will somehow find a way to read Aristarchus's treatise on the location of the Earth, for does not the mysterious geometry in which it is written suggest something of the *prisca sapientia*?)

A year or so after he has made this discovery, Aristarchus finds himself sometimes stumbling as he makes his way along the old, familar bypaths of this hill on the outskirts of Athens. This makes him uneasy; he knows these bypaths like the back of his hand, and in his youth he had often raced along them at night without a lantern. He stumbles once a night, and then twice a night, and then sometimes three. It is as if the plateau, which he has loved so much, from which he has gained so much, is exacting some revenge.

One night he realizes that the stumbling comes from his own fears. He does not believe in the gods, but increasingly he cannot push away from hs mind the knowledge that, for the Stoics, the Earth is the home of Zeus, and that Aristarchus, in making his discovery, has removed the home of the king of the gods from the center of the universe to a second-tier role in the cosmos. Now he fears that the Earth, which he indubitably knows moves, may move against him, may even throw him off, because he has sinned against it.

Aristarchus isn't a religious man, but so radically has he turned the tables of man's perception of the universe that he fears he may have sinned against the gods. For he has displaced Zeus himself from his position at the center of the universe. Have not the priests long worshipped Earth as the home, the hearth, the dwelling place of Zeus? It is precisely the Earth's centrality that has cemented that belief. Has he, Aristarchus, then committed the sin of impiety in destroying that centrality? Many of his fellow countrymen think so. They fear the gods will punish him—and them. Lately he has been threatened in the marketplace. There is pressure upon him to recant.

Suddenly he hears, ever so softly, the sound of singing. He recognizes both the hymn and the voice of the singer. He listens.

> *Thou, O Zeus, art praised above all gods:*
> *many are thy names and thine is all power for ever.*
> *The beginning of the world was from thee:*
> *and with law thou rulest over all things.*[13]

These are the first lines of the Hymn to Zeus. They are being sung by its composer, Cleanthes, the leader of the Stoics.

With growing irritation, Aristarchus hurries toward the source of the singing. He knows all about Cleanthes. He had come to Athens seventy years ago, a youthful prizefighter without a drachma in his pocket. By chance he heard Zeno, head of the Stoics, give a talk and was enchanted. He became a devotee of philosophy and Zeno's student.

As a student Cleanthes was so poor he took notes on oyster shells and the dried shoulder blades of oxen. He worked all night as a water bearer and studied all day with Zeno. When he was named Zeno's successor his classmates were astonished, because he was thought to be the dullest student in the school. Clearly he is still poor. He still draws water at night, and he must be more than ninety.

Aristarchus crests a hilltop and, looking down, sees in the circle of light cast by a lantern a very tall, very lean, extremely aged man drawing a bucket of water from the well. It is Cleanthes. Aristarchus listens as he sings:

> *Under thee may all flesh speak:*
> *for we are thy offspring.*
> *Therefore will I raise a hymn unto thee:*
> *and will ever sing of thy power.*

Rage overtakes Aristarchus. Surely Cleanthes has come singing to this well to taunt him! This is the man who is leading the charge against him—who has stated publicly that Aristarchus should be indicted for impiety—and now it occurs to the astronomer that this bullying priest has come here tonight to harass and to bedevil him.*

Aristarchus scrambles down the slope, and as he does so Cleanthes sings out strongly:

> *The whole order of the heavens obeyeth thy word:*
> *as it moveth around the earth:*
> *With little and great lights mixed together:*
> *how great art thou, King above all for ever.*[14]

Aristarchus stumbles and crashes to the bottom of the slope. He can no longer contain his fury. He shouts: "Cleanthes, the whole order of the heavens does *not* obey Zeus's word as it moves around the earth! The whole order of the heavens moves around the *sun* and *not* the earth!"

Cleanthes turns and stares haughtily at Aristarchus. Ten yards separate the two men. Aristarchus charges across those ten yards. The shade of Newton observes the numinosity of the *prisca sapientia* still glittering here and there in the thousand wrinkles of the face of Cleanthes, ninety years old if he's a day. The shade listens, shocked as Aristarchus cries out: "Old man! You, Cleanthes! *You* shall be indicted for impiety! You, old

---

*Plutarch writes of a certain Cleanthes (330–230 BC) who "thought it was the duty of Greeks to indict Aristarchus of Samos on the charge of impiety for putting in motion the Hearth of the Universe [the earth]" (qtd. in Heath, *Aristarchus of Samos*, 304). As far as we know, Aristarchus was never indicted; but thereafter he said nothing about a heliocentric universe.

man, not I! For blasphemy against the truth, blasphemy against reason!"

These are to be the last words the two men utter that night. They will fight; eventually they will separate from each other; and the shade of Newton, having seen as much as he wants to see, realizes that religion is already cutting into those powers of the *prisca sapientia* that still remain within men.

And he is appalled. One of his purposes in life was to chart the troubled descent, from the heights of Mount Ararat to the sunny uplands of humanity, of the purest form of human spirituality. And what a falling off he has found here. These two Greeks, one the greatest scientist of his time, the other the greatest religious leader—they both endure, they both seem immortal, and yet they cannot even speak to one another. In the mid-third century BC, the two components of the *prisca sapientia,* science and religion, are falling irrevocably further and further away from each other.

When, in the life of ancient Greece, did this falling away begin to be inevitable?

Perhaps in a prison cell in Periclean Athens after midnight on a late spring night in 432 BC, where an old man of amazing intelligence is hefting stones and feeding the rats.

It is to this prison cell that the shade of Isaac Newton now hastens.

Newton and his contemporaries called it the Golden Age of Pericles, and so do we. But today we know that Periclean Athens was also Silver. To some extent it was supported on the backs of tens of thousands of slaves who, laboring twenty-four hours a day, tore whole mountains of silver ore out of the mines of nearby Taurium and used huge carts to transport their glittering haul to Athens. The mines were owned by the ruling elite of the city; the ore, turned into bars, brought a hundred talents a year (one million dollars) into the coffers of Athens.

Pericles saw to it that the sweat of these tens of thousands of slaves produced public buildings of rare beauty. Pericles, who ruled Athens for close to thirty years during the period 467–428 BC, poured much of the earnings from the silver into a program that rid the city of unemployment and saw the construction, overseen by sculptors such as Phidias and Praxiteles, of buildings, like the Erecteum and the Acropolis, that were themselves works of art. The creativity spread; during this period Aeschylus wrote his last play, Aristophanes his first, and Euripides and Sophocles wrote between them a hundred masterpieces. Socrates flourished, Plato was born, Hippocrates organized medicine, and science continued to make inroads against religion.

The golden-tongued orator behind all this, Pericles, was not immune from attacks. Nor were his friends. One of these, named Anaxagoras, had been thrown in jail on trumped-up charges of atheism. There he had been languishing for several months. It was he whom the shade of Newton had come to see.

Within four sinister walls shutting out even the thought of liberty, and certainly all hope, Anaxagoras of Clazomenae (500–428 BC), ragged and gaunt, sprawls across the earthen floor. A stubby candle casts a wan glow on cakes, fish, and cheese, all uneaten. A rat darts by; Anaxagoras, suddenly alert, throws it a piece of cheese and watches closely to see how it eats it. He abruptly picks up a large stone—the cell has a dirt floor—and examines it with fierce concentration. Then he drops it and dejectedly lies back on the floor.

Born into wealth, Anaxagoras gave it all away while still a youth to devote his life to the study of the Earth, the sun, the moon, the stars—of all of the heavens. He is the first person to assert that the moon shines by the reflected light of the sun; the first to suggest that man became developmentally superior to other animals because "his erect posture freed his hands for grasping things"; the first to explain that it is the spring thaws and rains of Ethiopia that make the Nile overflow once a year; and much more.[15]

Astronomy is his greatest love. When a fiery meteorite crashed to the ground at Aegospotami in 468 BC, the Athenians, thinking it was a fallen star, enshrined it in a temple and worshipped it. It was said that Anaxagoras, already a legend, had prophesied the fall of this "star"; but it's more likely that the scientist from Clazomenae, gazing on the blackened stone, saw it as a most undivine example of the true nature of the heavens. Anaxagoras concluded that celestial bodies are not gods but glowing lumps of stone; that the Earth is a stone dusted with dirt; and that the sun is "a red-hot mass or a stone on fire" (many times larger than the Peloponnesian peninsula!); in short, that "the sun, the moon, and all the stars are stones on fire, which are carried round by the revolution of the aether."[16]

These declarations have gotten him into trouble, especially with those who worship the sun. Helios the sun god is a minor deity, but he commands enough respect for the Spartans to sacrifice horses to him every year and for the Rhodians to throw four horses and a chariot into the ocean annually to honor him. If the idea of a universe consisting mostly of rocks has offended the religious establishment, the notion that the sun is just another rock among rocks, albeit it a very hot one, has outraged them. Anaxagoras has been indicted for impiety and sent to prison, there to await trial and possibly execution.

(Some think he's an atheist, but it's hard to tell. He has put forth the concept of the *nous,* a divine intelligence that suffuses all and is the underlying cause of everything. But he's so adept at finding natural causes for things that he seems to have rendered the nous superfluous, and the people nickname him "Nous" to make fun of him and to applaud him. Some believe Anaxagoras uses the word *nous* to cover up his atheism; Aristotle will accuse him of using it to cloak his ignorance.[17])

Anaxagoras is an intimate friend of Pericles, and he might not have been indicted at all (religious observance being on the wane in Athens) were it not that Pericles's enemies, during a time when the statesman is not in power, are punishing him through his friends. Now, stretched out in his cell, the astronomer thinks bitterly about Pericles. This isn't because the statesman is the reason why he's here. It's because for many months now Pericles, fighting for his own survival, hasn't given

Anaxagoras the spiritual and material support that he has been wont to give him. Over the years, the scientist has passed on to the ruler everything he knows; now, depleted and starving, he needs Pericles to give back.

Suddenly the cell door swings open and a blazing torch, held aloft by the muscular arm of a slave, fills the cubicle with light. A veiled woman enters. She wears a scarlet cloak of such purity of coloration that it seems impossible it could coexist for a moment with the filthy confines of this cell. Anaxagoras gets slowly to his feet. The woman unfastens the jeweled clasps of her cloak and lets it slip to the floor; it spreads out swiftly in a wide circle as if determined to hide the rocky, earthen surface from her eyes. She steps forward, removes the veil—reveals a face that is beautiful when she wants it to be but is now contorted by an expression of wild despair. She pulls the fleecy white robe that she is wearing closer around her and throws herself at Anaxagoras's feet.

This is Pericles's mistress, Aspasia, who has lived with him ever since he and his wife parted ways in mutual agreement. Aristotle said that slaves do not have souls, that a woman has a soul but it is inoperative, and that only men have operative souls. But Aspasia's soul is surely operative; she has the knowledge and skills of a great man, even if, for her, the amorous arts substitute for the martial arts. (Plutarch wrote: "Her occupation was anything but creditable, her house being a home for courtesans."[18]) But Aspasia's house is a school too, where women learn all that men learn, and Pericles's consort holds symposia that even Socrates attends. All seek her out, for her skill in rhetoric, for her sinuous intelligence, for her beauty, for her connections, and because she seems capable of making anything possible.

A second torch flares up in the doorway, more than doubling the light in the cell. A tall man in a blue cloak slips in. He moves with certain freedom and yet with discipline. His skull is oddly elongated, in a shape that has earned him the nickname "Jupiter Long-pate."[19] He has the eyes of a visionary, an enabler, and a most unhappy man. He carries himself with an incomparable self-possession. His movements are so unobtrusively and yet so overwhelmingly commanding that the shade of Newton, watching, might be forgiven for thinking that he sees the rats in the cell bow humbly and retreat before the inimitable presence of this man.

He crosses the cell in one stride and, grasping Anaxagoras firmly by the hand, addresses him in words that are—as is always the case when this man speaks—striking both in their dignity and their high purposefulness.[20]

This is Pericles, whose rule over Athens Thucydides describes as "an aristocratical government, that went by the name of a democracy, but was, indeed, the supremacy of a single great man." His powers of persuasion are such that he can change the mind of a god or, almost, the trajectory of a stone, sometimes by the "thunder and lightning" of his oratory, other times by an Olympian detachment and a daunting self-assuredness that enable him to transform reality, first in the eyes of the beholder, and then—almost!—in the world itself. One of his rivals, when asked which of the two was, figuratively speaking, the better "wrestler," replied, "When I've thrown him and given him a fair fall, by insisting that he had no fall, he gets the better of me, and makes the bystanders, in spite of their own eyes, believe him."[21]

Pericles loves Aspasia with a "wonderful affection; every day, both as he went out and as he came in from the market-place, he saluted and kissed her."[22] He loves her, and he loves Anaxagoras, for it is Anaxagoras who has taught him the best of what he knows, and it is for the astronomer that Pericles entertains, says Plutarch, "an extraordinary esteem and admiration." Plutarch continues:

> But he that saw most of Pericles, and furnished him most especially with a weight and grandeur of sense, superior to all arts of popularity, and in general gave him his elevation and sublimity of purpose and of character, was Anaxagoras of Clazomenae; whom the men of these times called by the name of *Nous;* that is, mind, or intelligence. ... The style of speaking most consonant to his [Pericles's] form of

life and the dignity of his views he found, so to say, in the tones of that instrument with which Anaxagoras had furnished him; of his teaching he continually availed himself, and deepened the colors of rhetoric with the dye of natural science."[*23]

Pericles and Aspasia have come to rescue Anaxagoras. These three, Anaxagoras, Aspasia, and Pericles, talk together rapidly in the jail cell, and in so similar a manner that it's as if each has taught the other how to speak; they address each other with the captivating, high-minded, clear-speaking eloquence that Anaxagoras has taught Pericles and Pericles has passed along to Aspasia even as she is learning it from his teacher. To the shade of Newton they seem like Titans from before the Flood, beings hardly tainted with the sin of Adam and Eve. For the mathematician this is a conversation so vibrant and numinous that it opens out to him, as he had hoped and as he had expected, the hidden possibilities of races and lands over which hang the high mysteriousness of unrecorded ages—ages governed by the *prisca sapientia*.

Suddenly, as if such an electrifying fusion of human beings simply cannot last, a torch goes out, the light dims precipitately, there are muffled blows in the corridor—a beringed priest's arm reaches through the bars, is violently hauled back. There is a ringing silence; then light returns and the cell door opens. A soldier appears and beckons to Pericles. He sweeps his arms around the shoulders of Anaxagoras and Aspasia and ushers the two of them swiftly out.

Pericles will return to power soon. He has bribed some, called in favors from others, charmed many, threatened a few, to save Anaxagoras.

---

*Plutarch is here drawing upon Plato, as Socrates says in the *Phaedrus,* "All the great arts require discussion and high speculation about the truths of nature; hence come loftiness of thought and completeness of execution. And this, as I conceive, was the quality which in addition to his natural gifts, Pericles acquired from his intercourse with Anaxagoras, whom he happened to know. He was thus imbued with the higher philosophy, and attained the knowledge of Mind and the negative of Mind, which were favorite themes of Anaxagoras, and applied what suited his purpose to the art of speaking." (Socrates, *Dialogues of Plato,* 1:273.)

To protect him further, he banishes him to his birthplace, the coastal city of Lampsacus. The great astronomer and master of the nous will die there, in two years, from boredom and despair.

Aspasia will be accused of impiety. At her trial, attended by fifteen hundred people, Pericles defends her with all his ingenious wit. She is acquitted, but Pericles is a broken man. The plague that has overwhelmed Athens and killed his sister and two of his sons will claim him too.

There is a third port of call that the shade of Newton, though he wants to, cannot visit, because it never existed.

There existed—or so the living Newton thought—a pocket of *prisca sapientia* power a little more than two centuries further back in time than the Athens of Pericles.

The pocket was an illusion, but Newton and his contemporaries didn't know that. They studied this Shangri-la of high antiquity as if it were the realest physical place that had ever existed, mining it for fragments of the most ancient wisdom in the world.

The place was ancient Rome—but an imaginary ancient Rome. In 390 BC, the Gauls crushed the Roman army and demolished the great city, obliterating three hundred years of historical documents. This knowledge vacuum soon filled with myths. One myth was that of Numa Pompilius (715–672 BC), the second king of Rome, who had to be coaxed into becoming king, who imposed a strict religion on the peoples of Rome, and who kept the Eternal City at peace for forty years.

Numa had friends in unusually high places. The idea of "guardian angel"—*daimon* to the ancient Greeks—first emerged in fifth-century BC Athens. The philosopher Socrates had a daimon that unfailingly told him when *not* to do something, and the voice was always right (except, perhaps, when Socrates wondered if perhaps he shouldn't reconsider drinking the hemlock that was meant to kill him and go into exile instead, and the daimon answered, "No").

Numa had two major daimons and a host of minor ones. There are entities in ancient Greek literature that we might categorize as invisible daimons, but this same ancient literature often represents them as separate supernatural beings, outside the guidee and communicating with him as if they were real people. The most beguiling of Numa's two major spirit guides was the woodland goddess Egeria. Numa married her; in Plutarch's words, he was "admitted to celestial wedlock in the love and converse of Egeria" and thereby "attained to blessedness, and to a divine wisdom." (Plutarch remarks that the ancient Egyptian priests debated whether a man and goddess could really get married; they decided they could but were less certain about whether sex could actually take place between a supernatural and a natural being.)[24]

Numa's other major guardian angel who manifested as an actual mortal was the muse Tacita, who sometimes put Numa in touch with the other nine muses. Tacita means "the silent one" and was named, says Plutarch, "perhaps in imitation and honor of the Pythagorean silence."[25] The Pythagorean silence was the silence Pythagoras imposed on his students in asking them not to repeat or publish his teachings outside his school. What is curious here is that Pythagoras lived two hundred years after Numa Pompilius. We are certainly in the realm of myth, and myth that does everything it can to put Numa at the level of the incomparable fifth-century Greek scientist-seer Pythagoras. Plutarch's sources describe the second king of Rome as, variously, Pythagoras's pupil, Pythagoras's first cousin, and close enough to Pythagoras that he named one of his sons after him.

Numa introduced the worship of the goddess Vesta to ancient Rome, so the ancient legend goes. Egeria had come to Earth expressly to help him do this—that is, in the words of the philosopher Thomas Hobbes, to help him to "receive the ceremonies [of Vesta] he instituted among the Romans."[26] Hobbes, who didn't know history had invented Numa, thought Numa had invented Egeria to give his precepts divine authority.

Numa built a temple to Vesta, and with this the true nature of Tacita's role is revealed: she had come to supervise the temple's design.

What fascinated Newton about Numa's temple was that it was a picture-perfect copy of a pre-Flood prytaneum—a rational-religious temple that, worshipping God's creations, honored the wisdom of the *prisca sapientia*. It was if Numa had found in a time capsule the blueprint of an aboriginal prytaneum. Newton writes of "that wise king of the Romans, Numa Pompilius, who, as a symbol of the figure of the world with the Sun in the center, erected a temple in honor of Vesta, of a round form, and ordained perpetual fire to be kept in the middle of it."[27]

Vestal virgins danced in circles around the central fire in imitation of the orbiting planets. Newton further writes: "Plutarch records this instruction from Numa: You should twirl around as you worship the Gods, and sit down when you have worshiped them; then he [Plutarch] adds: The turning of the worshipers is said to be an image of the orbit of the world. . . . They implied that by turning about the central fire we men are revolving in the true system of the world."[28]

So the second king of Rome knew that the cosmos was heliocentric! Most striking of all for Newton, though, was that the God of Numa Pompilius—the God of the Temple of Vesta—was Newton's God. Plutarch writes that the second king of Rome "conceived the first principle of being [God] as transcending sense and passion, invisible and incorrupt, and only to be apprehended by abstract intelligence." Therefore, Plutarch continues, the design of the temple was simple and without images of God, since it was idolatrous to liken "the highest" to any earthly object. You could not know God "except by the pure act of the intellect";[29] therefore, the walls of the temple were absolutely without decoration or ornamentation.

William Blake would have regarded the spirit guides surrounding Numa Pompilius as representations of his expanded consciousness. (Today we might call them aspects of his genius.) Newton would likely have looked at them as the measure of the wisdom of the *prisca sapientia* still within Numa.

(The second king of Rome had occasional concourse with other divine and semidivine figures: Plutarch says he was able to draw to himself two demigods, Picus and Faunus, who "revealed to him many secrets and future events, and particularly a charm for thunder and lightning"[30] with which he was able to bring Zeus to Earth. Zeus was angry at being disturbed, but Numa figured out how to placate him, and a mollified king of the gods returned to heaven. Plutarch says all this, it seems, to complete his picture of the fullness of Numa's soul.)

In a remarkable passage, Plutarch tells us about the spirit guides of other great men and women of Classical times, and we get another picture of how the ancients described genius. The author of *Lives of Ancient Greeks and Romans* writes: "It is reported, also, that Pan became enamored of Pindar for his verses, and the divine power rendered honor to Hesiod and Archilochus after their death for the sake of the Muses; there is a statement, also, that Aesculapius sojourned with Sophocles in his lifetime, of which many proofs still exist, and that, when he was dead, another deity took care for his funeral rites."[31]

Newton was fascinated by Numa because he so perfectly exemplified Newton's notion of what an ancient priest of the *prisca sapientia* might be like, from the spirits representing his expanded being, who flock around him, to the rational perfection and unassuming worship of God in the temple that he constructed in Rome.

From Numa—who didn't exist but was cobbled together to represent certain prevailing trends of the sixth century BC—we move on to the greatest ancient master of the *prisca sapientia* of them all, Pythagoras. We'll introduce him through his disciple, Philolaus, who displaced the sun from the center of the cosmos several hundred years before Aristarchus did. In so doing, Philolaus seemed to Newton to provide overwhelming proof that the scientists of the ancient world had known everything that Newton and his contemporaries were just now learning.

CHAPTER SEVENTEEN

MASTERS OF THE *PRISCA SAPIENTIA*, PART 2

Philolaus, Pythagoras, Moses, Pauli

Around the year 450 BC, the Greek philosopher Plato made strenuous efforts to obtain a rare and valuable book whose very existence was a surprise to everybody.

Plato was at the court of Dionysius II in Syracuse, Sicily, helping the king implement the rules of ideal government that the philosopher sets forth in his *Republic* (man being the imperfect creature that he is, the attempt was unsuccessful). The book was called *On Nature;* it was written by Philolaus (470–385 BC), successor to the seer Pythagoras; and, two hundred years before Aristarchus, it displaced the sun from the center of the universe.

The book had been produced in the famed Pythagorean school at Croton in the heel of Italy. There are two versions of how Plato obtained it. One is that he bought it from Philolaus's relatives for the enormous sum of forty Alexandrian *minas* of silver, perhaps $1,000 today. The other is that, in a deliberate effort to procure this extraordinary volume, he persuaded Dionysius to free a disciple of Philolaus whom the king had thrown in prison.[1] A grateful Philolaus rewarded him with *On Nature*.

The book was extraordinary not only for its revolutionary displacement of the sun. It was well known at the time—or so legend tells us—that Pythagoras had forbade his disciples to even talk about his teachings outside his school, let alone publish them. His followers had devoutly honored his request. At the time of the seer's death, when he and his disciples had been attacked and were fleeing from Croton, a disciple named Timycha, trailing behind because she was pregnant, was caught and dragged before the tyrant Cylon. He threatened to torture her if she didn't tell him why Pythagoras forbade the eating of beans. Timycha ground her tongue with her teeth and bit it off, rather than reveal, because of torture, any of the master's secrets.[2]

And yet Philolaus had published *On Nature*. Perhaps Plato, finally reading the book in Syracuse, concluded he should not assume that (in the words of modern-day commentator Thomas Heath) "the absence of any written record of early Pythagorean doctrine [is] to be put down to any pledge of secrecy binding the school; there does not seem to have been any secrecy observed at all unless perhaps in matters of religion or ritual; the supposed secrecy seems to have been invented to explain the absence of any trace of documents before Philolaus."[3] The story of Timycha is only a poignant myth, an homage to Pythagoras.

Plato was impressed by *On Nature*. Though only very late in life did he come to believe—perhaps!—that the Earth is not at the center of the universe, he nevertheless incorporated many of Philolaus's best ideas in his own mystical-astronomical masterpiece, *Timaeus*.

Philolaus's *On Nature* has long been lost, but we know a great deal about it from Aristotle's *On the Heavens*. In *On Nature*, the second-generation Pythagorean does indeed displace the Earth from the center of the universe, setting it in an orbit around that center. But he doesn't replace it with the sun; that he leaves where it is, also orbiting the center, while he puts in the place of the Earth what the calls the "central fire."

The central fire is mysterious indeed. Philolaus arrived at this belief through a mystical-mathematical process of deduction. Aristotle describes the process in *On the Heavens;* some scholars think he does so just to make fun of Philolaus's mystical beliefs. Aristotle writes:

"Most people—all, in fact, who regard the whole heaven as finite—say it [the Earth] lies at the center. But the Italian philosophers known as Pythagoreans take the contrary view. At the center, they say, is fire, and the earth is one of the [wandering] stars, creating night and day by its circular motion about the center."[4]

Aristotle says the Pythagoreans regarded the center as the most precious of all possible places in the universe, and fire as the most precious of all possible things in the universe. Therefore, since the most precious thing belonged in the most precious place, fire must be at the center of the universe. And since what is most precious "should be most strictly guarded,"[5] and there can be no better guardian than a god, then it follows that the central fire is protected by Zeus or Apollo.

The Pythagoreans didn't know that the Earth rotates on its axis. And they believed only half of our world's surface was inhabited, the half that faced perpetually away from the central fire, which was why we never saw it. They believed that not just the Earth and the sun but also the moon, the five known planets, and the fixed stars (the stars functioning as a single unit) revolved around this central fire.

This makes nine celestial bodies. The Pythagoreans maintained that nine was a "limited" number—in fact a very second-rate number—and therefore it could not have played a role in the original and divine structuring of the universe. However, the number ten, being the sum of 1, 2, 3, and 4, was not only an "unlimited" number but a perfect number. So, to arrive at a total of ten celestial bodies, the Pythagoreans tacked on a tenth body, the "Counter-Earth."

Even if the Counter-Earth were farther away from the sun than the Earth, we wouldn't be able to see it any more than we can see the central fire. This is because it shares the same orbit as the earth and is directly opposite us on the other side of the central fire, and remains directly opposite us because it revolves around the central fire at the same rate of speed as we do.

It's not clear whether the Pythagoreans thought the central fire was really a fire at all; they may have regarded it as a "creative force which from the center gives life to all the earth and warms afresh that part of it which has cooled."[6] They believed the sun, orbiting the central fire, shone by the reflected light of that, perhaps only apparently, fiery body.

Philolaus professed to believe that the central fire had an organizing function vis-à-vis the universe; it was "the directing fire . . . which the Demiurge [the being created by God to put together the physical universe] has placed as a sort of keel to serve as a foundation to the sphere of the All."[7]

Isaac Newton was impressed by the Pythagorean concept of the central fire. He talks about it as if it were Pythagoras's idea, though we have no proof of this. Oddly, Newton conflates the central fire with the sun, speaking of the Pythagorean "sun" when he actually means Philolaus's central fire.

What truly fascinated Newton about the notion of a central fire was its nature as a unifying or organizing principle. This smacked of gravity to him. The Pythagoreans applied many epithets to this central fire that Newton regularly calls the sun: the Tower of Zeus; the Watchtower of Zeus; the Throne of Zeus; the House of Zeus; the Hearth of the Universe (the "fire" being "placed like a hearth round the center"); the Fire in the Middle; the Mother of the Gods; the Altar, Bond, and Measure of Nature; and more. For Newton these epithets had the feel of ancient hieroglyphs from the world before the Flood; he must have wondered if they were encoded descriptions of the laws of mathematics and physics that he was rediscovering.

Newton writes that Pythagoras (by whom he means the Pythagoreans), "on account of its [the central fire's] immense force of attraction, said that the sun was the prison of Zeus; that is, a body possessed of the greatest circuits. . . . [It was] the prison of Jupiter because he keeps the planets in their orbs."[8]

The "greatest circuits" must refer, Newton surely thought, to orbiting bodies held in place by the gravitational pull of the sun. He wrote that "the souls of the Sun and of all the Planets the more ancient philosophers held for one and the same divinity exercising its powers in all bodies whatsoever."[9] Here he seems to refer to that universal force of gravitation that, he believed, must, in the absence of any medium through which it can travel, be sustained by the power of God.

Usually when Newton thought about the theory of universal gravitation in ancient times, he thought about Philolaus's master, Pythagoras, who, Isaac Newton believed, knew everything he, Newton, knew about gravity and had veiled his knowledge in the legend of the music of the spheres.

Humanity has taken three giant leaps forward in the way it understands the world we live in. The second leap was made by Isaac Newton, when he invented calculus and formulated the laws of gravitation, motion, light, and more. The third giant leap forward was as recent as a hundred years ago, when Albert Einstein discovered relativity and a group of mathematicians and physicists that included Werner Heisenberg and Wolfgang Pauli developed quantum mechanics.

The first giant leap forward was taken in the fifth century BC by the Greek philosopher-seer Pythagoras (ca. 570–495 BC). Previous Greek philosophers had struggled to identify a single substance underlying all other substances: Thales suggested water, Anaximenes named air, Anaximander said breath and air, and Heraclitus posited fire. Pythagoras's unique genius was to discover in the chaos of the ever-changing world of sense phenomena the truth that Number is the essence of all, that it is the ultimate reality—that God, or the Universe, thinks in numbers.

Early biographers attribute Pythagoras with a vitality and plenitude of being that make him sound like a Titan roaming the Earth in antediluvian majesty. Diogenes Laertius says Pythagoras's students believed he was the god Apollo come back to Earth.[10] Plutarch tells us he taught an eagle to come at his command and swoop down to him in flight.[11] Pythagoras could talk to animals and once made a bear swear an oath that it would stop preying on the local countryside; the bear kept its word.[12] He could talk to the rivers: Iamblichus records that, "once, passing over the river Nessus along with many associates, he addressed the river, which, in a distinct and clear voice, in the hearing of all his associates, answered, 'Hail Pythagoras!'"[13]

The man who discovered Number could be in two places at once: his contemporaries report that, "during the same day he was present in Metapontum in Italy, and at Tauromenium in Sicily, discoursing with his disciples in both places, although these cities are separated, both by land and sea by many stadia, the traveling over which consumes many days."[14] He was personally captivating; the poet Timon speaks of him as "inclined to witching works and ways, / Man-snarer, fond of noble *periphrase* [juggling of solemn speech]."[15]

The tremendous efflorescence of being that was Pythagoras swept through a tumultuous lifetime that began on the island of Samos, in Ionia, Greece, in 570 BC, and ended in the Greek colony of Croton, in southern Italy, in 495 BC. Pythagoras spent nearly all of his twenties, thirties, and forties in Egypt, Chaldea, and Phoenicia, a foreign student abroad imbibing the essence of mystery religions so old (or so it was believed) that their initial rites had been performed by the gods themselves. He may have visited Mongolia, and perhaps Persia where he communed with Zoroaster. Pythagoras returned to Samos just in time to flee the advance of the Persian army into Asia Minor, ending up at Croton in southern Italy. Here he set up a school, or more accurately a mystical fellowship, that adhered to the rule of communal ownership and included women, not the least of them Pythagoras's wife, Theano, who was a philosopher in her own right.*

The power of Pythagoras's mind and soul were such that he was soon lecturing to as many as two thousand followers[16] (if "lecturing" is the word, since he spoke to his pupils from behind a curtain in the manner of the central Asian potentates of his day[17]). Pythagoras taught reincarnation and claimed to remember twenty-two of his past lives. He enjoined his pupils to harm nothing living, eat only vegetables, keep

---

*Theano was a philosopher, mathematician, physician, and administrator. She and Pythagoras had a daughter, Damo (ca. 535–475 BC), who may have published several of her father's treatises. Some fragments of Theano's writings have survived.

speech at a minimum, and strive for salvation through the assimilation to and knowledge of God.[18] He invented acoustics and turned it into a therapeutic tool, curing diseases and modifying emotions by singing and playing on his lyre in ways never known before.

Newton believed that Pythagoras, in the course of his travels, had tapped into many an ancient and still bubbling wellspring of the *prisca sapientia*. This was true both for his science and for the moral laws he promulgated. Newton writes that,

> Pythagoras, one of the oldest Philosophers in Europe, after he had traveled among the eastern nations for the sake of knowledge & conversed with their Priests & Judges & seen their manners, taught his scholars that all men should be friends to all men & even to brute Beasts & should conciliate the friendship of the Gods by piety, & his disciples were celebrated for loving one another. Th[is] religion . . . was therefore [called] the Moral Law of all nations put in execution by their courts of Justice until they corrupted themselves.[19]

Pythagoras's moral and spiritual teachings were part and parcel of his discovery of Number. Number was eternal. Number was God's structuring of the world. Number was sacred; it was to be mediated on, imbibed, understood—but never exploited. Philolaus wrote of Pythagoras's discovery that "Number became great, all powerful, all sufficing, the first principle and the guide in the life of gods, of heaven, of men. Without it, all is without limit, obscure, indiscernible."[20]

Future generations would agree. Johannes Kepler (1571–1630) wrote: "Geometry existed before the creation, is co-eternal with the mind of God, is God himself."[21] The twenty-first-century physicist Alex Vilenkin declares: "The Creator is obsessed with mathematics. Pythagoras, in the sixth century BC, was probably the first to suggest that mathematical relations were at the heart of all physical phenomena. His insight was confirmed by centuries of scientific research, and we now take it for granted that nature follows precise mathematical laws."[22] Pythagoras's second great discovery was that the musical scale depends on numerical proportions: the octave represents the proportion of 2:1; the fifth, 3:2; and the fourth, 4:3. This led to the idea of harmony. Pythagoras discovered that the most "agreeable" harmonies—those whose tones seem to be "in sync" with one other—are formed by the simplest kinds of mathematical ratios. If the vibrations of one tone are twice as fast as the vibrations of another, the two tones will be an octave apart thus making a unity. The separate constituents of this musical marriage oscillate in the proportion of 2 to 1. This is a very basic ratio. But only through the simplest kind of proportion does pleasing harmony arise.

The dynamic seer of Croton, whose presence drew beauty and harmony out of all that surrounded him, first detected the numerical proportions of music in the soot and clamor of a blacksmith's shop. Striding by a smithy one day, he saw five blacksmiths striking five anvils with hammers of five different weights. Beautiful harmonies irregularly cut through the cacophony. Pythagoras recorded the weights of the five hammers and when at home hung from his ceiling on stretched sinews of oxen and sheep's intestines five equivalent weights.

He randomly plucked the strings in pairs, and discovered that bell-like harmonies sounded when the weights were in the proportions of 2:1, 3:2, and 4:3. The discovery that music arises from unchangeable mathematical relationships had profound implications. Since mathematical relationships were the groundwork of all being, it followed that ultimate reality expressed itself in music. Pythagoras and the Pythagoreans had come to believe, as Aristotle succinctly put it, that "the whole cosmos is a scale and a number."[23]

All of this suggested to Newton the deeper mysteries of gravitation. His tortuous interpretation of the lyre of Apollo, the pipes of Pan, and the harmony of the spheres rests on the belief that the "true system of the world" was known to the ancients but had been turned into a "great mystery" that only the initiates could penetrate. His thoughts ran thus:

since all bodies moving in space produce sounds whose pitch depends on the size and speed of the body, then each planet in its orbit about the Earth (or this was what Newton believed Pythagoras thought) makes a sound proportional to its speed which is a function of its distance from the Earth; and these diverse notes constitute a harmony or "music of the spheres" that we never notice because we hear it all the time.

According to Pythagoras, God, in placing the moon, the planets, and the fixed stars in motion around the Earth, had varied their sizes, their orbital velocities, and their distances from us (this last determining how high or low the pitch was) in such a way that the distinct and different sounds that each of the nine celestial bodies produced created when sounded together a harmonious whole.

Let's imagine that each of these celestial bodies is attached to the Earth by a taut violin string. God touches each string, and that is what produces the sound. Let's further imagine that these bodies, revolving around the Earth, are analogous to anvil hammers whose weights as they hang from the end of strings differ in the proportions first discovered by Pythagoras.

Newton believed Pythagoras knew that when varying the weights attached to the ends of the strings, rather than string length, the proportions had to be squared and inverted. To Newton, the Pythagoras story was hidden science; the "music of the spheres" was a myth meant to conceal the true scientific fact that "the weights of the planets towards the sun were reciprocally as the squares of their distances from the sun."[24] This was nothing less than an early iteration of Newton's principle of universal gravitation, which states, in modern terms, that the force of gravity varies inversely as the square of the distance; that is, this force grows smaller as the distance grows larger; and if the distance increases by a ratio of x, the force increases by a ratio of x^2.

Newton was able to make this application because in his time Marin Mersenne, the French theologian, philosopher, mathematician, and music theorist, often referred to as the "father of acoustics," and Vincenzo Galilei, the Italian lutenist, composer, and music theorist who was the father of Galileo Galilei, had discovered a quantitative relation in which two equally thick strings stretched by suspended weights would be in unison when the weights were reciprocally as the lengths of the strings. Newton contended that this must have been the quantitative relation that Pythagoras applied to the heavens, that he must have recognized that the harmony of the spheres required the force of the sun to act upon the planets in that harmonic ratio of distance by which the force of tension acts upon strings of different length—that is, inversely as the square of the distance.

So Mersenne and Galileo were only rediscovering what Pythagoras has discovered! David Gregory tells us that "Pythagoras afterwards applied the [much later Mersenne/Galileo] proposition he had thus found by experiments, to the heavens, and thus learned the harmony of the spheres."

Newton's own explanation of how Pythagoras made this application, which may be found in the draft Scholium to Proposition VIII, is, for most modern readers, very difficult to understand.

> Pythagoras discovered by experiment an inverse-square relation in the vibrations of strings (unison of two strings when tensions are reciprocally as the squares of lengths); that he extended such a relation to the weights and distances of the planets from the sun; and this this true knowledge, expressed esoterically, was lost through the misunderstanding of later generations.[25]

By way of further explanation, we don't find ourselves much better off with the explanation of Newton's most brilliant pupil, Colin Maclaurin.

> A musical chord gives the same notes as one double in length, while the tension or force with which the latter is stretched is quadruple of the gravity of a planet at a double distance. In general, that any musical chord may become unison to a lesser chord of the same kind, its tension must be increased in the same proportion as the square

of its length is greater; and that the gravity of a planet may become equal to the gravity of another planet nearer to the sun, it must be increased in proportion as the square of its distance from the sun is greater. If therefore we should suppose musical chords extended from the sun to each planet, that all these chords might become unison, it would be requisite to decrease or diminish their tensions in the same proportions as would be sufficient to render the gravities of the planets equal. And from the similitude of those proportions the celebrated doctrine of the harmony of the spheres is supposed to have been derived.[26]

Maclaurin goes on to say that

these doctrines of the Pythagoreans, concerning the diurnal and annual motions of the earth, the revolutions of the comets ... and the harmony of the spheres, are very remote from the the suggestions of sense, and opposite to vulgar prejudices; so we cannot but suppose that they who first discovered them must have made a very considerable progress in astronomy and natural philosophy.[27]

The average reader does not come away from this enlightened; it is almost as if Newton and Maclaurin (and perhaps David Gregory) were themselves wrapping up deep truths in arcane language to keep we common people from learning things we won't be able to handle.

Newton was no great listener of music as a form of art to be appreciated. He hardly ever listened to music at all. Newton told William Stukeley that he "never was at more than one Opera. The first Act, he heard with pleasure, the 2nd stretched his patience, at the 3rd he ran away." On another occasion, emerging from a concert at which Handel had played the harpsichord, the only thing he could think of to comment on was the elasticity of the great composer's fingers.[28]

But Newton was fiercely interested in music in the abstract. He wondered if the spacing of the seven colors of the spectrum (which Newton had discovered) imitated "the proportional 'distances' between the tones in a musical scale."[29] Had nature created laws for melody and harmony in the same way that it had created hues of light? Strangely enough, writes Stuart Isacoff, Newton decided that man "is not inherently musical . . .[, that] natural singing is the sole property of birds. In contrast to our feathered friends, humans perform and understand only what they are taught. . . . Still, he acknowledged, the proportions 'which the God of Nature has fitted' were now impractical. A system in tune with both man and heaven was needed."[30] But, wait! Before we go on, we need to visit with Pythagoras one last time.

During this visit, Pythagoras will introduce us, if indirectly, to perhaps the only master of the *prisca sapientia* of high antiquity who was his equal.

This was the prophet Moses.

Let's travel to the port of Sidon, in Phoenicia, in about 530 BC. Pythagoras is on the dock. He has just arrived from Croton. He wears white robes and pants and a gold tiara, the standard outfit for a seer of his high rank. Probably an eagle is circling obediently overhead.

If any animals have wandered down from the adjoining fields he's probably talking to them. They listen with ears pricked wildly to the master. (Pythagoras may be exchanging a word or two with the lapping waves beneath the dock.) Behind him, in a large trireme, the crew sits huddled on the huge deck, watching the master's back. They gaze at him like a farmer stares at the sky, looking for signs of wind or rain or sun.

A procession is advancing slowly along the dock toward Pythagoras. It seems to be one family; they look alike, and they are dressed in similarly brightly colored robes. The man in front is a high priest. Pythagoras knows these are Phoenician Jews, and they have come here to perform a high office. The high priest carries before him a cedar chest the size of a large book. He reaches Pythagoras, bows; explains that he and his family are descendants of the prophet Moses; that their illustrious ancestor had bequeathed this chest to his family; that it was not to be opened by any of them, but that, probably several centuries in the future, an eagle would swoop down from high in the sky, perch on the chest, and, speaking to them in Hebrew, tell them for whom this cedar chest was for and how to find him and give it to him.

All had transpired as Moses had foretold. And now the priest passed the chest to Pythagoras, who opened it—just as the eagle swooped down, perched on his arm, and peered down into the chest. It contained a book in rich gold binding; Pythagoras opened it and for several minutes could not read the strange-looking words at all. Then he began to realize he was looking at a textbook on atomic physics.

Isaac Newton believed that many of the ancient Greek philosophers who had tapped into the wisdom of the *prisca* had learned all of the secrets of the atom, not just what Lucretius and Democritus knew, but everything Isaac Newton knew. These Greek physicist-seers included, along with Lucretius and Democritus, Epicurus and Ecphantus and Empedocles and Zenocrates and Heraclites and Aesculapius and Diodorus and Metrodorus of Chios—and many more.*[31]

Newton could back up his assertions with a chestful of citations. For him the greatest atomic physicist among the ancients was Moses. Moses's accumulation of knowledge was so important that the prophet had somehow arranged for it to be preserved and passed along to Pythagoras.

Here is Newton summarizing the Cambridge Platonist Ralph Cudworth:

> Posidonius, an ancient & learned Philosopher, did (as Strabo & Empiricus tell us) avouch it for an old tradition, that the first inventor of atomical Philosophy was one Moschus [Moses] a Phoenician, who as Strabo notes, lived before the Trojan war. Perhaps this Moschus was

---

*For a lengthy excerpt from Newton's discussion of the achievements of the scientists of the ancient world, see appendix E.

then that Mochus a Phoenician Physiologer in Iamblichus with whose successors Priests & Prophets he affirms that Pythagoras sometimes sojourning at Sidon (his native city) had conversed.[32]

Would an atomic physicist, even of the most primitive sort, ever say that Earth was created in six days? Here, Newton is as fertile with answers as was scientist-theologian William Whiston, his surrogate apologist for Moses. Newton argued that the entire creation could have taken place in six twenty-four-hour periods, because God had started the world rotating on its axis only on the fourth or fifth day; at first it had rotated very slowly, or not at all. The Deity could make the days as long or short as he wished so as to accomplish in a given time period anything that he wanted.

Why didn't Moses simply say this in the Hebrew Bible? Frank Manuel responds: "Moses knew the whole of the scientific truth—of this Newton was certain—but he was speaking to ordinary Israelites, not delivering a paper to the Royal Society, and he popularized the narrative without falsifying it."[33]

Newton Project director Rob Iliffe adds that in the early post-Flood era religious leaders "concealed these [scientific] truths from the vulgar. At this time, there was a 'sacred' philosophy—communicated only to the *cognoscenti*—and a 'vulgar' version, promulgated openly to the common people."[34]

Newton seems to have believed that the ancients withheld great truths from the "vulgar" for reasons that were noble. In the beginning, primal truth had shone in the souls of humankind. We were infants of genius—but we were infants. We couldn't handle the knowledge; it gave us power we couldn't help wielding often for the wrong reasons. We harmed the planet, and eventually God felt he had no choice but to wipe us out and start over with a clean slate.

So Newton felt that it must be true that the last of the sages of the antediluvian period—the "remnant," led by Noah—who had, however incompletely, preserved the *prisca* from the Flood, had felt it was necessary to obscure the deepest truths from the vulgar; they had learned, at the expense of the loss of a world, that ordinary men and women couldn't handle them. This wasn't because we were stupid, Newton thought. It was because we forgot to keep worshipping God; the old devil idolatry kept slipping and luring us once again down the path of moral corruption.

Today we believe Newton came up with his discoveries on his own. Did his immersion in ancient scientific doctrines, however disguised in myth, help him make these discoveries? Some critics wonder if notions like the pipes of Pan and the music of the spheres, lengthily and zealously pondered by Newton, didn't decisively jog his mind so far outside the box as to actually play an indispensable role in enabling him to come up with his amazingly innovative theories. Did the prolonged meditation of Pythagoras on the mystical number of seven provide Newton with an indispensable impetus pushing him to the discover that light is composed of seven colors? Did Pythagoras's seminal notion of the music of the spheres not anticipate the theory of universal gravitation so much as inspire it?

If the ancient sages knew so much, what happened to all that knowledge? Why didn't we build steadily on the discoveries of Archimedes and Aristarchus; if we had, wouldn't we be just as advanced today as our descendants will be in the twenty-third or twenty-fourth century?

How did it happen that, in the words of mathematician Alfred North Whitehead, "by the year 1500 Europe knew less than Archimedes, who died in 212 BC"?[35]

Scholar Piyo Rattansi explains that "it was the Greeks who first inaugurated the search for truth, unfettered by authority. But Aristotle had dimmed that light."[36] Aristotle rejected many of the most daring ideas of his time, and his dominance became such that no other daring ideas could break through his eminence as the ultimate authority on everything. "The light of truth," continues Rattansi, "was almost wholly extinguished after the Gothic invasions [of Italy]. Christianity had prospered among the ruins of Rome, and the world returned to credulity and superstition."[37]

So complete was the loss of the "sacred philosophy" of the ancients that it did not reemerge until after the fall of Constantinople in 1453, when it was rediscovered in ancient texts, many thought lost forever, many never known. Much of this lost knowledge was fairly quickly restored, and to the philosophers of the Italian Renaissance as well as the natural philosophers of Newton's century, its superiority over modern ideas was such that it was very easy to believe there was a *prisca sapientia* of which this newfound knowledge was a part.

The French mathematician René Descartes (1596–1650), who tried to bring mathematical precision to philosophy, had been forced to do so by a series of terrifying, irrational visions that beset him when he was twenty-three. Descartes believed that "certain primary seeds of truth [had been] implanted by nature in our human minds." But their growth had been "stifled owing to our reading and hearing, day by day, so many diverse errors." Descartes hated the ancients for concealing those truths in esoteric symbols. He suspected their motives, declaring cynically that "they have perhaps feared that their method being so very easy and simple, would if made public, diminish, not increase public esteem."[38]

Might it not be true, mused Newton, that the sages of old—the members of the "remnant" who had restored the *prisca* after the Flood—had felt compelled to obscure its deeper meanings from the vulgar simply because the knew that ordinary men couldn't handle it, not because mankind were stupid, but because we tend so easily to moral corruption?

We've already spoken of Wolfgang Pauli as one of the group of twentieth-century scientists, masterfully led by Albert Einstein, who are responsible for mankind's third giant leap forward in terms of our understanding of the physical universe.

Wolfgang Pauli (1910–1958), a native of Vienna, Austria, won the Nobel Prize in Physics in 1945 for his discovery of the exclusion principle. Pauli was first a patient, then a friend, and finally a collaborator of the famed analytical psychologist Carl Jung. One of the most brilliant and curmudgeonly hardheaded scientists who ever lived, Pauli came to embrace Jung's notions of the collective unconscious of mankind and the universe of archetypes that lies behind the visible world.

Pauli was haunted all his life by a spectral otherness for which he could never account. His presence triggered (somehow psychokinetically, because he was never near any of the apparatus) equipment failure. He visited the observatory in Bergesdorf, Germany, and a major accident rendered the telescope useless. The train taking him to Denmark stopped in Göttingen, Germany, and a massive equipment meltdown paralyzed a laboratory at the University of Göttingen.

One night Pauli sat down in a lecture hall and "two dignified-looking ladies simultaneously and symmetrically collapsed with their chairs on either side." Once, when he was on a train, the rear cars accidentally decoupled and stayed behind while Pauli continued on his way in a front car.

So there was in Pauli already an unasked-for opening to other realms of reality, when, one day in 1932, on the point of complete collapse, he appeared on Carl Jung's doorstep in Zurich, Switzerland. Jung quickly took him on as a patient. The famed psychologist, in 1936, said that Pauli "has a most remarkable mind and is famous for it. He is no ordinary person. . . . It unfortunately happens that such intellectual people pay no attention to their feeling life and so they lose contact with the world that feels."[39]

To coax feeling back into Pauli, Jung encouraged him to search out in his dreams the archetypes and alchemical symbols that, Jung believed, embodied the basic configurations of human emotion. Pauli increasingly had dreams and fantasies in which "terms and concepts from physics appeared in a quantitative and figurative—i.e., symbolic sense." Pauli decided the "scientific" symbols these personal dreams and fantasies contained were not arbitrary, or subjective, but rather they were proof that "background physics" was archetypal in nature.

Pauli was now collaborating with Jung, and this man who had

an incomparable grasp of modern physics decided, astonishingly, that modern science had "come to a dead end." He wondered if the only way to break through and develop new insights was "to take a radically different approach and return to science's alchemical roots." He wrote that, "beginning with Kepler, modern scientists deliberately excluded the *anima* [inner feminine part of the male personality] as they tried to mechanize the world, partially guided, perhaps, by the image of the Trinity, which they saw in the three dimensions of space."[40]

Pauli wanted to get back to the moment when, for the first time, mysticism and alchemy clashed with the new, rational, scientific thinking. He wished to return to the place where Newton was when he toiled simultaneously over his alchemical crucibles and the *Principia* and believed the two could be reconciled. Pauli didn't believe mysticism would be resurrected in its old form. He believed, rather, that "the natural sciences will out of themselves bring forth a counter pole in their adherents, which connects with the old mystic elements." Jungian archetypes would function "as the long sought-for bridge between the sense perceptions and the ideas, and are, accordingly, a necessary presupposition even for evolving a scientific theory of nature."[41]

So Pauli realized that the despiritualization of science had led to an impasse—and he may, like Newton, have looked gloomily upon the proliferation of atomic bombs around the world in an age when science had all but lost any connection with loftier realms of being. He embraced that concept of archetypes whose reality may (though Pauli doesn't say so) have accounted for Newton's intense intimation that his own discoveries had been known in other places and other times in the life of man. (Does the idea of archetypes account for the whole notion of the *prisca sapientia*?) Pauli, like Newton, looked back to Pythagoras—and also, like Newton, to Archimedes—and we will see, in the next and final chapter of this book, just how powerful these influences were on Isaac Newton, and just how much they may have accounted for the dismay and anger he felt as he entered his final years.

CHAPTER EIGHTEEN

SON OF ARCHIMEDES

When a physicist becomes entranced with the symbolic and aesthetic qualities of the quantum theory and begins to see it as a forum for contemplation, rather than a problem to be solved in order to make a bigger bomb, then he begins to see that there is a higher level of hieroglyphic knowledge in which art, religion and science reconverge.

WILLIAM IRWIN THOMPSON,
DARKNESS AND SCATTERED LIGHT

In 1694, word got around the scientific community that Isaac Newton had gone mad.

Christiaan Huygens wrote in his journal that he was told on May 29

that Isaac Newton, the celebrated mathematician, eighteen months previously, had become deranged in his mind, either from too great application to his studies, or from excessive grief at having lost, by fire, his chemical laboratory and some papers. Having made observations before the alienation of his intellect, he was taken care of by his friends, and being confined to his house, remedies were applied, by means of which he has lately so far recovered his health as to begin to again understand his own *Principia*.[1]

Huygens took it for granted that Newton's career was ended. In early 1695, two different sources told John Flamsteed that Newton was dead. That summer, the philosopher Johann Christoph Sturm informed the mathematician John Wallis that Newton's house and all the books in it had been destroyed in a fire and that Newton himself was "so disturbed in mind thereupon, as to be reduced to very ill circumstances."[2]

Wallis knew this wasn't so. He knew Newton had had a nervous breakdown, and told Sturm this. Two other Englishmen, both very distinguished, had known about the breakdown from the start. They were Samuel Pepys, chief administrator of the Royal Navy and diarist, and the philosopher John Locke. On September 13, 1693, Pepys had received a letter from Newton that read in part: "I am extremely troubled by the embroilment I am in, and have neither ate nor slept well this twelve month, nor have my former consistency of mind. I never designed to get anything by your interest, nor by King James's favor, but am now sensible that I must withdraw from your acquaintance, and see neither you nor the rest of my friends anymore."[3]

Three days later, John Locke received a similar letter. One paragraph read: "Being of opinion that you endeavored to embroil me with women and by other means I was so much affected with it as that when one told me you were sickly and would not live I answered 'twere better you were dead. I desire you to forgive me this uncharitableness."

Newton confessed that in the presence of others he had accused Locke of trying to set him up with prostitutes. He apologized for harboring evil thoughts that Locke was an immoral man and a follower of the atheist Thomas Hobbes. Newton signed the letter: "Your most humble and most unfortunate servant, Is. Newton."[4]

The two very distinguished Englishmen responded calmly and compassionately to these letters. Richard Westfall writes: "One cannot sufficiently admire Pepys and Locke. Confronted without warning with such letters, neither gave thought to taking offense. Rather both assumed at once that Newton was ill and acted accordingly."[5]

Pepys made discreet inquiries at Cambridge and received

reassurances that Newton was perfectly sane, though he had been indisposed. Two months later, the diarist sent the mathematician a dignified letter in which, rather than inquire about Newton's health, he asked him a complex question about gambling odds; Pepys needed the information for a lottery investment he was thinking of making, but he also wanted to see how well Newton could still reason. (Newton solved the problem overnight and mailed the answer the next day.)

Locke waited two weeks, then replied with a letter that was a simple expression of "humanity and forgiveness, a tender witness of his 'love and esteem.'"[6]

By the end of 1693, Newton was apparently his old self again. He sent apologies to both men, explaining that for a year he had been in a state of exhaustion and anxiety that culminated in his not being able to sleep for more than an hour a night for two weeks and not being able to sleep at all for five days. His letters, he assured them, were solely the product of his a sadly disordered state of mind.

What caused Newton's nervous breakdown? It's easy to say it was the immense labor of writing the *Principia Mathematica*. But the book had been completed six years earlier.

Was mercury poisoning a factor? Newton had been doing alchemical experiments almost every day for thirty years, and almost every day he had tested mercury on the tip of his tongue. Chronic mercury poisoning can cause sleeplessness and paranoid delusions, but there are other significant symptoms that Newton did not exhibit.

We can't discount the role Newton's young protégé Nicholas Fatio de Duillier might have played in his breakdown. The two met when Newton was forty-five and the Swiss mathematical genius twenty-three, and took to each other immediately.

Early in 1693, the charming and volatile Fatio seemed to be on the point of moving from London to Cambridge to be near Newton. The great mathematician was very willing that this should happen. Then Fatio abruptly returned to the Continent. This was just weeks before

Newton's nervous breakdown. Isaac Newton had no involvements with women during his life and died a virgin; he probably felt more love for the engaging young Swiss polymath than he did for any other person in his life. Frank Manuel speculates that Newton's feelings for Fatio had a sexual aspect and "reached a high pitch" early in 1693, "creating a demand for repression that was an element in the breakdown."*7

But only an element. Perhaps there is a more fundamental reason why Isaac Newton had a nervous breakdown.

There is something else. At the latter end of the century Newton became master of the London Mint. Today we marvel at how successful Newton was when he exchanged the dreaming spires of Cambridge for the grimy smokestacks of London. At Cambridge he was an absent-minded professor living in an ivory tower. In London he was a world-class administrator, striding purposefully through the corridors of a long, narrow building attached to the Tower of London that housed nine clattering coin presses. He held this position for twenty years and proved a most able public servant. Richard Westfall extols him for having been able to "pick up the threads of a practical pursuit and to perform it with distinction."[8]

Newton had his nervous breakdown at about the same time as he first decided to leave Cambridge and look for a position in London. He didn't actually move to London until 1695, but the decision was made in 1693, and it may be that it unleashed in Newton a storm of conflicting psychic forces that immobilized him for many weeks and sent the rumor flying about Europe that Isaac Newton had gone mad.

---

*Fatio, who had once been seen as the heir apparent to Newton, abandoned his career in mathematics bit by bit. In 1707, he surfaced in London as a "French Prophet"—part of a band of Protestant enthusiasts, newly exiled from France, who "ranted in the streets and conducted wild séances during which frenzied men and women prophesied the imminent coming of Judgment Day." The authorities put two of these charismatic prophets in the stocks. One was the recording secretary of the Prophets: Nicholas Fatio de Duillier. A hysterical streak in Fatio had gradually taken hold; he never got back to mathematics and died alone at eighty-two. (Manuel, *A Portrait*, 288.)

Up until early 1687, when he was completing book 3 of the *Principia Mathematica,* Newton was not much interested in the world beyond the gates of Trinity College and relatively innocent of the evil that flared up in the bloody arena that was Augustan political life. What little he knew of evil he'd gleaned from the ancient tomes he'd studied periodically so he could put together his various nonscientific writings—in particular, that somber thread, running through many of these, that we've been calling his "History of the Corruption of the Soul of Man."

But this secondhand exposure to evil wasn't enough to make Newton heed the warning of Moses and Pythagoras that the highest knowledge should be hidden from the commonality of mankind, since the ordinary run of man and woman couldn't be expected to do more than misunderstand and abuse such knowledge and might well turn it to evil purposes. He took this seriously—but not seriously enough to make him say no to Edmund Halley when the latter begged him to write a book that would reveal Newton's own highest knowledge to the world. Instead, he went on to write the *Principia Mathematica.*

Something was about to happen that would make Newton regret that decision.

He was about to have a personal encounter with evil.

You can find scattered here and there in Barbados today white men who've never intermarried with the blacks and who call themselves "red legs." They are the descendants of twelve hundred rebels whom the British government sold as slaves to the island plantations in 1685. The white slaves were the mark of George Jeffries, one day to be Lord Chief Justice of England and the man behind the deportation—a man, in the words of one historian, "whose brutal nature, savage partisanship and high professional gifts made him the perfect instrument of judicial murder."[9]

Charles II had died early in 1685, and the Catholic king James II succeeded him to the throne. A small army led by the protestant pretender to the throne the Duke of Monmouth took the field against

James but was roundly defeated after a short and bloody campaign.

Lord George Jeffries presided over the tribunal set up to punish Monmouth and the army. History calls it the "Bloody Assise." Jeffries hung three hundred men, sold many more into slavery, and personally sold hundreds of pardons—at illegal and exorbitant rates—to the captive soldiers. He made sure the imprisoned daughters of well-to-do families who had taken part in the rebellion were set free only after their families paid enormous ransoms.

From the moment James II came to the throne, he favored England's Catholics and promoted them to high office. In 1687, James ordered Cambridge to award a master of arts degree to a Benedictine monk named Alban Francis. The king stipulated that Father Francis should not be required to take the usual oaths of loyalty to the Anglican Church.

The regents of the solidly Protestant university of Cambridge did not dare defy James. Isaac Newton, his loathing of the papacy aroused, leaped into the fray. He rallied the university to stand up to the Catholic king. He warned of the dangers of setting a precedent if it bowed to James's demands. He assembled documents and statutes to prove James was legally in the wrong. In general Newton stiffened the spines of his unworldly colleagues. In April, Cambridge vice chancellor John Peachell, better known for his drinking prowess than his scholarly achievements, was summoned to appear with representatives of the university before the Court of Ecclesiastic Commission in London. Newton was one of the delegation of eight.

Presiding over the court was the notorious Lord George Jeffries.

The eight dons knew Jeffries had been the mastermind of the Bloody Assise, and they knew that Londoners called him the Hanging Judge. What met their eyes when they entered the court didn't belie what they had heard. Judge Jeffries's voluminous black robes of a presiding magistrate failed to hide his obesity; and, since he was hunched forward over the bench in an almost predatory fashion, his notoriety, combined with his appearance, made the dons think of a huge, overfed vulture; only the skewed white periwig on his head made him look vaguely human. His face was hollowed, ridged, and pockmarked above a jagged fringe of gray beard. His tiny eyes seemed blank and unseeing, unstamped coins stuck in his head. Jeffries was a man of crude yet formidable power, but he was also a ghoul, a vampire, a bloodsucker. His voice was hoarse, breathless, and low; apart from his words, which struck them like hammer blows, that voice seemed not to resonate so much as to suck the vitality out of them. From one harangue to the next, he exhausted them.

During the first two sessions the dons made their case; Jeffries responded with mockery, contempt, and invective. The arbitrariness of his words dismayed them; they could find in them little connection to justice. At the third session, Jeffries asked Peachell point-blank why he had disobeyed the king's order. Perhaps because he had been in his cups earlier that day, the vice chancellor couldn't come up with a satisfactory answer; then and there, Jeffries stripped him of his position along with his salary.

It would have been at this point that a sense of the evil in the world entered Newton's bones. In confronting James, the dons had put their careers on the line, Newton's no less than any. And now it struck him that if there were any justice in this courtroom, it was largely accidental. His livelihood could be cut off at any moment. The case against James he had so meticulously set forth was no guarantee against the arbitrary cruelty of this judge.

But at the final session the eight dons were surprised and relieved when the mastermind of the Bloody Assise did no more than dismiss them (albeit with scorn and contempt), admonishing the dons to "Go your way, and sin no more, lest a worse thing come to you" (John 5:14). Near the end, a dread thought must have passed through Newton's mind. There were Jeffries everywhere. Many ruled the land. Newton would have been devastated by this realization. They would not be expected to apply Newton's equations to the world in an ethical manner. Newton suddenly felt a profound sense of guilt: Halley was wrong, and Pythagoras and Moses were right: the world was not good enough for its finest creation. He had made a great mistake in writing the *Principia* and releasing it to the world.

It was in a state of exaltation that the dons left the Court of Ecclesiastic Commission. Newton shared their enthusiasm, but it was overshadowed by the guilt he now felt at having betrayed his equations.

But he was proud of the *Principia,* and so he repressed his guilt.

It lingered, and festered.

Father Francis never received his degree—"not least," writes Richard Westfall, "because Newton had refused to be frightened."[10] James II was overthrown near the end of 1688, and the protestant Dutch sovereign William of Orange was placed on the throne of England. Jeffries was hurried off to the Tower of London, where he died after five miserable years.

Newton had demonstrated leadership, personal bravery, and moral integrity during the standoff with James, and he was given much of the credit for its success. He had shown that he could act with great effectiveness in the greater world beyond Cambridge University; perhaps he hadn't known that himself when he climbed down from his ivory tower to help his fellow dons.

Everybody knew what Newton had done; doors were opened to him that hadn't been open before, and it would have been impolitic for him not to walk through a few. Cambridge asked him to run for one of two seats as a university representative in the Commission Parliament scheduled to convene in January 1688 to ratify William's succession. Newton accepted, and was elected. He arrived in London early in the new year and began a one-year stay in the city by dining with William of Orange and two others on January 17.

It's likely Newton attended every session of the Commission Parliament and the regular Parliament that followed it. He spoke only once, and that one time, or so the story goes, was to ask an usher to close a window because he felt a draft. Newton watched over the interests of

Cambridge assiduously. He watched everything; he heard everything; he saw great men honorably advancing the interests of England, and he saw these same great men equivocate, lie, and deceive. He witnessed men of undoubted nobility at work and men not unlike Judge George Jeffries. He saw much to admire, and much to abhor. Men were two-faced creatures; both faces, it seemed, the reprehensible and the saintly, were needed to conduct the business of this nation.

England was at war with France in 1690; vast sums of money were requisitioned regularly, almost arbitrarily. Detailed reports of battles, prepared for Parliament only, were read aloud; this information was kept from the public. Newton was appalled by the number of casualties—and yet such slaughter was surely necessary if it were a question of crushing the Catholic king Louis XIV. It may have been at this time that he began to feel that the application of science to warfare did not, as he would later disingenuously put it, "serve the interests of science." Some years later, he cautioned David Gregory's son to have nothing to do with a new cannon his father (who was an early expert on Newton's *Principia*) was developing. In later years Newton would profess that he had no interest in the practical applications of science; sometimes he expressed hostility to the very idea of technological progress. He would always believe, writes Manuel, that the practical use of his equations should be "controlled by the two fundamentals of religion, love of neighbor and love of God as set forth in Scripture."[11] And all through his life he would see that this rarely happened. Newton must have come away from his year as a parliamentarian with the growing conviction that mankind, always needing to defend itself, usually desperate, could not be trusted—not even its best-intentioned members—to apply Newton's equations to the world in a way that could be beneficial to all.

Back behind the timeless walls of Trinity, Newton threw himself into his studies with as much vigor as before. He plunged into a new attempt to understand why gravity happened, carefully examining those achievements of the scientists of the ancient world that seemed to have bearing on his universal law of gravitation. He tackled anew a problem he had left unresolved in his treatment of optics: whether the forces responsible for optical effects operate not only on the macro level but on the micro level as well. The solution he came up with was so outlandish, so "far out" even by Newton's standards, that he decided to drop this line of inquiry lest that solution "should be accounted an extravagant freak."[12]

Newton continued to work furiously on theology, church history, and chronology, and it was at this time that he coaxed out into the open (but still hid) the corrupted texts of the New Testament. He continued to hide his Arianism and, to some extent, his alchemy. And (or so we are suggesting) he continued to hide from himself his growing sense of guilt at having revealed his equations to the unappreciative and perhaps dangerous masses.

Newton's yearlong exposure to the vibrant life of London had changed him: he was becoming more interested in people. Much shouldn't be made of this; Isaac Newton remained aloof, autocratic, and dangerously touchy. But he became less elusive. You could run after him on his brisk walks through the Trinity gardens, catch up with him, stop him, talk to him. More and more people were getting to see the inside of his apartment with its gleaming columns of mint guinea pieces stacked on the windowsills and its impeccably arranged and very eclectic furniture. Newton served rather inexpensive wine at these tête-à-têtes but made up for it with the startling richness of some of his observations.

Beyond the confines of Cambridge University, Newton was becoming the most famous natural philosopher in the world. The rich, the famous, and the powerful sought him out. He was drawn to power, this man who dreaded what those in power might do to his equations. In about 1691—we're not sure when—he began to think about leaving Cambridge for a nonacademic position in London.

Why? The creative power of even the greatest of mathematicians begins to wane when they are in their late fifties, and Newton was no exception. But his mind, his thirst for knowledge, his inquisitiveness, his sheer level of energy, were as powerful as ever, and he began to cast about restlessly for new worlds to conquer. And he may have felt the need for a richer intellectual community than that offered by Cambridge.

But a move to London presented not only practical but psychological difficulties for Newton.

Many of the great writers of the world have looked with a jaundiced eye on the great cities of the world. Kóstya Lévin, the country-estate-owning protagonist of Leo Tolstoy's novel *Anna Karenina*, regarded rural life as wholly good and urban life as wholly bad. He even (as did Tolstoy) condemned the lavish opera productions of the city of Moscow as contrived and artificial—fundamentally deceitful. Joseph Conrad called London "the cruel devourer of the world's light," while Charles Dickens hit out at the British capital as "a place of great energy, but also of constant deceit, of others and of oneself."

The puritanical child in Newton who saw London as Babylon; the pious child in Newton who, as a man confronting Jeffries, felt the life's blood drained out of him by that raucous voice and realized men like this could be expected to apply his equations without the slightest reverence for God's will, this combined with what he'd seen of certain parliamentarians whose type was everywhere and who could be expected to apply his equations only with ruthless pragmatism—all of this (which he had immediately repressed beause of the grief and guilt he now felt at having published the *Principia*) must have come welling up into his conscious mind. But not quite into his conscious mind. He could not afford to feel any of this, and he quickly repressed it.

But the decision to go to London would have stirred up this buried knowledge. This repressed material would have risen almost, but not quite, to the forefront of his consciousness; it would have been vaguely sensed and not fully understood by Newton. His divided mind would have feared that he himself, once in London and surrounded by the city's corrupting practices, would gradually forget his good intentions vis-à-vis his equations and even, however inadvertently, aid and abet those who sought to profit from them without thought of the welfare of humanity.

We know that for Newton the Puritan, London was a Babylon, and the great but vulnerable mathematician feared its allurements. This comes out in his letters to Locke and Pepys: Locke foisting prostitutes on him and Pepys forcing him to pull strings for high preferment were merely the mirror of his terror that the corruption of the world he was about to enter would be too powerful for him to resist; that the world itself, and not his friends, would impose these horrors on him.

Some time in mid-1693, in the weeks before he made his final decision to go to London, all this erupted chaotically into Newton's mind. He may not have understood it exactly, but somehow over the next two years he worked through it sufficiently that late 1695 found him the warden of the London Mint.

As has been noted, behind Newton's pessimism lay the conviction, shared by many of his peers, that the universe had been becoming steadily more corrupt and the human race was now on its last ropes. A model for this was the decline and fall of four empires, each one less powerful than the last, expressed in the four visions of Daniel. This model itself recapitulated the concept, given voice by the Greek poet Hesiod in the ninth century BC, of the four ever-declining ages of mankind: the Gold, the Silver, the Lead, and the Clay. The Golden Age began in the Garden of Eden; mankind now lives in pain and suffering in the twilight years of the Clay Age.

But another, quite different current of thought was making its way through England at the time. Its advocates did not believe the world was dying; they believed it was changing for the better and that science could help it immeasurably. It held out the hope that man was perfectible and it contained the seeds of the modern-day conception of progress. Its great proponent was Francis Bacon (1561–1626), a judge and Lord Chancellor of England. Bacon scholar Hugh G. Dick explains:

> When he [Bacon] looked about him, he came to realize (as others had done) that three recent discoveries—printing, gunpowder, and

the compass—had done more to transform the world in which he lived than had any political theory or school of philosophy. "For these three have changed the whole face and state of things throughout the world; first in literature, the second in warfare, the third in navigation; whence have followed innumerable changes; insomuch that no empire, sect, no star seems to have exerted greater power and influence in human affairs than these mechanical discoveries." Awakened by these realizations, Bacon felt compelled to review the whole intellectual history of the western world to see why philosophy had been so productive of words but so barren of fruit for the "benefit and use of life."[13]

Bacon believed mankind would develop new tools at the same time as he systematically acquired greater knowledge of himself; in such a way would a necessary balance be maintained. But Bacon overestimated mankind. In her 2014 bestseller *This Changes Everything,* Naomi Klein puts him at the beginning of a chain of events that culminated in global warming. She writes:

> If the modern-day extractive [fossil fuel–producing] economy has a patron saint, the honor should probably go to Francis Bacon. The English philosopher, scientist, and statesman is credited with convincing Britain's elites to abandon, once and for all, pagan notions of the earth as a life-giving mother figure to whom we owe respect and reverence (and more than a little fear) and accept the role as her dungeon warden.[14]

Isaac Newton was no son of Francis Bacon. He did not regard the earth as a life-giving mother but as a giant riddle that if unraveled properly would reveal God's benevolent activities in the world. We should not exploit the fruits of these activities; rather, we should respect, reverence, and adore God for having so acted. It was not for us to command the Earth, but to adore it; and the worldly applications of

Newton's equations (which were descriptions of those activities of God in the world) must be conducted in the same spirit. Newton's ideas, his inclinations—the promptings of his soul—had little affinity with Baconism. They went back to the third century BC.

He was the son of Archimedes.

The Roman general Marcus Claudius Marcellus, who besieged the Carthage-controlled city of Syracuse in Sicily from AD 210 to 212, had a reputation for personal bravery; he had twice fought the enemy in single man-to-man combat, emerging victorious and earning the right to spare his army a battle.

But what he saw from the swaying deck of his trireme on the first day of the siege of Syracuse was enough to make even the bravest soldier want to turn and flee. From the heights of the fortified walls of the city, weapons he'd never seen before rained terror down on the Roman fleet. Plutarch tells us that

> huge poles thrust out from the walls over the ships sunk some by the great weights which they let down from on high upon them; others they lifted up into the air by an iron hand or beak like a crane's beak and, when they had drawn them up by the prow, and set them on end upon the poop, they plunged them to the bottom of the sea; or else the ships, drawn by engines within, and whirled about, were dashed against steep rocks that stood jutting out under the walls, with great destruction of the soldiers that were aboard them. A ship was frequently lifted up to a great height in the air (a dreadful thing to behold), and rolled to and fro, and kept swinging, until the mariners were all thrown out, when at length it was dashed against the rocks, or let fall.[15]

The Romans had brought with them a floating siege tower with grappling hooks; it rested on a bridge of planks stretched across eight ships. Earlier in the day, Marcellus had watched as three giant boulders, plummeting down the face of wall, shattered the siege tower and sent it flying in pieces into the turbulent water. Roman battleships scraped against each other as they maneuvered to escape the falling rocks. Their crews were so unnerved that when the slightest object, like a rope or a beam of wood, suddenly protruded from the city walls, they turned their vessels around and tried to get away.

The siege of Syracuse lasted for two years. The city's defenders didn't weaken, and the uncanny war machines, continually replaced or repaired, never stopped raining destruction on the Roman fleet. Only a ruse, based on a weakness Marcellus detected in Syracuse's defenses while he was conducting negotiations, enabled his men to overwhelm the city and slaughter or take prisoner most of its defenders.

These extraordinary war machines had never been seen before. They would rarely be seen again. They were the creation of the illustrious Greek mathematician, physicist, engineer, inventor, and astronomer Archimedes (ca. 287–212 BC). This was the man who, according to legend, ran naked through the streets of Athens crying, "*Eureka!*" ("I have found it!") when he discovered, while sitting in his bath, that a sinking object displaces its volume in fluid; it was the same Archimedes who understood the principles of levers so well that he could boast, "Give me a place to stand on, and I will move the world!"*

Archimedes had been reluctant to build these war machines. He had done so only at the request of King Hiero II of Syracuse, who was his friend and near relation—and besides, it was never wise to disobey a king. But perhaps it would have been wise in this case; Archimedes was accidentally killed when the Roman soldiers swarmed across the city.

He left no blueprints of his brilliantly destructive war machines. Archimedes disdained all engineering as "the mere corruption and annihilation of the one good of geometry"; he believed, like Plato, that applied science shamefully turned its back "upon the unembodied

---

*Plutarch renders this line as, "if there were another earth, by going into it he could remove this." (Plutarch, *Lives of the Ancient Greeks and Romans*, 376.)

objects of pure intelligence, [so as] to recur to sensation and to ask help (not to be obtained without base supervisions and deprivation) from matter."[16]

Plutarch writes:

> Yet Archimedes possessed so high a spirit, so profound a soul, and such treasures of scientific knowledge, that though these inventions had now obtained him the renown of more than human sagacity, he yet would not deign to leave behind him any commentary or writing on these subjects; but, repudiating as sordid or ignoble the whole trade of engineering, and every sort of art that lends itself to mere use and profit, he placed his whole affection and ambition in those purer speculations where there can be no reference to the vulgar needs of life; studies, the superiority of which to all others is unquestioned, and in which the only doubt can be whether the beauty and grandeur of the subjects examined, of the precision and cogency of the methods and means of proof, most deserve our attention.[17]

Isaac Newton possessed an equally high spirit and an equally profound soul. His sometime Cambridge roommate, William Wickins, reminiscing to his son many years later about the time he spent with Newton, declared:

> When he has sometimes taken a turn or two [he] has made a sudden stand, turned himself about, run up the stairs like another Archimedes, with an εὔρηκαευρrjKa [*Eureka*], and fall[en] to write at his desk standing, without giving himself the leisure to draw a chair to sit down on. At some seldom times when he designed to dine in the hall, [he] would turn to the left hand and go out into the street, when making a stop when he found his mistake, would hastily turn back, and then sometimes instead of going into the hall, would return to his chamber again.[18]

Plutarch adds of Archimedes:

> His familiar and domestic Siren made him forget his food and neglect his person, to that degree that when he was occasionally carried by absolute violence to bathe or have his body anointed, he used to trace geometrical figures in the ashes of the fire, and diagrams in the oil on his body, being in a state of entire preoccupation, and, in the truest sense, divine possession with his love and delight in science.[19]

Behind Newton and Archimedes looms the godlike figure of Pythagoras. Pythagoras and the Pythagoreans maintained that mankind's duty was to "honor first the Immortal Gods—themselves Numbers—before embarking on any enterprise."[20] Numbers were the supreme reality, the ground of all other realities (which by comparison were not real). For Pythagoras, even to *discuss* geometry, let alone apply it to the world, was to sully the immortal form of Number. Isaac Newton tells us that Pythagoras believed the contemplation of pure Number could bring us to the knowledge of God himself. Newton wrote: "Hence the Pythagorean harmony is between God acting harmoniously and matter reacting harmoniously, and the geometrical elements, which are composed of numbers and their symmetry, are the whole of nature. Hence in Lucian, Pythagoras says *you will come to know God himself, who is number and harmony.*"[21]

Despite the chasm of more than two thousand years separating them, Newton had striking affinities with Archimedes and Pythagoras. As has been mentioned earlier, if Newton had been left to his own devices it's possible he would never have made his discoveries known to the world. He would probably have retired to his mother's estate in Woolsthorpe. The scholar Derek Gjertsen writes that, had Newton done this,

> no doubt his scientific writing, in manuscript, would continue to grow, and would be shown to a few friends and colleagues. While

a few might have suspected the full range of his genius, none could have been quite sure. His results would probably have leaked out piecemeal, or have been independently discovered over the years by other scholars. Eventually, long after his death, his papers would come to light and would show that Newton had anticipated much of the science of a later age.[22]

Johann Sebastian Bach signed his musical compositions with the letters "SDG": *Soli Deo Gloria*—"to the Glory of God Only." Newton's deepest impulse was to do his mathematics for the glory of God only. If he had been left to himself, he would likely have followed that path. But Edmund Halley's intervention made him famous, enriched our world—and probably divided Newton's soul.

When he was writing the *Principia,* Newton believed his work would help exorcise the demons of atheism. As has been noted earlier, he wrote to Richard Bentley on December 10, 1692: "When I wrote my treatise about our System [of the world] I had an eye upon such Principles as might work with considering men for the belief of a Deity & nothing can rejoice me more then to find it useful for that purpose."[23]

Frank Manuel tells us Newton's scrutiny of nature was directed almost exclusively "to the knowledge of God and not the increase of sensate pleasure or comfort. Science was pursued for what it could teach men about God, not for easement or commodiousness.... To be constantly engaged in studying and probing into God's actions was true worship and fulfillment of the commandments of a warden."[24]

As Newton himself put it near the end of the *Optics:*

And if natural philosophy in all its parts, by pursuing this method, shall at length be perfected, the bounds of moral philosophy will also be enlarged. For so far as we can know by Natural Philosophy what is the First Cause, what power He has over us, and what benefits we receive from Him: so far our duty toward Him, as well as towards one another, will appear to us by the light of Nature.[25]

In *The Clockwork Universe,* modern-day author Edward Dolnick sums it up beautifully.

> Newton had ambitions for his discoveries that stretched far beyond science. He believed that his findings were not merely technical observations but insights that could transform men's minds. The transformation he had in mind was not the usual sort. He had little interest in flying machines or labor-saving devices. Nor did he share the view, which would take hold later, that a new era of scientific investigation would put an end to superstition and set men's minds free. Newton's intent in all his work was to make men more pious and devout, more reverent in the face of God's creation. His aim was not that men rise to their feet in freedom but that they fall to their knees in awe.[26]

They should fall to their knees in recognition of the fact, among so many others, that it was impossible not to see in the design of the universe the handiwork of a designer God. This "argument by design" was proof of God's existence, and it seemed to Newton and his contemporaries that every day the new sciences revealed more examples of the order, beauty, symmetry, and purpose apparent all around us. Newton wrote in "A Short Schem of the True Religion":

> Atheism is so senseless & odious to mankind that it never had many professors. Can it be by accident that all birds beasts & men have their right side & left side alike shaped (except in their bowels) & just two eyes & no more on either side [of] the face & just two ears on either side [of] the head & a nose with two holes & no more between the eyes & one mouth under the nose & either two forelegs or two wings or two arms on the shoulders & two legs on the hips one on either side & no more?

The bilateral symmetry of birds, animals, and humanity alike could only have arisen from "the counsel & contrivance of an Author."[27] And the presence of a designer God was as obvious beyond the Earth as it was upon the Earth; Newton wrote in the *Principia* that "this most beautiful System of the Sun, Planets, and Comets could only proceed from the counsel and dominion of an intelligent and powerful being."[28] Astronomers were discovering that the stars were vastly more distant from each other than the planets were from each other. Flamsteed and Hooke put the nearest star at 648 billion miles away; Huygens thought the distance was 2.25 trillion miles.[29] Newton regarded these vast stretches of space as a further argument from design for God's existence; he explained that "lest the systems of the fixed stars should, by their gravity, fall on each other, He hath placed those systems at immense distances from one another."*[30]

For two centuries scholars believed that Newton was a Deist. The Deists believed that, having created the universe and set it in motion, God withdrew from his creation for all eternity; God was, in the mocking words of the philosopher Leibniz, an absentee landlord.

Today we know Newton believed God intervenes from time to time in his creation to make adjustments. Newton and his peers called this divine interventionalism special providence, while calling the Deist notion of a caretaker God general providence. It's not inaccurate to say that Newton believed in miracles; God's sending a comet to Earth to trigger the Noachic Flood and remake mankind was one such miracle. This divine flexibility seemed to Newton to be yet another argument from design for God's existence.

With the coming of the Age of Enlightenment, the argument from design began to crumble. The complexities and contradictions of the universe became more and more apparent to physicists, mathematicians, and geologists, and that process continues. Today, scientists don't believe a design factor is woven into the fabric of the universe. They believe we live in an optimal world—that is, the best physically and

---

*The nearest star, Proxima Centauri, is 4.24 light-years, or 2.511×10^{13} (21 trillion) miles from Earth, about nine times farther than Huygens estimated.

mentally possible for us—among many possible worlds, all formed randomly. Life on Earth is a freak phenomenon, the result of a sequence of chemical actions so rare that it is unlikely they have happened twice in the observable universe. Impeccable design is the result of a near-infinite series of random attempts and failures, one finally taking hold because it indispensably and effectively furthers the survival of our species. (God is not necessarily excluded from this process; Spinoza and Einstein believed God works through blind chance to achieve incredible fertility of form and invention, these being achievable only through many degrees of freedom.)[31]

Gone is Newton's sacramental view of the universe, which was itself only the final flickering expression of that mighty power of creation which men call God and that helped sustain them through the first few millennia of their existence.

If Isaac Newton had returned to Earth in 2017, what would he have found? How would he have thought?

He would have been shocked and horrified to discover that the human race had devised not one, but two means of destroying itself, and that, in this very year of 2017, both these means, having been held in check so far by man, were on the brink of becoming uncontrollable.

On July 16, 1945, the world's first atomic bomb was detonated at the Manhattan Project's Almogordo bombing site in New Mexico. It lit up the night sky with a fireball four times hotter than the center of the sun. Watching, project supervisor Robert Oppenheimer recalled with a pang the words of Shiva in the Bhagavad Gita: "I am become death, the destroyer of worlds."

The successful test was followed by the dropping of A-bombs on the Japanese cities of Hiroshima and Nagasaki in August 1945. This ended World War II. But the horrific death toll that resulted—80,000 in Hiroshima and 40,000 in Nagasaki, with thousands more deaths to come by radiation poisoning—sparked a heated debate over the use of atomic weapons that has not died down to this day. In 1949, the Soviet

Union detonated its own atomic bomb. Both the United States and the USSR were soon stockpiling hydrogen bombs. As of 2017, thirty-four nations had laid claim to the possession of a nuclear arsenal. Some of these countries don't seem entirely responsible; the world's greatest fear today is that a rogue state with unstable leadership, such as North Korea under Kim Jong-un, will accidentally or intentionally launch an A-bomb and trigger a worldwide nuclear holocaust.

As early as 1960, scientists began to notice that waste materials were collecting in the world's oceans. By century's end, it was apparent that the burning of fossil fuels like coal and oil had also clogged our planet's atmosphere, and Earth's temperature was slowly but surely rising. "Climate change" and "global warming," though not accepted by everyone, had become household words.

By December 2016, global warming had taken on such menacing proportions that Noam Chomsky, the eminent political dissident, linguist, author, and professor emeritus at MIT, warned today's young people that they will soon be "facing problems that have never arisen in the 200,000 years of human history—hard, demanding problems. You in particular, and all the rest of us, will be in there struggling hard to save the human species from a pretty grim fate."[32]

Chomsky's words were prompted by the United States' refusal last fall to sign off on the COP 22 climate-change agreement that would commit the global community to a unified assault on global warming.

Isaac Newton had expected the world to end, and to us he and his millenarian colleagues seem to have looked forward to the end with almost the same unseemly eagerness. But Newton, touring the polluted cities of the world beneath the murky glimmering of a sun growing hotter every day, surely hadn't expected that mankind, no matter how corrupt, was fated to die by its own hand. It would have been inconsistent for him not to believe that it was the failure to read prophecy properly that had brought us to this sorry pass—which prophecy we will leave for Newton scholars to guess at. Newton would have realized that neither a comet's flight past the Earth nor a comet's plunge into the sun would be the cause of the upcoming apocalypse, but, rather, the technologies born, albeit with many steps in between, from his own science. He would have thought that the benefits that those same technologies had bestowed upon the Earth—antibiotics, space flight, computers, a thousand more—hardly compensated for the destruction of all those who were now enjoying them.

All this would have been devastating for Isaac Newton. It would have seemed to him that his whole life had been a travesty, not to say a tragedy. He had wanted his equations to bring man to God; instead, they had brought man to the brink of destruction. Writing the penultimate chapter of his "History of the Corruption of the Soul of Man," (the last chapter would be devoted to the Apocalypse), he would have ascribed man's present-day predicament to a general, overarching cause, idolatry—our persistent tendency to put other gods between ourselves and God. It was the same cause he had ascribed to all the other fallings away of man from God he had recorded in his "History."

We live in an age that is rapidly abandoning religion, and today idolatry seems to us a trifling thing. We don't understand why it's such a problem for God. Why does He need us to worship him? Doesn't he exist whether we believe in him or not? Isn't he, being God, sufficient unto himself?

It would take a cathedral full of theologians to answer this question. Suffice it to say that for Newton, Christianity consisted basically in worshipping God and loving our fellow man and woman. But mankind finds it difficult to worship God; our deity is without earthly attributes, is unknowable—is so other, so beyond, that we can't get our imaginations around him for very long and quickly tire of the effort. Jesus, on the other hand, is easy to worship. He is one of us, and we can identify with his trials and tribulations. And that gives rise very quickly, says Newton to the idolatry that is Christ worship.

But worship is essential, if for no other reason than that the world does not have to end tomorrow and perhaps not for a very long time. Pursuing his heartbreaking trek through the world of 2017, deploring mankind's folly, he would have believed that the apocalypse still had to come. But he would also have believed, even in that eleventh hour, that in the short term mankind could save itself from self-destruction. He proposed a method of worship that would make it easier for ourselves. God should be worshipped for his activities in themselves, and not for himself alone—for his activities we can feel and see and hear and taste and touch.

Dobbs tells us that, in a sermon against idolatry in all its forms, Newton argued that

> it is for His *actions* that God wants to be worshiped, His actions in "creating, preserving, and governing all things." To celebrate God for His essence is "very pious" and indeed it is "the duty of every creature to do it according to his capacity." But those attributes spring not from the freedom of God's will "but from the necessity of His nature." And so he must be worshiped for his *actions* in the world.[33]

We shouldn't worship God for being that which he had no choice but to be, but for what he has chosen to do. From the point of view of humans, God's greatest activity is to have created the Earth and sustained it in its creaturehood.

We'll recall that Newton's God was a God of special providence. He can intervene and make changes in his divine plan even as it is unfolding. In the remarkable document in which he explains God's "Preamble to the Prophetic Visions" in Revelation, Newton seems to be implying that the course of God's future history as set down in the Book of Revelation can be changed. He writes: "It [God's preamble] was done in prosecution of the main design of the Apocalypse [John's Revelation], which was to describe & obviate the great Apostasy."[34]

Obviate means "to remove." It's difficult to understand what Newton is saying here, but it seems to be that the Great Apostasy can be ended—that we can undo the corrupting effects of the doctrine of the Trinity on our lives—prevent the Apocalypse from happening, if only we will worship God, for to truly worship God alone is to overcome the doctrine of the Trinity.

It would be wonderful to believe that Newton intended his equations to raise the consciousness of mankind, that he wanted to expand our souls. It would be wonderful to believe that he wanted to harness within us the energy the Pythagoreans called the music of the spheres; that he saw man as potentially a being of great power and that it was that potential power that he wished to harness.

And in fact his goal was indeed to restore man and the world to that state of paradisaical being that man had enjoyed in the world before the Flood.

But that worship had to be sustained as the world grew older, and Newton thought it had been sustained through the system of temples that we have talked about so often in this book. The key here is that these antediluvian houses of worship embodied in their design God's primal activity of creating the solar system; they are meld of planetarium and cathedral. The ancient Egyptians remembered them and "built temples in the form of the solar system and derived the names of their gods from the order of the planets. As such, the ancient religion was modeled on the understanding of the heavens and Newton occasionally referred to the Ancients' astronomical theology."[35] All his life Newton searched for the lost religion of the prytaneum. Newton sometimes called this ancient lost religion "astronomical theology."

Perhaps Newton, returned to Earth and looking around at a world on the verge of extinction, would have suggested we build temples that, embodying in their design God's primal activities, somehow resembling our planet itself in their construction, were dedicated to worshipping God through his activities in nature. It may seem to us moderns that just worshipping God will do little to staunch the ecological disaster in which we find ourselves. But Newton would have thought differently: that the true worship of God would draw out of us right action, action conducive to the proper healing of our planet—action like a united front against global warming.

Whatever the case may be, we may be certain that Newton, returned to our world, would have issued a statement somewhat like the following:

Beware the anger of the Earth. You are violating that which sustains you. It cannot continue.

That is Isaac Newton's warning to mankind.

APPENDIX A

Newton's Prophetic Hieroglyphs

1. Beasts = armies whereby kingdoms are usually founded and upheld (armies being wild beasts by definition, called beasts because of their pugnaciousness)
2. Court = parcel of ground exempted from the general possessions of the people for sacred purposes
3. Dens and rocks = buildings (and ruins of buildings)
4. Dragon = hostile king
5. Earth = inferior people (commoners)
6. Eating the flesh of a scorpion = receiving the wealth of an enemy
7. "Eating the flesh of my people" and flaying skins = military exaction and oppression
8. Eye = knowledge
9. Heads = successive parts of a kingdom
10. Heaven = the throne, court, honors
11. Herbs and other vegetables = men
12. Horn = more than one horn denotes "collateral parts of a Kingdom or of that head of it upon which they grow. . . . If the Horns Heads & Body of a Beast be considered strictly according to their relation one among another, the horns as being the

most exalted member will chiefly respect Kings or successions of them, the Heads the Nobles & great men & the Body the rest of the Kingdom. But becaus a King is sometimes put for his Kingdom, thence it is that a horn is also sometimes used in the same signification."

13. Idols = men (idolaters)
14. Insects (for example, locusts) that destroy herbs and other vegetables = men
15. Mountain = a city and more especially the head city, as Jerusalem or Babylon
16. Snake = smaller-rank persons (this from Achmed)
17. Stars, great = nobles
18. Stars, rest of the = whole world
19. Teeth = great men and soldiers
20. Wild beasts, because they prey upon men = armies
21. Wilderness = an Abomination (such as the destruction or desecration of the Temple at Jerusalem) spreading over Christian world
22. Woman = church

APPENDIX B

FURTHER CORRUPTIONS FOUND IN THE NEW TESTAMENT

Note: The following is taken from "Two Notable Corruptions of Scripture" (part 4: ff. 70–83), "The Third Letter," Newton Project, www.newtonproject.ox.ac.uk/view/texts/diplomatic/THEM00263.

1) **John 3:6:** "That which is born of the flesh is flesh, and that which is born of the spirit is spirit." But Ambrose cites several contemporary texts: "That which is born of the flesh is flesh because it is born of flesh and that which is born of the spirit is spirit because the spirit is God"; this reading is found neither in all the Greek manuscripts nor in all the versions both ancient and modern.

2) **Philippians 3:3:** "For we are the true circumcision, who worship God in spirit, and glory in Christ Jesus, and put no confidence in the flesh." But Ambrose, in *De Spiritu Sancto,* book 2, has "worship God the holy spirit" for "worship God in the spirit." Augustine admitted both versions could be found but said he favored the latter. Numerous Greek mss. and early Syriac, Ethiopian, and Arabic versions contain the former.

3) **1 John 5:20:** This passage has virtually disappeared from the

modern Bible. But Hilary in *De Trinitate,* and Basil, Cyrill, Ambrose, and others quote scripture saying here that Christ is "the true God."

4) **Luke 19:41:** "And when he [Christ] drew near and saw the city [Jerusalem] he wept over it." Epiphanius reveals, in *Anacorato,* chap. 31, that near the beginning of the Arian controversy the Catholic Church struck this passage from the Bible, fearing it because it shows Christ's human weaknesses, and claiming that it was a corruption. But both Irenaeus and Origen had [in the previous century] commented on this passage.

5) **Luke 22:43–44:** "And there appeared to him an angel from heaven, strengthening him. And being in an agony he prayed more earnestly; and his sweat became like great drops of blood falling down upon the ground." The Catholics also sought to strike this passage from the Bible, for the same reasons given in example 4. Sir Isaac writes: "These words are now found in all Greek & Latin manuscripts & in all versions, but Hilary tells us that in his age, they wanted in very many copies both Greek & Latin, & Jerome that they were only extant in some." Even today, this text appears only as a footnote to the Revised Standard Version.

6) **Matthew 19:16–17:** "And he said to him 'Why do you ask me about what is good? Over there is one who is good. If you would enter life, keep the commandments.'" Sir Isaac notes that many ancient versions contained, "Why askest thou me of a good one? There is one who is good."

7) **Matthew 24:36:** "But of that day and hour no one knows, not even the angels of heaven, nor the Son, but the Father only." Origen, Chysostom, Theophylact, Hilary, Augustine, and Cyrill all testify to the existence of this passage. Mark copied it in toto into his Gospel, and it appears in some ancient Greek and Latin copies and in the Ethiopic versions. But Ambrose (and others) testify to the fact that at the time of the Eusebian controversy

(early fourth century) the Catholics struck out the phrase "nor the son" so as not to bring attention to the difference in knowledge (and therefore in being) of God and Christ.

8) **Ephesians 3:14–15:** "For this cause I bow my knees unto the Father of our Lord Jesus Christ, of whom the whole family in heaven and earth is named." The words, "of our Lord Jesus Christ," were added after.

9) **Ephesians 3:9:** "and to make all men see what is the fellowship of the mystery, which from the beginning of the world hath been hid in God, who created all things." In Sir Isaac's time, the reading generally received was the added, "Who created all things by Jesus Christ." Sir Isaac writes: "But the last words *by Jesus Christ* have been added by the Greeks. For they are still wanting in the oldest Greek MSS the Alexandrin & the Claromontan, gr & lat. That of S. Germ. One of M. Colbert & in the Syriac Latin & Ethiopick Versions: nor did Tertullian nor Jerome nor Ambrose read them."

10) **Apocalypse 1:8:** "'I am the Alpha and the Omega,' says the Lord God, who is and who was and who is to come, the Almighty." To prove the omnipotence of Christ, Ambrose substituted "the Lord Jesus" for "the Lord God."

11) **Corinthians 10:9:** "We must not put the Lord to the test, as some of them did and were destroyed by serpents." The Greek mss. and most of the old versions read, "We must not put Christ to the test." Yet *the Lord* was in "Theodorets MSS & is still conserved in the MS of Lincoln College in Oxford & in one of Dr. Covels MSS. In the Alexandrine MS & Ethiopick Version 'tis, *"Neither let us tempt God."*

12) **Jude 1:5:** "having saved the people out of the land of Egypt, afterward destroyed them that believed not." The Alexandrine mss. and some others, and the Latin and Arabic versions, had, "Jesus having saved the people." Almost all the early manuscripts, and the Syriack and Arabic versions, had, "The Lord having saved the people."

13) **1 John 4:3:** "and every spirit that confesseth not that Jesus Christ is come in the flesh is not of God," was, "Every spirit that separates Jesus is not of God."

14) **John 19:40:** "They took the body of Jesus, and bound it in linen clothes with the spices, as is the burial custom of the Jews." In the Alexandrine ms., the reading is, "Then they took the body of God." (Changes of only one or two characters in the Greek text can effect these changes.)

15) **Acts 13:41:** "Behold, you despisers, and wonder, and perish; for I work a work in your days, a work which you shall in no wise believe, though a man declare it unto you." In a ms. of New College, "because God is crucified" is inserted after "a work in your days."

16) **2 Thessalonians 1:9:** "Who shall be punished with everlasting destruction from the presence of the Lord and the glory of his power." In a ms. in Lincoln College, Oxford, the word "God" has been placed after the word "Lord."

17) **Acts 20:28:** "Take heed therefore unto yourselves, and to all the flock, over which the Holy Spirit hath made you overseers, to feed the Church of God which he hath purchased with his own blood of his own son." Footnotes in the Revised Standard Version: (1) Other ancient authorities read "of the Lord"; (2) Greek: "with the blood of his Own or with his Own blood" (or "one church of Christ").

18) **1 John 3:16:** "By this we know love, that he has laid down his life for us." In the Vulgar Latin Version, someone has inserted the word *Dei* [of God] after *love,* thereby changing the sense to: "Here[by we know] the love of God because he [that is, God] laid down his life for us." All the ancient Greek mss. have the former Sir Isaac: "By this and other instances it appears that the Spanish Divines in their edition of the Bible have corrected the Greek Testament by the Vulgar Latin."

19) **Jude 1:4:** "For admission has been secretly gained by some who

long ago were designated for this condemnation, ungodly persons who pervert the grace of our God into licentiousness and deny our only Master and Lord, Jesus Christ." Newton cites ten important ancient manuscripts leave out "ungodly persons who pervert the grace of our God into licentiousness," thus suggesting to the unwary reader that God and Christ are one. Other important ancient texts make the sense ambiguous by saying, "denying the only God and master even our Lord Jesus Christ." The Ethiopic version has, "denying the only God Jesus Christ."

20) **Philippians 4:13:** "I can do all things through him who strengtheneth me." "Him" means "God." This reading is found in most of the important texts, but some have added one word in Greek that changes the sense to, "through Christ who strengtheneth me." This seems to conflate Christ and God.

21) **Romans 15:32:** "So that by God's will I may come to you with joy and be refreshed in your company." Sir Isaac says: "Some have changed the will of God into the will of Christ Jesus."

22) **Colossians 3:15:** "And let the peace of God rule your hearts, to the which also you were called in one body; and be thankful." Newton remarks, "Some have changed the peace of God into the peace of Christ."

23) **Apocalypse 1:11:** "Saying, I am Alpha and Omega, the first and the last." Newton says in "Apoc. 1:11, the words of the son of man, *I am Alpha & Omega the first & the last* have crept erroneously into some few Greek MSS. . . . God is called ye first & ye last to signify not his eternity but that it is he who sits upon the throne in the beginning & end of the prophecy: which some not understanding have applied here to Christ to prove his Eternity."

24) **2 Peter 3:18:** "But grow in grace and in the knowledge of our Lord and Savior Jesus Christ, and of God the Father. To him be glory both now and for ever. Amen." Sir Isaac says: "[Some] MSS & versions have left out the words *& of God the father*, that the Doxology may refer to Christ."

25) **Romans 9:5:** "Whose are the fathers, and of whom as concerning the flesh Christ came, who is over all, God blessed for ever. Amen." The ancient "Syriack version" had "Who is God over all to whom be praises and blessings forever. Amen." Newton thought the Syriack originally had "to him who is God over all be praises."

26) **Hebrews 2:9:** "But we see Jesus, who was made a little lower than the angels for the suffering of death, crowned with glory and honour, that he by the grace of God should taste death for every man." Newton says that "[The corrupted Syriac] version now hath *For God himself by his Grace tasted death for all men* corruptly for *That He by the Grace of God should taste death for all men.*"

27) **Philippians 2:6:** "Who, being in the form of God, thought it not robbery to be equal with God." Newton says that this passage "was anciently understood in another sense . . . which makes me suspect some of the other versions have been tampered with."

APPENDIX C

Newton's Twenty-Three Queries Concerning the Word ὁμοούσιος

Query 1. Whether Christ sent his apostles to preach metaphysics to the unlearned common people, and to their wives and children.

Query 2. Whether the word ὁμοούσιος ever was in any creed before the Nicene; or any creed was produced by any one bishop at the Council of Nice [Nicaea] for authorizing the use of that word.

Query 3. Whether the introducing the use of that word is not contrary to the Apostles' rule of holding fast the form of sound words.

Query 4. Whether the use of that word was not pressed upon the Council of Nice against the inclination of the major part of the Council.

Query 5. Whether it was not pressed upon them by the emperor Constantine the Great, a catechumen not yet baptized, and no member of the Council.

Query 6. Whether it was not agreed by the Council that that word should, when applied to the Word of God, signify nothing more than that Christ was the express image of the Father, and whether many of the bishops, in pursuance of that interpretation of the word allowed by the Council, did not, in their subscriptions, by way of caution, add τουτ' ἐστιν ὁμοιούσιος?

Query 7. Whether Hosius (or whoever translated that Creed into Latin) did not impose upon the Western Churches by translating ὁμοούσιος by the words *unius substantiæ,* instead of *consubstantialis,* and whether by that translation the Latin Churches were not drawn into an opinion that the Father and Son had one common substance, called by the Greeks Hypostasis, and whether they did not thereby give occasion to the Eastern Churches to cry out, presently after the Council of Sardica, that the Western Churches were become Sabellian.

Query 8. Whether the Greeks, in opposition to this notion and language, did not use the language of three hypostases, and whether in those days the word *hypostasis* did not signify a substance.

Query 9. Whether the Latins did not at that time accuse all those of Arianism who used the language of three hypostases, and thereby charge Arianism upon the Council of Nice, without knowing the true meaning of the Nicene Creed.

Query 10. Whether the Latins were not convinced, in the Council of Ariminum, that the Council of Nice, by the word ὁμοούσιος, understood nothing more than that the son was the express image of the father;—the acts of the Council of Nice were not produced for convincing them. And whether, upon producing the acts of that Council for proving this, the Macedonians, and some others, did not accuse the bishops of hypocrisy, who, in subscribing these acts, had interpreted them by the word ὁμοιούσιος in their subscriptions.

Query 11. Whether Athanasius, Hilary, and in general the Greeks and Latins, did not, from the time of the reign of Julian the Apostate, acknowledge the Father, Son, and Holy Ghost to be three substances, and continue to do so till the school men changed the signification of the word *hypostasis,* and brought in the notion of three persons in one single substance.

Query 12. Whether the opinion of the equality of the three substances was not first set on foot in the reign of Julian the Apostate, by Athanasius, Hilary, etc.

Query 13. Whether the worship of the Holy Ghost was not first set on foot presently after the Council of Sardica?

Query 14. Whether the Council of Sardica was not the first Council which declared for the doctrine of the consubstantial Trinity and whether the Council did not affirm that there was but one hypostasis of the Father, Son, and Holy Ghost.

Query 15. Whether the Bishop of Rome, five years after the death of Constantine the Great, AD 341, did not receive appeals from the Greek Councils, and thereby begin to usurp the universal bishopric.

Query 16. Whether the Bishop of Rome, in absolving the appellants from excommunication, and communicating with them, did not excommunicate himself, and begin a quarrel with the Greek Church.

Query 17. Whether the Bishop of Rome, in summoning all the bishops of the Greek Church to appear at the next Council of Rome, AD 342, did not challenge dominion over them, and begin to make war upon them for obtaining it.

Query 18. Whether that Council of Rome, in receiving the appellants into communion, did not excommunicate themselves, and support the Bishop of Rome in claiming appeals from all the world.

Query 19. Whether the Council of Sardica, in receiving the appellants into communion, and decreeing appeals from all the churches to the Bishop of Rome, did not excommunicate themselves, and become guilty of the schism which followed thereupon, and set up Popery in all the West.

Query 20. Whether the emperor Constantius did not, by calling the Council of Millain and Aquileia, AD 365, abolish Popery and whether Hilary, Lucifer, were not banished for adhering to the authority of the Pope to receive appeals from the Greek Councils.

Query 21. Whether the emperor Gratian, AD 379, did not, by his edict, restore the universal bishopric of Rome over all the west and whether this authority of the Bishop of Rome hath not continued ever since.

Query 22. Whether Hosius, Saint Athanasius, Saint Hilary, Saint Ambrose, Saint Hierome, Saint Austin were not Papists.

Query 23. Whether the Western Bishops upon being convinced that the Council of Nice by the word ὁμοούσιος did.

APPENDIX D

Newton on God's Anti-Trinitarian Introduction to John's Book of Revelation

Note: From Newton's "Untitled Treatise on Revelation" (section 1.4), Yahuda MS. 1.4, National Library of Israel, Jerusalem, Israel. Newton Project, www.newtonproject.sussex.ac.uk/view/texts/normalized/THEM00182. Text begins under the heading 156r, "Of the Lambs taking the Book."

Having thus framed a conception of the heavenly court, you are in the next place to consider what was done there by way of preparation to these visions: & this was the Lamb's taking a sealed book out of the hand of him that sat on the throne, which none in heaven or earth or under the earth besides the Lamb was found worthy to open & read; & upon his doing this he is celebrated for his worthiness, first by a double commemoration of it as if he became worthy by the merits of his death, & then by a doxology hereupon given him together with him that sat on the Throne, & this is followed with a higher degree of worship given to him only which sate upon the throne.

The Book you may conceive rolled up & sealed in such a manner that the opening of every seal may undo some of the leaves so that more &

more of the book may be opened by steps till the whole be open. And the contents of it you must conceive of so transcendent excellency that they were fit to be communicated to none but the Lamb. You are not to imagine that this is the book of the Apocalypse written by Saint John, but rather a book representing that plenary revelation which the great God imparted to our Saviour after his resurrection & to none but him. For first it was a book written within & on the backside, that is a book containing the knowledge of things past as well as to come whereas the Apocalypse contained only things to come, Apoc. 1.1, 3 & 22.6, 10; & accordingly the visions thereof were represented concomitant to the opening of the seals for the Lamb to look on the inside after he had viewed the backside, as you may conceive. Secondly there is nothing in the Apocalypse which can be pretended to be a transcript of this book: for there is nothing set down there but certain visions which Saint John saw concomitant to the opening of the Seals, & those too such as by the motions of some & voices of others, & also by Saint John's being called by the four beasts to go from place to place to see them, were manifestly no flat pictures in the book, but appearances to the life, such as (like those made to Daniel in his visions) had the full proportions, dimensions & gestures of the things they were a shew of, as if they had been those real things. And Thirdly it is expressly said, not only that none but the Lamb was worthy to take & open the book, but also that none but he were worthy to read it or to look thereon: & if so, then Saint John himself was not worthy to read it, & much less was the world worthy to whom the full contents of this book should be made public to be read by all, bad as well as good. I might add also, that the great emphasis laid upon this book, first by the solemn declaration with a loud voice that none in heaven nor in earth nor under the earth were found worthy to open it, so as to make Saint John weep thereat; & then by the following celebrations of the Lamb for his worthiness to receive wisdom & all other perfections, after he appeared to take & open it: is an intimation plain enough that this book signifies one of the greatest treasures that he who sat upon the Throne ever conferred upon the Lamb, & consequently nothing less then all that fullness of knowledge of things past & to come which God gave him after his resurrection. This is certain, that it signifies such knowledge as the Lamb had not received before the Apocalypse it self being a new revelation to him Apoc. 1.1: & why it should signify less then all that knowledge communicated to him at that time when he received this knowledge, I see no reason. But further this book . . . This book is an emblem of the Revelation as it was made by God to the Lamb, & it cannot be thought that God would give him a revelation in obscure types & figures such as the Apocalypse consists of. The Apocalypse is called indeed the revelation of Jesus Christ which God gave him: but it is not to be supposed that it is all the revelation which God gave him, or that God gave it him in those obscure terms in which we have it, but rather that God gave him a full & clear revelation, & that he gave us only so much of that revelation as was fit for us to have, & that too wrapped up in obscurity. Wherefore since the sealed book signifies the revelation as it was given by God to the Lamb, & not as it was given by the Lamb to us (for God gave it to the Lamb but the Lamb gave it not to Saint John) it must signify a full & perspicuous revelation, such an one as eminently contains the revelation made to us. And therefore you are to conceive that the Lamb opened the book for his own perusal only & that the concomitant visions which appeared to Saint John were but general & dark emblems of what was particularly & perspicuously revealed to the Lamb in this book.

But then you will say: Why were we told of this book if it contained a revelation for the Lamb only, & not for us? To which I answer that it was done in prosecution of the main design of the Apocalypse, which was to describe & obviate the great Apostasy. That Apostasy was to begin by corrupting the truth about the relation of the Son to the Father in putting them equal, & therefore God began this prophesy with a demonstration of the true relation: shewing the Son's subordination, & that by an essential character, his having the knowledge of futurities only so far as the father communicates it to him. And least you should think he had this knowledge given him from all eternity, the book was represented in the hand of God alone sealed at first. Yea it was represented sealed in his hand when there were beings in heaven & earth which could not open it, that is after the creation of the world: & consequently was not given to the Lamb at his first generation but since his resurrection; he meriting it by his obedience to death: which you need not wonder at if you consider his declaration which he made before his death concerning the day of judgment[:] *Of that day & hour knoweth no man, no, not the angels which are in heaven, neither the Son, but the Father* Mark 13.32—*but the Father only,* Matth. 24.36. Yea here, upon the Lamb's taking the book he is celebrated by this song of the Saints. *Thou art worthy to take the book & to open the seals thereof: for thou wast slain & hast redeemed us to God by thy blood,* which is as much as to say that he merited this dignity by his death to take & read the book & consequently that the book continued sealed up in the hand of the father till after his resurrection. ~ ~ ~ ~ ~ ~ ~ And to inculcate this further there immediately follows another song of angels & saints together, saying: *worthy is the Lamb that was slain to receive power & riches & wisdom & strength & honor & glory & blessing.* Which is as much as to say, the Lamb which was slain became worthy thereby to receive at the hand of the father, not only the wisdom of this book but power & honor & other perfections. And to this purpose speak other places of scripture also. To the Son he saith, Thy throne, O God, is for ever—*thou hast loved righteousness & hated iniquity, therefore, O God, thy God hath anointed thee with the oyle of gladness above thy fellows.* Heb. 1.8, 9. We see Jesus who was made a little lower then the Angels, *for the suffering of death crowned with glory & honor:* for it became him for whom are all things & by whom are all things, in bringing many sons unto glory to make the captain of their salvation *perfect through sufferings.* Heb. 2.9. Let us run with patience the race that is set before us; looking unto Jesus the Author & finisher of our faith; *who for the joy that was set before him endured the cross,* despising the shame, & is set down at the right hand of the throne of God. Heb. 12.2. He emptied himself & took on him the form of a servant & was made in the likeness of men, & being found in fashion as a man he humbled himself & became obedient to death even the death of the cross; wherefore God hath highly exalted him, & given him a name which is above every name, that at the name of Jesus every knee should bow of things in heaven & things in earth & things under the earth, & that every tongue should confess that Jesus Christ is Lord to the glory of God the Father Philip 2.8, 9. See also Acts 2.36. Rom 8.17. Heb 5.8, 9. Apoc. 3.21.

But further because the Apostates were to deceive themselves with a sophistical distinction, saying that these things were spoken of the Lamb in respect of his humane nature & not as he was God: there is also care taken in this vision to obviate this distinction & that by a threefold insinuation: first by representing the book in his hand . . . first by representing the book in his hand alone that sate upon the throne, with a solemn declaration added that there was none in heaven or earth or under the earth besides the Lamb worthy to take & open it or to look thereon. Here was an universal assembly of all beings from the great God that sate upon the throne down to the lowest of the creatures, & in this assembly the two supreme (who were therefore afterward worshipped together by all the rest,) were God that gave the book & the Lamb that received it; & the Lamb till he received it was as absolutely represented without it as any of the other beings: whereas if the λογὸς had known it before, the Lamb was as much a possessor of the . . . book from the beginning as he was that sate upon the Throne, & ought to have been represented so, & not to have received it from another, but only the humane nature to have received it from the divine. For to what purpose was the humane soul hypostatically united to the Λογὸς if the Λογὸς communicated not with it but left it to receive knowledge from another hand? Or how could the Lamb as he was the Lamb (which is as much as to say, the Λογὸς incarnate) be represented at first without this book & afterward receiving it, if the Λογὸς had it from the beginning? And also since the communication of the wisdom of this book is called unsealing it, & consequently it's being sealed in his hand only who sate upon the Throne must denote its being shut up in his breast till then uncommunicated, how could it be properly represented sealed in his hand if he had communicated it to the Son or any other person before? We must therefore, unless we will do violence to the vision, affirm that this was the first communication of this book by the Father, & that he communicated it to the Lamb absolutely & properly so-called without any ambiguity, that is, to the Λογὸς incarnate. And indeed what else can we affirm if we consider our Saviour's own confession which he made concerning the day of judgment which he made before this book was given him. *Of that day & hour knoweth none, no, not the Angels in heaven nor the Son, but the father.* He first asserts in general, *none but the father;* & then to take away all suspicion of further exception, he instances in the chief of those none, the Angels & the Son.

Secondly the said distinction is obviated by making the Lamb, as he was worthy to take & open the book. &c. by making the Lamb, as he was worthy to take & open the book to be the object of worship[;] for he is here worshiped both alone & together with him that sate upon the throne: the first by the four beasts & 24 Elders, falling down before him & singing a new song saying thou art worthy to take the Book & to open the seals thereof for thou was slain & hast redeemed us to God by thy blood out of every kindred & tongue & people & nation: the last by the whole creation saying, Blessing & honor & glory & power to him that sitteth upon the throne & unto the Lamb for ever & ever. Now this worship was given to the Lamb as he was a God without all doubt, Divinity & worship being relative terms, & yet it was given to him as he was worthy to take & open the Book for at the falling down of the four Beasts & 24 Elders before him to worship him, the very act of their worship was to celebrate him for his worthiness to take & open the book. The Lamb therefore as he was a God was worshiped for his worthiness to take & open the book & therefore took & opened the Book as he was the object of worship, that is a God. But to make all this plainer you may compare it with Philip: 2.9 where tis expressly said, that for his obedience to death God gave him a name above every name that at the name of Jesus every knee should bow &c. that is that all the creation should worship him which is as much as to say that he should be ισα θεω as a God over the creation: for Deity & worship are relative terms & infer one another.

Thirdly the said distinction whereby Christ is made equal to the father as he is God though inferior as man, is obviated by the difference put between God & the Lamb in their worship & that in a double respect: first in that the Lamb, while he was celebrated together with the great God in a Doxology by the whole creation, did not sit upon a Throne as God did but only stood by the Throne which God sate upon; for what else is meant by his sitting upon the Throne but to signify that he was king over all that did not sit upon the throne & consequently over the Lamb too who as a God was worshipped together with him: And secondly in that after the Doxology given to God & the Lamb together, there followed a higher degree of worship given to God alone without the Lamb. I call it a higher degree of worship, for so I gather it to be first by the falling down of the worshippers which it is not said they did in saying the doxology: secondly by the gradation in worship which began with a celebration peculiar to the Lamb, & then proceeded higher to a Doxology common to the Lamb & God, & ended in the worship of God alone: Thirdly by calling this last absolutely worship, as if it were an act distinct from those that went before, to which the name of worship specially belongs. But further to make the stronger impression of this difference of worship, you may see it repeated in the seventh chapter, where there is also first a doxology given to God & the Lamb together by a multitude standing, & then the Angels fall down before the Throne & worship only God. Now why the Lamb should not be joined with God in this supreme worship as well as in the precedent Doxology, I think there can be no reason given but this, that it was a worship peculiar to God: for otherwise he could not have been omitted seeing he as well as God was in the middle before the worshippers & the design of the vision here was to celebrate him As the solemnity began with the celebration of the Lamb for his worthiness to receive the book & other blessings at the hand of God, which was too low a worship for God & therefore given to the Lamb alone; & then proceeded to a Doxology which agreed both to God & the Lamb & therefore was given to both together: so ascending still higher to the supreme worship, worship properly & absolutely so called, it argues that this agreed to none but God because given only to him; & consequently the Lamb must be a God inferior to the great God that sate upon the throne. [F]or a close I might produce the whole strain of scripture to confirm this, but doing that in another place I shall content my self here with the first Chapter to the Hebrews: where you may see the son all along described by things agreeable only to the Λογὸς, as the worlds being made by him, his upholding all things, his being worshipped by the Angels, his being called God, & his founding the earth & making the heavens: & yet in the middest of this career, even in the same sentence where he is once at least if not twice called God, he is said to have a God above him & to be anointed by this his God with the oyle of gladness above his fellows, & that because he loved righteousness & hated iniquity.

You have now had a view of the Preamble to the Prophetic visions, & by what has been said, I hope you conceive this is no insignificant ceremony but a very weighty passage, a system of the Christian religion, showing the relation of the father & Son, & how they are to be worshipped in a general Assembly of the Church & of the whole creation. The father the supreme King upon the Throne, the fountain of prescience & of all perfections. The Lamb the next in dignity, the only being worthy to receive full communications at the hand of the father. No Holy Ghost, no Angels, no Saints worshipped here: none worshipped but God & the Lamb, & these worshipped by all the rest. None but God upon the Throne worshipped with the supreme worship; none with any other degree of worship but the Lamb; & he worshipped not on the account of what he had by nature, but as he was slain, as he became thereby worthy to be exalted & endowed with perfections by the father. This was the religion to be corrupted by the Apostasy[.] This therefore was very pertinently shadowed out in the exordium to the Prophesy of that Apostasy. Which having explained, I proceed now to consider the Prophesy it self, & first the four Horsemen which appeared at the opening of the first four seals.

APPENDIX E

NEWTON ON ANCIENT SCIENCE

Newton on the Scientific Achievements of the Philosophers of the Ancient World, published in *De Mundi Systemate* (*The System of the World*), 1728.

It was the ancient opinion of not a few, in the earliest ages of philosophy, that the fixed stars stood immovable in the highest parts of the world; that under the fixed stars the planets were carried about the sun; that the earth, as one of the planets, described an annual course about the sun, while by a diurnal motion it was in the meantime revolved about its own axis; and that the sun, as the common fire which served to warm the whole, was fixed in the center of the universe.

This was the philosophy taught of old by Philolaus, Aristarchus of Samos, Plato in his riper years, and the whole sect of the Pythagoreans. And this was the judgment of Anaximander, more ancient than any of them, and of that wise king of the Romans, Numa Pompilius, who as a symbol of the figure of the world with the sun in the center, erected a temple in honor of Vesta, of a round form, and ordained perpetual fire to be kept in the middle of it.

The Egyptians were early observers of the heavens. And from them probably this philosophy was spread abroad among other nations. For from them it was, and the nations about them, that the Greeks, a people of themselves more addicted to the study of philology than of nature, derived their first as well as soundest notions of philosophy. And in the vestal ceremonies we may yet trace the ancient spirit of the Egyptians. For it was their way to deliver their mysteries, that is their philosophy of things above the vulgar way of thinking, under the veil of religious rites and hieroglyphic symbols.

It is not to be denied that Anaxagoras, Democritus, and others did now and then start up, who would have it that the earth possessed the center of the world, and that the stars of all sorts were revolved towards the west, about the earth quiescent in the center, some at a swifter, others at a slower rate.

However, it was agreed on both sides that the motions of the celestial bodies were performed in spaces altogether free, and void of resistance. The whim of solid orbs was of a later date, introduced by Eudoxus, Calippus, and Aristotle; when the ancient philosophy began to decline, and to give place to the new prevailing fictions of the Greeks.

But, above all things, the phenomena of comets can by no means consist with the notion of solid orbs. The Chaldeans, the most learned astronomers of their time, looked upon the comets (which of ancient times before had been numbered among the celestial bodies) as a particular sort of planets, which describing very eccentric orbits, presented themselves to our view only by turns, viz. once in a revolution when they descended into the lower parts of their orbits.

And as it was the unavoidable consequence of the hypothesis of solid orbs, while it prevailed, that the comets should be thrust down below the moon; so no sooner had the late observations of astronomers restored the comets to their ancient places in the higher heavens, but these celestial spaces were at once cleared of the encumbrance of solid orbs, which by these observations were broke into pieces and discarded for ever.

APPENDIX F

NEWTON'S TRANSLATION OF THE EMERALD TABLET

Note: Translation by Isaac Newton, circa 1690 (see B. J. Dobbs, "Newton's Commentary on the Emerald Tablet of Hermes Trismegistus," 183–84; or view online at www.sacred-texts.com).

1) Tis true without lying, certain & most true.
2) That wch is below is like that wch is above & that wch is above is like yt wch is below to do ye miracles of one only thing.
3) And as all things have been & arose from one by ye mediation of one: so all things have their birth from this one thing by adaptation.
4) The Sun is its father, the moon its mother,
5) the wind hath carried it in its belly, the earth its nourse.
6) The father of all perfection in ye whole world is here.
7) Its force or power is entire if it be converted into earth.
7a) Seperate thou ye earth from ye fire, ye subtile from the gross sweetly wth great indoustry.
8) It ascends from ye earth to ye heaven & again it desends to ye earth and receives ye force of things superior & inferior.

9) By this means you shall have ye glory of ye whole world & thereby all obscurity shall fly from you.
10) Its force is above all force. ffor it vanquishes every subtile thing & penetrates every solid thing.
11a) So was ye world created.
12) From this are & do come admirable adaptaions whereof ye means (Or process) is here in this.
13) Hence I am called Hermes Trismegist, having the three parts of ye philosophy of ye whole world.
14) That wch I have said of ye operation of ye Sun is accomplished & ended.

NOTES

CHAPTER ONE.
"A HISTORY OF THE CORRUPTION OF THE SOUL OF MAN"

1. Westfall, *Never at Rest*, 473.
2. Gjertsen, "Newton's Success," 29.
3. Rattansi, "Newton and the Wisdom of the Ancients," 187.
4. Ferris, "Measuring an Intellect."
5. See www.newtonproject.ox.ac.uk.
6. Snobelen, "Isaac Newton (1642–1727): Natural Philosopher," 6.
7. Force and Popkin, *Newton and Religion*, x.
8. Manual, *Religion*, 17.
9. Gleick, *Isaac Newton*, 90.
10. Christianson, *Isaac Newton and the Scientific Revolution*, 30.
11. Ferris, "Measuring an Intellect."
12. Stukeley, "Memoir of Newton."
13. Stukeley, "Memoir of Newton."
14. Stukeley, "Memoir of Newton."
15. Snobelen, "Isaac Newton . . . Natural Philosopher," 1.
16. Snobelen, "Isaac Newton . . . Natural Philosopher," 5.
17. Dolnick, *The Clockwork Universe*, 289.
18. Popkin, "Plans for Publishing," 2.
19. Rattansi, "Newton and the Wisdom of the Ancients," 200.
20. Qtd. in Dry, *The Newton Papers*, 158.
21. Qtd. in Dry, *The Newton Papers*, 160, 162.

CHAPTER TWO. THE NEWTON CODE

1. Snobelen, "'A Time and Times,'" 539–40.
2. Snobelen, "'A Time and Times,'" 541.
3. National Oceanic and Atmospheric Administration (NOAA), "For 11th straight month, the globe was record warm." April 19, 2016, www.noaa.gov/11th-straight-month-globe-was-record-warm.
4. Klein, *This Changes Everything*, 31.
5. Newton, *Observations upon the Prophecies of St. John*, chap. 1, "Introduction, Concerning the Time When the Apocalypse Was Written."
6. Bournis, "St. John on Patmos," 42–44.
7. Ebon, "In the Grotto of St. John," 3–4.
8. Durant, *Caesar and Christ*, 594.
9. Newton, *Observations upon the Prophecies of St. John*, chap. 1.
10. Pagels, *Revelations*, 163.
11. Pagels, *Revelations*, 11–13.
12. Pagels, *Revelations*, 1.
13. Pagels, *Revelations*, 2.
14. Grosso, *The Millennium Myth*, 23.
15. Grosso, *The Millennium Myth*, 23.
16. Jung, "The Dark Side of God," 89–100.
17. Grosso, *The Millennium Myth*, 19.
18. Ingermanson, *Who Wrote the Bible Code?* 26.
19. Clouse, "The Apocalyptic Interpretation of Thomas Brightman and Joseph Mede," 181.
20. Clouse, "The Apocalyptic Interpretation of Thomas Brightman and Joseph Mede," 183.
21. Buchwald and Feingold, *Newton and the Origin of Civilization*, 132.
22. Manuel, *Religion of Isaac Newton*, 224.
23. Westfall, *Never at Rest*, 349.
24. Manuel, *Isaac Newton, Historian*, 141.
25. Force, *Whiston*, 114.
26. Qtd. in Goff, 2.
27. Apollonius, *The Voyage of Argo*, 82 (*Argonautica*, 2:314–16).
28. Qtd. in Goff, "The Millennial Scientist," 2.
29. Newton, *Observations upon the Prophecies of Daniel*, www.newtonproject.sussex.ac.uk/view/texts/normalized/THEM00209.

30. Westfall, *Never at Rest*, 327–28.
31. Murrin, "Newton's Apocalypse," 210.
32. Newton, "Two Incomplete Treatises on Prophecy," www.newtonproject.sussex.ac.uk/view/texts/normalized/THEM00005.
33. BBC, *A History of the World,* www.bbc.co.uk/ahistoryoftheworld/objects/GOe8Mt6vRdSNcg-yeivrEA.
34. Qtd. in Ben Uzziel, *The Chaldee Paraphrase,* vi.
35. Newton, "Untitled Treatise on Revelation," section 1.1a, www.newtonproject.ox.ac.uk/view/texts/normalized/THEM00136.
36. Newton, "Extract from the Untitled Treatise on Revelation," section 1.1. www.newtonproject.sussex.ac.uk/view/texts/normalized/THEM00135.
37. "Rules for Methodizing the Apocalypse," Rule 9, qtd. in Manuel, *Religion of Isaac Newton,* 49.
38. Newton, "Two Incomplete Treatises on Prophecy," bk. 1, chap. 1, www.newtonproject.sussex.ac.uk/view/texts/normalized/THEM00005.
39. Newton, "An Account of the Empires of the Babylonians, Medes, Persians, Greeks, and Romans, According to the Descriptions Given of Them by Daniel," in "Four Draft Chapters on Prophecy," www.newtonproject.sussex.ac.uk/view/texts/normalized/THEM00367.
40. Manuel, *Religion of Isaac Newton,* 95.
41. Newton, "Words for Interpreting the Rules and Language in Scripture," and "The Proof," in "Untitled Treatise on Revelation" section 1.1, www.newtonproject.ox.ac.uk/view/texts/normalized/THEM00135.
42. Newton, "The Proof," in "Untitled Treatise on Revelation," www.newtonproject.ox.ac.uk/view/texts/normalized/THEM00135.
43. Manuel, *Religion of Isaac Newton,* 87.
44. Flaubert, *Temptation of Saint Anthony,* 27.
45. Frye, *The Great Code,* 176–77.
46. Newton, "Of the Vision of the Four Beasts," chap. 4 in *Observations upon the Prophecies of Daniel,* www.newtonproject.sussex.ac.uk/view/texts/normalized/THEM00198.
47. Smoley, "2012 and the Annoying Persistence of Time."
48. Frye, *The Great Code,* 223.
49. Goff, "The Millennial Scientist."
50. Newton, *Observations,* chap. 1, www.newtonproject.sussex.ac.uk/view/texts/normalized/THEM00209.

51. Westfall, *Never at Rest*, 328–29.
52. Westfall, *Never at Rest*, 329.

CHAPTER THREE. NEWTON'S GOD

1. Ellis, *Jesus, King of Edessa*.
2. Haycock, *William Stukeley*, 196–97; see also MacCulloch, *Christianity*, 748.
3. Haycock, *William Stukeley*, 196–97.
4. Snobelen, "Isaac Newton, Heretic," 393.
5. Westfall, *Never at Rest*, 235.
6. Westfall, *Never at Rest*, 313.
7. Champion, "'Acceptable to Inquisitive Men,'" 82.
8. Newton, *Two Notable Corruptions of Scripture*, part 1: ff. 1–41.
9. "Newton to a Friend," in *Correspondence*, vol. 3, 87–88; also *Two Notable Corruptions of Scripture*, part 1: ff. 1–41.
10. Snobelen, "Isaac Newton, Heretic," 405.
11. Erasmus, *The Praise of Folly*, 205.
12. Erasmus, *The Praise of Folly*, 83.
13. Erasmus, *Responsio ad Annotationes Eduardi Lei* of May 1520, www.e-rara.ch/bau_1/content/structure/1096927.
14. Voltaire, *Philosophical Dictionary*, "Arianism," http://ebooks.adelaide.edu.au/v/voltaire/dictionary/chapter46.html.
15. Qtd. in Westfall, *Never at Rest*, 534.
16. Newton, *Correspondence*, 3:129, www.newtonproject.ox.ac.uk/view/texts/diplomatic/THEM00263.
17. Küng, *Christianity*, 95.
18. Küng, *Christianity*, 95.
19. Ehrman, *The Orthodox Corruption*, 275.
20. Qtd. in Force, *Whiston*, 138.
21. Qtd. in Manuel, *Portrait of Isaac Newton*, 124.
22. Newton, "General Scholium from the *Mathematical Principles of Natural Philosophy*," www.newtonproject.sussex.ac.uk/view/texts/normalized/NATP00056.
23. Newton, "General Scholium from the *Mathematical Principles of Natural Philosophy*," www.newtonproject.sussex.ac.uk/view/texts/normalized/NATP00056.

24. Newton, "General Scholium from the *Mathematical Principles of Natural Philosophy*," www.newtonproject.sussex.ac.uk/view/texts/normalized/NATP00056.
25. Iliffe, "Prosecuting Athanasius."
26. Dobbs, *The Janus Faces*, 83.
27. Castillejo, *Expanding Force in Newton's Cosmos*, 60.
28. Newton, "Irenicum."
29. MacCulloch, *Christianity*, 782.
30. MacCulloch, *Christianity*, 258.

CHAPTER FOUR.
BLOODBATH IN A BOGHOUSE:
MURDER IN THE FOURTH CENTURY AD, PART I

1. Qtd. in Manuel, *Isaac Newton, Historian*, 5.
2. Westfall, *Never at Rest*, 344.
3. Westfall, *Never at Rest*, 312.
4. Newton, "Paradoxical Questions," question 18.
5. Anatolios, *Athanasius*, 3.
6. MacCulloch, *Christianity*, 216.
7. MacCulloch, *Christianity*, 216.
8. Gibbon, *The Decline and Fall of the Roman Empire*, 1:612.
9. Johnson, *Christianity*, 87.
10. MacCulloch, *Christianity*, 213.
11. Qtd. in Iliffe, "Prosecuting Athanasius," 129.
12. Eusebius, *The Life of the Blessed Emperor Constantine*, 23.
13. MacCulloch, *Christianity*, 191.
14. Becker, *Eagle in Flight*, 32–33.
15. Voltaire, *Philosophical Dictionary*, "Arianism," http://ebooks.adelaide.edu.au/v/voltaire/dictionary/chapter46.html; see also Eusebius, *Life of the Blessed Emperor Constantine*, 89; and Becker, *Eagle in Flight*, 32–33.
16. Gibbon, *The Decline and Fall of the Roman Empire*, 1:588.
17. Eusebius, *Life of the Blessed Emperor Constantine*, 110.
18. Johnson, *Christianity*, 68.
19. Gibbon, *The Decline and Fall of the Roman Empire*, 1:588.
20. Levitt, "The Model for Messianic Community."
21. Newton, "Twenty-Three Queries," query 1.

22. Newton, "Irenicum."
23. Newton, "Twenty-Three Queries," query 2.
24. Newton, "Two Incomplete Treatises on Prophecy," no. 8, first book, www.newtonproject.sussex.ac.uk/view/texts/normalized/THEM00005.
25. Manuel, *Isaac Netwon, Historian,* 158.
26. Koloski-Ostrow, "Raising a Really Big Stink," 43.
27. Newton, "Paradoxical Questions."
28. Newton, "Paradoxical Questions."
29. Gibbon, *The Decline and Fall of the Roman Empire,* 1:608, 624–25.
30. MacCulloch, *Christianity,* 428.
31. Gibbon, *The Decline and Fall of the Roman Empire,* 1:395.

CHAPTER FIVE.
THE SEVERED HAND:
MURDER IN THE FOURTH CENTURY AD, PART 2

1. Newton, "Untitled Treatise on Revelation," section 1.4.
2. Gibbon, *The Decline and Fall of the Roman Empire,* 1:450, 451.
3. Epiphanius, *Panarion,* 317.
4. Epiphanius, *Panarion,* 317.
5. Newton, "Paradoxical Questions," question 2.
6. Epiphanius, *Panarion,* 317.
7. Gibbon, *The Decline and Fall of the Roman Empire,* 1:450.
8. Küng, *Catholic Church,* 35.
9. Newton, "Paradoxical Questions," question 12.
10. Newton, "Paradoxical Questions," question 12.
11. Voltaire, *Philosophical Dictionary,* "Alexandria."
12. Dio Chrysostom, qtd. in Durant, *Life of Greece.*
13. Voltaire, *Philosophical Dictionary,* "Alexandria."
14. Newton, "Paradoxical Questions," question 12.
15. Anatolios, *Athanasius,* 227.
16. Anatolios, *Retrieving Nicaea,* 20.
17. Newton, "Paradoxical Questions," question 2.
18. Qtd. in MacCulloch, *Christianity,* 218.
19. Qtd. in MacCulloch, *Christianity,* 222.
20. Johnson, *Christianity,* 54.
21. Tichenor, *Creed of Constantine,* 89.

22. Cochrane, *Christianity and Classical Culture*, 198.
23. Newton, "Paradoxical Questions," question 3.
24. Newton, "Paradoxical Questions," question 13.
25. Theodoret, *Ecclesiastical History*, chap. 28, www.sacred-texts.com/chr/ecf/203/2030049.htm.
26. "The Canons of the Council of Nicea," canon 1, www.christian-history.org/council-of-nicea-canons.html.
27. Theodoret, *Ecclesiastical History*, chap. 28.
28. Newton, "Paradoxical Questions," question 10.
29. From Cave, *Ecclesiastici*, qtd. in Iliffe, "Prosecuting Athanasius," 134.
30. Sozomen, *Ecclesiastical History*, chap. 26, http://biblehub.com/library/sozomen/the_ecclesiastical_history_of_sozomenus/chapter_xxvi_erection_of_a_temple.
31. Sozomen, *Ecclesiastical History*, chap. 26, http://biblehub.com/library/sozomen/the_ecclesiastical_history_of_sozomenus/chapter_xxvi_erection_of_a_temple.
32. Qtd. in Iliffe, "Prosecuting Athanasius," 139.
33. Durant, *Caesar and Christ*, 498.
34. Johnson, *Christianity*, 94–95.
35. Newton, "Paradoxical Questions," question 7.
36. Iliffe, "Prosecuting Athanasius," 144.
37. Newton, "Paradoxical Questions," question 5.
38. Newton, "Paradoxical Questions," question 7.
39. Vidal, *Julian*, 46, 52, 178.
40. Gibbon, *The Decline and Fall of the Roman Empire*, 1:611.
41. Newton, "Paradoxical Questions," question 15.
42. Newton, "Paradoxical Questions," question 14.
43. Newton, "Paradoxical Questions," question 14.
44. Newton, "Paradoxical Questions," question 14.
45. Gibbon, *The Decline and Fall of the Roman Empire*, 1:623.
46. Gibbon, *The Decline and Fall of the Roman Empire*, 1:696.

CHAPTER SIX.
THE TEMPTATION OF SAINT ANTHONY

1. Newton, "Paradoxical Questions," question 18.
2. E. Bisland, introduction to Flaubert, *The Temptation of Saint Anthony*, 4.

3. E. Bisland, foreword by Marshall C. Olds, *The Temptation of Saint Anthony*, xx.
4. Newton, "Paradoxical Questions," question 18.
5. Westfall, *Never at Rest*, 502.
6. Qtd. in Westfall, *Never at Rest*, 502, from the *Diary of Abraham de la Pryme*, ed. Charles Jackson (Durham, 1870), 42.
7. Westfall, *Never at Rest*, 345.
8. Newton, "Paradoxical Questions," question 19. *Observations upon the Prophecies of Daniel, and the Apocalypse of St. John, Mahuzzims,* www.newtonproject.sussex.ac.uk/view/texts/normalized/THEM00208.
9. Newton, "Paradoxical Questions," questions 17, 19.
10. Newton, "Paradoxical Questions," questions 17, 19.
11. Newton, "Paradoxical Questions," questions 17, 19.
12. Newton, "Paradoxical Questions," question 19.
13. Newton, "Paradoxical Questions," question 19.
14. Newton, *Observations upon the Prophecies of Daniel*, chap. 14, "Of the Mahuzzims," www.newtonproject.sussex.ac.uk/view/texts/normalized/THEM00208.
15. Newton, "Paradoxical Questions," question 19.
16. Newton, "Paradoxical Questions," question 19.
17. Pagels, *Revelations*, 221.
18. Pagels, *Revelations*, 158.
19. Athanasius, *39th Festal Letter,* Christian Classics Ethereal Library, www.ccel.org/ccel/schaff/npnf204.xxv.iii.iii.xxv.html.
20. Pagels, *Revelations*, 165.

CHAPTER SEVEN. THE GREAT APOSTASY

1. Sachar, *A History of the Jews*, 41.
2. Manuel, *Isaac Newton, Historian*, 162.
3. Newton, "A Dissertation upon the Sacred Cubit of the Jews and the Cubits of the Several Nations," www.newtonproject.sussex.ac.uk/view/texts/normalized/THEM00276.
4. Force, *William Whiston: Honest Newtonian*, 129.
5. Josephus, *Antiquities*, 3.7.7.
6. Goldish, *Judaism in the Theology of Sir Isaac Newton*, 96–97.
7. John Spencer, qtd. in Manuel, *Isaac Newton, Historian*, 121.
8. Chambers, *Victor Hugo's Conversations*, 279–80.

9. Merrill, *Recitative*, 68.
10. Newton, "Untitled Treatise on Revelation," section 1.4.
11. Newton, "Untitled Treatise on Revelation," section 1.4.
12. Newton, "Untitled Treatise on Revelation," section 1.4.
13. Churchill, *Great Contemporaries*, 268. This is Winston Churchill on Lloyd George, but his words apply equally well to Isaac Newton.
14. Newton, *Observations upon the Prophecies of Daniel*, part 2, chap. 2, www.newtonproject.sussex.ac.uk/view/texts/normalized/THEM00210.
15. Newton, www.newtonproject.sussex.ac.uk/view/texts/normalized/THEM00270.
16. Conrad, *Heart of Darkness*, 13, 14.
17. Goonetilleke, *Joseph Conrad's "Heart of Darkness,"* 123.
18. Conrad, *Heart of Darkness*, 9, 10.
19. Newton, *Observations upon the Prophecies of Daniel*, part 2, chap. 2.
20. Newton, *Observations upon the Prophecies of Daniel*, part 2, chap. 2.
21. Newton, *Observations upon the Prophecies of Daniel*, part 2, chap. 2.
22. Newton, *Observations upon the Prophecies of Daniel*, part 2, chap. 2.
23. Newton, *Observations upon the Prophecies of Daniel*, part 2, chap. 2.
24. Gibbon, *The Rise and Fall of the Roman Empire*, vol I, 95–97.
25. Newton, "Untitled Treatise on Revelation," section 1.4.
26. Conrad, *Heart of Darkness*, 17.
27. Conrad, *Heart of Darkness*, 20–21.
28. Newton, *Observations upon the Prophecies of Daniel*, part 1, chap. 2, www.newtonproject.sussex.ac.uk/view/texts/normalized/THEM00196.
29. Newton, *Observations upon the Prophecies of Daniel*, part 1, chap. 2, www.newtonproject.sussex.ac.uk/view/texts/normalized/THEM00196.
30. Newton, *Observations upon the Prophecies of Daniel*, part 1, chap. 2, www.newtonproject.sussex.ac.uk/view/texts/normalized/THEM00196.
31. Castillejo, *Expanding Force in Newton's Cosmos*, 33.
32. Conrad, *Heart of Darkness*, 19.
33. Iliffe, *Newton: A Very Short Introduction*, 80.

CHAPTER EIGHT. APOCALYPSE 2060?

1. Newton, "The Synchronisms of the Three Parts of the Prophetick Interpretation," www.newtonproject.sussex.ac.uk/view/texts/normalized/THEM00049.

2. Newton, "The Synchronisms of the Three Parts of the Prophetick Interpretation," www.newtonproject.sussex.ac.uk/view/texts/normalized/THEM00049.
3. Newton, "The Synchronisms of the Three Parts of the Prophetick Interpretation," www.newtonproject.sussex.ac.uk/view/texts/normalized/THEM00049.
4. Newton, "The Synchronisms of the Three Parts of the Prophetick Interpretation," www.newtonproject.sussex.ac.uk/view/texts/normalized/THEM00049.
5. Smolinski, "Logic of Millennial Thought," 263.
6. Smolinski, "Logic of Millennial Thought," 263.
7. Iliffe, *Newton: A Very Short Introduction*, 81.
8. Newton, "Untitled Treatise on Revelation," section 1.6, www.newtonproject.sussex.ac.uk/view/texts/normalized/THEM00213.
9. Newton, *Observations upon the Prophecies of Daniel*, part 2, chap. 3, www.newtonproject.ox.ac.uk/view/texts/normalized/THEM00211.
10. Newton, "Untitled Treatise on Revelation," section 1.6.
11. Iliffe, *Newton: A Very Short Introduction*, 84.
12. Newton, in "Two Incomplete Treatises on Prophecy," and *Observations upon the Prophecies of Daniel*, part 3, chap. 3.
13. Iliffe, *Newton: A Very Short Introduction*, 63.
14. Conrad, *Heart of Darkness*, 60.
15. Conrad, *Heart of Darkness*, 75.
16. Newton, "Three Draft Chapters on Prophecy," www.newtonproject.sussex.ac.uk/view/texts/normalized/THEM00368.
17. Newton, "Three Draft Chapters on Prophecy," www.newtonproject.sussex.ac.uk/view/texts/normalized/THEM00368.
18. Newton, *Observations upon the Prophecies of Daniel*, part 1, chap. 7, www.newtonproject.sussex.ac.uk/view/texts/normalized/THEM00201.
19. Newton, *Observations upon the Prophecies of Daniel*, part 1, chap. 7, www.newtonproject.sussex.ac.uk/view/texts/normalized/THEM00201.
20. Newton, "Untitled Treatise on Revelation," section 1.1, www.newtonproject.ox.ac.uk/view/texts/normalized/THEM00135.
21. Newton, Yahuda MS 7, 3g, fol. 13v; qtd. in Snobelen, "A Time and Times and the Dividing of Time," 12, https://isaacnewtonstheology.files.wordpress.com/2013/06/newton-the-apocalypse-and-2060-ad.pdf.
22. Newton, Yahuda MS 7, 3g, fol. 13v; qtd. in Snobelen, "A Time and

Times and the Dividing of Time," 12, https://isaacnewtonstheology.files.wordpress.com/2013/06/newton-the-apocalypse-and-2060-ad.pdf.
23. Newton, Yahuda MS 7, 3g, fol. 13v; qtd. in Snobelen, "A Time and Times and the Dividing of Time," 12, https://isaacnewtonstheology.files.wordpress.com/2013/06/newton-the-apocalypse-and-2060-ad.pdf; and *Observations upon the Prophecies of Daniel,* part 1, chap. 7.
24. Manuel, *A Portrait of Isaac Newton,* 378–79.
25. Newton, Yahuda MS 7, 2a, fol. 13r; qtd. in Snobelen, "A Time and the Dividing of Time," 12.
26. Qtd. in Manuel, *A Portrait of Isaac Newton,* 379.
27. Qtd. in Manuel, *A Portrait of Isaac Newton,* 379.

CHAPTER NINE.
THE CONVERSION OF THE JEWS

1. Amichai, *Poems of Jerusalem,* www.pij.org/details.php?id=1009.
2. Lamartine, *A Pilgrimage to the Holy Land,* 2:40.
3. Hammer, "What Is Beneath the Temple Mount?" www.smithsonianmag.com/history/what-is-beneath-the-temple-mount-920764.
4. Chandler, "Hal Lindsey: Prophet of the 'Terminal Generation,'" in *Doomsday,* 53–61.
5. Qtd. in Snobelen, *Restitution,* 95, from Newton, "Of ye . . . Day of Judgmt & World to come," Jewish National and University Library (Jerusalem), Yahuda MS 6, fol. 12r.
6. MacCulloch, *Christianity,* 7.
7. Burton, *The Pageant of Stuart England,* 270–71.
8. MacCulloch, *Christianity,* 773.
9. Buchwald and Feingold, *Newton and the Origin of Civilization,* 381.
10. Johnson, *Christianity,* 276; and see Dimont, *Jews, God, and History,* 293.
11. Snobelen, "Mystery of This Restitution," 104.
12. Keynes, "Newton the Man," 316.
13. Qtd. in Dry, *The Newton Papers,* 162.
14. Qtd. in Dry, *The Newton Papers,* 163.
15. Qtd. in Snobelen, "Mystery of This Restitution," 102–3.
16. Newton, "Untitled Treatise on Revelation," section 1.4, www.newtonproject.ox.ac.uk/view/texts/normalized/THEM00182.
17. Snobelen, "The Mystery of This Restitution," 99.

18. Snobelen, "The Mystery of This Restitution," 99.
19. Newton, "Untitled Treatise on Revelation," section 1.1.
20. Goldish, *Judaism and the Theology of Isaac Newton*, 66.
21. Newton, "Untitled Treatise on Revelation," section 1.1.
22. Newton, "Untitled Treatise on Revelation," section 1.1.
23. Snobelen, "The Mystery of This Restitution," 109.
24. Newton, "Treatise on Revelation," section 1, www.newtonproject.sussex.ac.uk/view/texts/normalized/THEM00216.
25. Newton, "Treatise on Revelation," section 1, www.newtonproject.sussex.ac.uk/view/texts/normalized/THEM00216.
26. Whiston, *New Theory*, "Hypotheses, Book 2," 91–98.
27. Empedocles, qtd. in Whiston in *New Theory of the Earth*, "Hypotheses, Book 2," 98.
28. Castillejo, *Expanding Force in Newton's Cosmos*, 34.
29. Newton, *Of the Prophesy of the Seventy Weeks* (3 drafts), www.newtonproject.sussex.ac.uk/view/texts/normalized/THEM00369.
30. Dimont, *Jews, God, and History*, 394.
31. Sachar, *A History of the Jews*, 357.
32. Newton, *Observations upon the Prophecies of Daniel*, part 1, chap. 10, www.newtonproject.sussex.ac.uk/view/texts/normalized/THEM00204.
33. Qtd. in Sachar, *A History of the Jews*, 357.
34. Johnson, *History of the Jews*, 521.
35. Castillejo, *Expanding Force in Newton's Cosmos*, 34.
36. Snobelen, "The Mystery of This Restitution," 95.
37. Qtd. in "Beauties of the Truth," 8, www.beautiesofthetruth.org/Archive/Library/Doctrine/Mags/Bot/90s/BOTAUG07.PDF.
38. Qtd. in Chambers, "Did Newton Predict the State of Israel?"
39. Snobelen, "The Mystery of This Restitution," 110.

CHAPTER TEN.
WITH NOAH ON THE MOUNTAINTOP

1. Halley, "Some Considerations about the Cause of the Universal Deluge," R.S.S. no. 383, 118, http://archive.org/details/philtrans08240252.
2. Glassie, *A Man of Misconceptions*, 231–32.
3. Haycock, *William Stukeley*, 79.
4. *Gilgamesh*, Gardner and Maier, 226.

5. Qtd. in Tuval, "The Role of Noah," 177–78.
6. Johnson, *History of the Jews*, 9–10.
7. Berlitz, *The Lost Ship*, 73–74.
8. Newton, "Miscellaneous draft portions of *Theologiæ Gentilis Origines Philosophicæ*," www.newtonproject.ox.ac.uk/view/translation/TRAN00010.
9. Newton, "Miscellaneous draft portions of *Theologiæ Gentilis Origines Philosophicæ*," www.newtonproject.ox.ac.uk/view/translation/TRAN00010.
10. *Travels of John Mandeville*, ll. 1433–43, www.lib.rochester.edu/camelot/teams/tkfrm.htm.
11. Berlitz, *The Lost Ship*, 20–21, 23.
12. Berlitz, *The Lost Ship*, 31–42; 88–97; 101.
13. Jung, *Psychology and Alchemy*, 460.
14. Vavra, *Unicorns I Have Known*, 56.
15. Noorbergen, *Secrets of the Lost Races*, 121.
16. Newton, "Draft Chapters of a Treatise on the Origin of Religion and Its Corruption," www.newtonproject.sussex.ac.uk/view/texts/normalized/THEM00077.
17. Newton, "Draft Chapters of a Treatise on the Origin of Religion and Its Corruption," www.newtonproject.sussex.ac.uk/view/texts/normalized/THEM00077.
18. *Gilgamesh*, Gardner and Maier, notes to tablet 11, 233.
19. Newton, dimensions of Noah's ark, in *Chronology of Ancient Kingdoms*.
20. Minov, "Noah and the Flood in Gnosticism," in *Noah and His Book(s)*, ed. Stone et al., 215.
21. Newton, "Of the Church," in the Bodmer MS, qtd. in Goldish, *Judaism in the Theology of Newton*, 41
22. Newton, "Irenicum, or Ecclesiastical Polyty Tending to Peace."
23. Plato, *Timaeus*, 2:8 (sec. 22c).
24. Hawking, qtd. in www.businessinsider.com/stephen-hawking-predictions-about-the-end-of-the-world-2016-1.
25. From the movie description, www.imdb.com/title/tt0816692.
26. From the description here: www.amazon.com/Worlds-Collide-Bison-Frontiers-Imagination/dp/0803298145.
27. Newton, "Treatise on Revelation," section 2, www.newtonproject.ox.ac.uk/view/texts/normalized/THEM00270.
28. Newton, "Untitled Treatise on Revelation," section 1.1, www.newtonproject.ox.ac.uk/view/texts/normalized/THEM00135.

29. Newton, *Observations upon the Prophecies of Daniel*, part 1, chap. 3, www.newtonproject.ox.ac.uk/view/texts/normalized/THEM00197.
30. Westfall, *Never at Rest*, 325.
31. Amihay and Machiela, "Traditions of the Birth of Noah," in Stone, Amihay, and Hillel, *Noah and His Book(s)*, 65, 66.
32. Amihay and Machiela, "Traditions of the Birth of Noah," in Stone, Amihay, and Hillel, *Noah and His Book(s)*, 63.
33. Machiela, "Genesis Apocryphon," 183.
34. Amihay and Machiela, "Traditions of the Birth of Noah," 57.
35. Machiela, "Genesis Apocryphon," 102.
36. Amihay and Machiela, "Traditions of the Birth of Noah," 104.
37. Cranston and Williams, *Reincarnation: A New Horizon in Science, Religion, and Society*, 188.
38. Westfall, *Never at Rest*, 352.
39. Newton, "Draft Chapters of a Treatise on the Origin of Religion and Its Corruption," chap. 2, www.newtonproject.ox.ac.uk/view/texts/normalized/THEM00077.
40. Newton, "Miscellaneous Draft Portions of 'Theologiæ Gentilis Origines Philosophicæ,'" www.newtonproject.sussex.ac.uk/view/translation/TRAN00010.
41. Newton, "Miscellaneous Draft Portions of 'Theologiæ Gentilis Origines Philosophicæ,'" www.newtonproject.sussex.ac.uk/view/translation/TRAN00010.
42. Newton, "Miscellaneous Draft Portions of 'Theologiæ Gentilis Origines Philosophicæ,'" www.newtonproject.sussex.ac.uk/view/translation/TRAN00010.
43. Newton, "Draft Chapters of a Treatise on the Origin of Religion and Its Corruption," chap. 2.
44. Newton, "Draft Chapters of a Treatise on the Origin of Religion and Its Corruption," chap. 2.
45. Markley, "Newton, Corruption, and the Tradition of Universal History," 138.
46. Newton, "Theologiae Gentiles Origines Philosophicae," www.newtonproject.sussex.ac.uk/view/texts/normalized/THEM00260.
47. Newton, "Draft Chapters of a Treatise on the Origin of Religion and Its Corruption," chap. 2.
48. Newton, "Draft Chapters of a Treatise on the Origin of Religion and Its Corruption," chap. 2.

428   Notes

49. Newton, "Draft Chapters of a Treatise on the Origin of Religion and Its Corruption," chap. 2.
50. Newton, "Miscellaneous Draft Portions of 'Theologiæ Gentilis Origines Philosophicæ.'"
51. Knoespel, "Interpretive Strategies," 183.

CHAPTER ELEVEN. IN THE DAYS OF THE COMET

1. Flaste et al., *New York Times Guide to the Return of Halley's Comet*, 75–76.
2. Stach, *Kafka*, 16.
3. Flaste et al., *New York Times Guide to the Return of Halley's Comet*, 186.
4. Flaste et al., *New York Times Guide to the Return of Halley's Comet*, 50.
5. Yeomans, *Comets*, 20.
6. Qtd. in Yeomans, *Comets*, 20.
7. Flammarion, *Omega*, 149–50.
8. Ackroyd, *Newton*, 72.
9. Flammarion, *Omega*, 150.
10. Qtd. in Westfall, *Never at Rest*, 391.
11. Manuel, *Portrait*, 295.
12. Dobbs, *The Janus Faces of Genius*, 236.
13. Dobbs, *The Janus Faces of Genius*, 236.
14. Blake, "A Vision of the Last Judgment," in G. Keynes, *Poetry and Prose*, 652.
15. Blake, "Jerusalem" (15:15–18), in G. Keynes, *Poetry and Prose*, 449.
16. Dobbs, *The Janus Faces of Genius*, 234.
17. Newton, *Mathematical Principles of Natural Philosophy* (hereafter cited as *Principia*), 372. (See also *Principia*, 359.)
18. Dobbs, *The Janus Faces of Genius*, 83.
19. Dobbs, *The Janus Faces of Genius*, 83.
20. Dobbs, *The Janus Faces of Genius*, 240.
21. Qtd. in Haycock, *William Stukeley*, 79.
22. Newton, *Principia*, 368.
23. Dobbs, *The Janus Faces of Genius*, 24.
24. Gibbon, *The Decline and Fall of the Roman Empire*, 2:1426.
25. Qtd. in Force, *Whiston*, 129.

26. Stukeley, qtd. in Westfall, *Never at Rest,* 194.
27. Newton, *Optics,* book 3, part 1, 543.
28. Goldsmith, *The Vicar of Wakefield,* 30.
29. Donnelly, Meehan, "Whiston's Flood," http://web.stanford.edu/~meehan/donnelly/whiston.html.
30. Force, *Whiston,* xiii.
31. Milton, *Paradise Lost,* 10.668–80.
32. Whiston, *New Theory,* book 2, "Hypothesis," 111.
33. Milton, *Paradise Lost,* 12, 627–29, 632–34.
34. Whiston, *A New Theory,* 13, 471.
35. Force, *Whiston,* 32.
36. Force, *Whiston,* 13.
37. Manuel, *Isaac Newton, Historian,* 143.
38. Force, *Whiston,* 29.
39. Force, *Whiston,* 20.
40. Force, *Whiston,* 133–37.
41. Halley, "Saltiness," *Notes and Records,* 296.
42. Voltaire, *Le Siècle de Louis XIV,* vol. 2, 378, (author's translation).
43. Abraham de Moivre, qtd. in Westfall, *Never at Rest,* 403.
44. Westfall, *Never at Rest,* 404.
45. Halley, "Saltiness," *Notes and Records,* 296.
46. Newton, *Principia [III],* 367.
47. Newton, *Principia [III],* 354.
48. Qtd. in Manuel, *A Portrait,* 252.
49. Conduitt, "Account of a conversation between Newton and Conduitt," March 7, 1724, www.newtonproject.sussex.ac.uk/view/texts/normalized/THEM00173.
50. Conduitt, "Account of a conversation between Newton and Conduitt," March 7, 1724, www.newtonproject.sussex.ac.uk/view/texts/normalized/THEM00173.
51. Conduitt, "Account of a conversation between Newton and Conduitt," March 7, 1724, www.newtonproject.sussex.ac.uk/view/texts/normalized/THEM00173.
52. Force, *Whiston,* 134.
53. Force, *Whiston,* 135.

CHAPTER TWELVE.
DECONSTRUCTING TIME

1. Krawcewicz et al., "Investigation of the Correctness of the Historical Dating," http://wayback.archive.org/web/20060209081746/http://www.revisedhistory.org:80/investigation-historical-dating.htm.
2. "History: Fiction or Science," description of Fomenko et al.'s *History: Fiction or Science,* on CreateSpace website, www.createspace.com/6007705.
3. "Classical History Is a Lie Created to Control You," www.rexdeus.com/wp/secret-societies/history-lie-created-control.
4. Westfall, *Never at Rest,* 815.
5. Chazelle, interview, "Discovering the Cosmology of Bach," American Public Media, https://onbeing.org/programs/bernard-chazelle-discovering-the-cosmology-of-bach.
6. Castillejo, *Expanding Force in Newton's Cosmos,* 82.
7. Castillejo, *Expanding Force in Newton's Cosmos,* 91, 81.
8. Frye, *Anatomy of Criticism,* 354.
9. Cress, "Review," *Newton's Revised History of Ancient Kingdoms: A Complete Chronology,* Amazon.com, July 30, 2009.
10. Dry, *The Newton Papers,* 160, 162.
11. Manuel, *Religion,* 98.
12. Pierce, "Chronology Wars," www.answersingenesis.org/articles/am/v6/n1/chronology-wars.
13. Manuel, *Isaac Newton, Historian,* 90.
14. Newton, *Chronology,* 43.
15. Manuel, *Isaac Newton, Historian,* 30.
16. Castillejo, *Expanding Force in Newton's Cosmos,* 22.
17. Qtd. in Westfall, *Never at Rest,* 154.
18. Newton, *Chronology,* 151.
19. Castillejo, *Expanding Force in Newton's Cosmos,* 81.
20. Newton, *Chronology,* 21, 26.
21. Manuel, *Isaac Newton, Historian,* 57.
22. Voltaire [François Marie Arouet], "Letters on Newton" from the *Letters on the English or Lettres Philosophiques,* ca. 1778, www.fordham.edu/halsall/mod/1778voltaire-newton.asp.
23. Castillejo, *Expanding Force in Newton's Cosmos,* 43, 44.
24. Manuel, *Isaac Newton, Historian,* 128.

25. Newton, "The Original of Monarchies," www.newtonproject.sussex.ac.uk/view/texts/normalized/THEM00040.
26. Gibbon, *Decline and Fall*, 1:170.
27. Gibbon, *Decline and Fall*, 1:170.
28. Buchwald and Feingold, *Newton and the Origin of Civilization*, 166.
29. Buchwald and Feingold, *Newton and the Origin of Civilization*, 167, 168.
30. Buchwald and Feingold, *Newton and the Origin of Civilization*, 168, 169.
31. Qtd. in Buchwald and Feingold, *Newton and the Origin of Civilization*, 215.
32. Buchwald and Feingold, *Newton and the Origin of Civilization*, 220.
33. Raleigh and Wallis qtd. in Buchwald and Feingold, *Newton and the Origin of Civilization*, 213.
34. Buchwald and Feingold, *Newton and the Origin of Civilization*, 214.
35. Johnson, *History of the Jews*, 120.
36. Manuel, *Isaac Newton, Historian*, 137.
37. Manuel, *Isaac Newton, Historian*, 138.
38. Buchwald and Feingold, *Newton and the Origin of Civilization*, 306.
39. Reé, "I Tooke a Bodkin," review of *Newton and the Origin of Civilization* by J. Z. Buchwald and M. Feingold, *London Review of Books* 35, no. 19 (October 10, 2013): 16–18, www.lrb.co.uk/v35/n19/jonathan-ree/i-tooke-a-bodkine.

CHAPTER THIRTEEN.
CHIRON AND THE STAR GLOBE

1. Qtd. in Allen, *Star Names*, 149.
2. Greaves, Zaller, and Roberts, *Civilizations of the West*, 62–63.
3. Rieu, introduction to Apollonius, *Voyage of the Argo*, 21.
4. Newton, *Chronology*, chap. 1, "Of the Chronology of the First Ages of the Greeks," www.newtonproject.sussex.ac.uk/view/texts/normalized/THEM00186.
5. Newton, *Chronology*, chap. 1, "Of the Chronology of the First Ages of the Greeks," www.newtonproject.sussex.ac.uk/view/texts/normalized/THEM00186.
6. Twain, *Innocents Abroad*, 410.
7. Asimov, *Double Planet*, 106–8.
8. Aratos, *Phaenomena*, 155, ll. 1122–24.
9. Qtd. in Allen, *Star Names*, 113.

10. Allen, *Star Names*, 17.
11. Newton, "Notes on Ancient Religions," www.newtonproject.sussex.ac.uk/view/translation/TRAN00012.
12. Buchwald and Feingold, *Newton and the Origin of Civilization*, 324.
13. Newton, "Drafts on Chronology: Section 2d," www.newtonproject.ox.ac.uk/view/texts/normalized/THEM00402.
14. Buchwald and Feingold, *Newton and the Origin of Civilization*, 293–94.

CHAPTER FOURTEEN.
A GLITTER OF ATLANTIS

1. Qtd. in Mifsud et al., *Malta*, 54.
2. Glassie, *Man of Misconceptions*, 88–89.
3. De Santillana and von Dechend, *Hamlet's Mill*, 209.
4. De Santillana and von Dechend, *Hamlet's Mill*, 209.
5. Mifsud et al., *Malta*, 56.
6. Plutarch, *The Lives of the Ancient Greeks and Romans*, 132.
7. Plutarch, *The Lives of the Ancient Greeks and Romans*, 104, 132.
8. Herodotus, *Histories*, sec. 29.
9. Plato, *Timaeus*, in *Dialogues*, 2:8.
10. Pierce, "Chronology Wars," www.answersingenesis.org/articles/am/v6/n1/chronology-wars.
11. Newton, *Chronology*, 43.
12. Newton, *Chronology*, 231.
13. Newton, *Chronology*, 231.
14. Luce, *The End of Atlantis*, 38.
15. Luce, *The End of Atlantis*, 136.
16. Newton, *Chronology*, 45, 133, 135.
17. Newton, *Chronology*, 181.
18. Newton, *Chronology*, 229.
19. Mifsud et al., *Malta*, 54, 42.
20. Adams, "My Quest for Atlantis," March 20, 2015.
21. Newton, *Chronology*, 231–32.
22. Newton, *Chronology*, 234.
23. Qtd. in Buchwald and Feingold, *Newton and the Origin of Civilization*, 218.
24. Newton, *Chronology*, 233–34.
25. Newton, *Chronology*, 233.

26. Qtd. in Luce, *The End of Atlantis*, 13.
27. Qtd. in Luce, *The End of Atlantis*, 13.
28. Jowett's introduction to Plato's *Critias*, www.gutenberg.org/files/1571/1571-h/1571-h.htm#link2H_INTR.

CHAPTER FIFTEEN. THE SECRET OF LIFE

1. Dobbs, *The Janus Faces of Genius*, 55.
2. White, *Isaac Newton*, 137.
3. Dobbs, *The Janus Faces of Genius*, 13.
4. Westfall, *Never at Rest*, 286.
5. Qtd. in Brewster, *Life of Newton*, 2:98, www.newtonproject.sussex.ac.uk/view/texts/normalized/OTHE00077.
6. White, *Isaac Newton*, 137.
7. White, *Isaac Newton*, 137; Bosveld, "Isaac Newton, World's Most Famous Alchemist," *Discover*, July-Aug, 2010; and *NOVA*, "Newton's Dark Secrets," www.pbs.org/wgbh/nova/physics/newton-dark-secrets.html.
8. Lequeuvre, "Le jardin secret de Newton," 54.
9. Needham, *Science and Civilization in China*, 2:82.
10. Manuel, *A Portrait*, 163.
11. Qtd. in Brewster, *Life of Newton*, 2:94, www.newtonproject.sussex.ac.uk/view/texts/normalized/OTHE00077.
12. Castillejo, *Expanding Force in Newton's Cosmos*, 17, 21–22.
13. Needham, *Science and Civilization in Ancient China*, 4:455.
14. Manuel, *A Portrait*, 178.
15. Ovid, *Metamorphoses*, 107.
16. "William Newman Project," www.indiana.edu/~college/William Newman Project.shtml, quoted in Dr. Faustall, *Pataphysica 4*, 172.
17. Santillana, *Hamlet's Mill*, 178.
18. Lovelock, *The Ages of Gaia*.
19. Thompson, *Imaginary Landscapes*, 37.
20. Principe and Newman, "Newton's Dark Secrets," *NOVA*, www.pbs.org/wgbh/nova/physics/newton-dark-secrets.html.
21. Manuel, *A Portrait*, 169.
22. John Wickins, qtd. in Brewster, *Life of Newton*, 7.
23. Manuel, *A Portrait*, 169.

24. Qtd. in Manuel, *A Portrait*, 172.
25. Manuel, *A Portrait*, 173.
26. Golinski, "The Secret Life of an Alchemist," 160.
27. Manuel, *A Portrait*, 175–76.
28. Qtd. in Dobbs, *The Janus Faces of Genius*, 23.
29. Newton, "Letter from Newton to Henry Oldenburg," 26 April 1676, www.newtonproject.sussex.ac.uk/view/texts/normalized/NATP00268.
30. Needham, *Science and Civilization in Ancient China*, 4:408.
31. Boyle, *A Historical Account of the Degradation of Gold by an Anti-Elixir*, www.levity.com/alchemy/boyle.html.
32. Bosveld, "Isaac Newton: World's Most Famous Alchemist."
33. Brewster, *Life of Newton*, 121–22.
34. Brewster, *Life of Newton*, 123.
35. Brewster, *Life of Newton*, 123.
36. Newman, "Isaac Newton, World's Most Famous Alchemist," *Discover* (July–August 2010); "Newton's Dark Secrets," *NOVA*.
37. Newman, "Isaac Newton, World's Most Famous Alchemist," *Discover* (July–August 2010); "Newton's Dark Secrets," *NOVA*.
38. White, *Isaac Newton*, 145.
39. Dobbs, *Janus Faces of Genius*, 13.

CHAPTER SIXTEEN.
MASTERS OF THE *PRISCA SAPIENTIA*, PART I

1. Lawrence, *Fantasia*, 54–55.
2. Allen, *Star Names*, 17–18.
3. Qtd. in Noorbergen, *Secrets of the Lost Races*, 130–31.
4. McGuire and Rattansi, "Newton and the 'Pipes of Pan,'" 109 (translation by author).
5. McGuire and Rattansi, "Newton and the 'Pipes of Pan,'" 110.
6. Gregory, *Elementa astronomiae*, https://archive.org/details/elementsofastron00greg.
7. Dolnick, *The Clockwork Universe*, 36–37.
8. McGuire and Rattansi, "Newton and the 'Pipes of Pan,'" 113.
9. Qtd. in Heath, *Aristarchus of Samos*, back cover.
10. Durant, *Life of Greece*, 169.
11. Qtd. in Heath, *Aristarchus of Samos*, 299.

12. Archimedes, qtd. in Heath, *Aristarchus of Samos*, 302.
13. Qtd. in Durant, *Life of Greece*, 653.
14. Durant, *Life of Greece*, 654.
15. Durant, *Life of Greece*, 340, 341.
16. Durant, *Life of Greece*, 654.
17. Diogenes Laertius, *Lives of Eminent Philosophers*, 2.3.
18. Plutarch, *Lives of Ancient Greeks and Romans*, 200.
19. Laertius, *Lives of Eminent Philosophers*, 3.8.
20. Copleston, *History of Philosophy*, 1.2.32.
21. Plutarch, *Lives of Ancient Greeks and Romans*, 188.
22. Plutarch, *Lives of Ancient Greeks and Romans*, 201.
23. Plutarch, *Lives of Ancient Greeks and Romans*, 185, 187.
24. Plutarch, *Lives of Ancient Greeks and Romans*, 76–77.
25. Plutarch, *Lives of Ancient Greeks and Romans*, 80–81.
26. Hobbes qtd. in Durant, *Louis XIV*, 558.
27. Qtd. in Dobbs, *Janus Faces of Genius*, 187.
28. Newton, "Notes on Ancient Religions," www.newtonproject.sussex.ac.uk/view/translation/TRAN00012.
29. Plutarch, *Lives and Ancient Greeks and Romans*, 81.
30. Plutarch, *Lives of Ancient Greeks and Romans*, 83.
31. Plutarch, *Lives of Ancient Greeks and Romans*, 87.

CHAPTER SEVENTEEN.
SAPIENTIA, PART 2

1. Diogenes Laertius, "Pythagoras," www.perseus.tufts.edu/hopper/text?doc=Perseus%3Atext%3A1999.01.0258%3Abook%3D8%3Achapter%3D1>; see also Guthrie, *Pythagorean Sourcebook*, 150.
2. Iamblichus, in Guthrie, *Pythagorean Sourcebook*, 103–4.
3. Heath, *Aristarchus of Samos*, 57.
4. Aristotle, *On the Heavens*, 2.13.1.384.
5. Aristotle, *On the Heavens*, 385.
6. Simplicius qtd. in Heath, *Aristarchus of Samos*, 96.
7. Heath, *Aristarchus of Samos*, 96.
8. Newton, "Draft Scholium to Proposition 9, *Principia*," qtd. in McGuire and Rattansi, "Newton and the 'Pipes of Pan,'" 119.
9. Dreyer, *History of the Planetary Systems*, 141.

10. Guthrie, *Pythagorean Sourcebook*, 144.
11. Plutarch, *Lives of Ancient Greeks and Romans*, 80.
12. Guthrie, *Pythagorean Sourcebook*, 127.
13. Iamblichus, qtd. in Guthrie, *Pythagorean Sourcebook*, 90.
14. Guthrie, *Pythagorean Sourcebook*, 90.
15. Diogenes Laertius, "Pythagoras," www.perseus.tufts.edu/hopper/text?doc=Perseus%3Atext%3A1999.01.0258%3Abook%3D8%3Achapter%3D1; see also Guthrie, *Pythagorean Sourcebook*, 150.
16. Guthrie, *Pythagorean Sourcebook*, 126–27.
17. Kingsley, *A Story Waiting to Pierce You*, 154.
18. Zeller, *Outlines*, 47–48.
19. Newton, "Irenicum," www.newtonproject.sussex.ac.uk/view/texts/normalized/THEM00003.
20. Guthrie, *Pythagorean Sourcebook*, 168.
21. Vilenkin, *Many Worlds in One*, 200.
22. Vilenkin, *Many Worlds in One*, 200–201.
23. Aristotle, *Metaphysics*, 1.5.504.
24. Plutarch, *Plutarch's Lives*, 121.
25. Qtd. in McGuire and Rattansi, "Newton and the 'Pipes of Pan,'" 117.
26. Maclaurin, qtd. in McGuire and Rattansi, "Newton and the 'Pipes of Pan,'" 117.
27. McGuire and Rattansi, "Newton and the 'Pipes of Pan,'" 117–18.
28. Gouk, "The Harmonic Roots of Newtonian Science," 101.
29. Isacoff, *Temperament*, 32.
30. Isacoff, *Temperament*, 32.
31. Qtd. in McGuire and Rattansi, "Newton and the 'Pipes of Pan,'" 119.
32. Newton, "Out of Cudworth," www.newtonproject.sussex.ac.uk/view/texts/normalized/THEM00118.
33. Manuel, *A Portrait*, 364–65.
34. Iliffe, *Newton: A Very Short Introduction*, 21.
35. Qtd. in Vilenkin, *Many Worlds in One*, 85.
36. Rattansi, "Newton and the Wisdom of the Ancients," 185.
37. Rattansi, "Newton and the Wisdom of the Ancients," 185.
38. MathPages, "Prisca Sapientia," www.mathpages.com/home/kmath066/kmath066.htm.
39. Miller, *137: Jung, Pauli, and the Pursuit*, 126–27.

40. Miller, *137: Jung, Pauli, and the Pursuit*, 198, 201, xxi, 177–78.
41. Miller, *137: Jung, Pauli, and the Pursuit*, xx, 179.

CHAPTER EIGHTEEN. SON OF ARCHIMEDES

1. Qtd. in Biot, *Life of Sir Isaac Newton*, www.newtonproject.sussex.ac.uk/view/texts/normalized/OTHE00089.
2. Qtd. in Westfall, *Never at Rest*, 535.
3. Westfall, *Never at Rest*, 534.
4. Westfall, *Never at Rest*, 534.
5. Westfall, *Never at Rest*, 535.
6. Westfall, *Never at Rest*, 535.
7. Manuel, *A Portrait*, 111.
8. Westfall, *Never at Rest*, 617.
9. https://wikilivres.ca/wiki/Marlborough:_His_Life_and_Times,_Book_One.
10. Westfall, *Never at Rest*, 479.
11. Manuel, *The Religion of Isaac Newton*.
12. Qtd. in Iliffe, *Newton: A Very Short Introduction*.
13. Dick, *Francis Bacon*, xv.
14. Klein, *This Changes Everything*, 170.
15. "Marcellus," in Plutarch, *Lives of the Ancient Greeks and Romans*, 376.
16. "Marcellus," in Plutarch, *Lives of the Ancient Greeks and Romans*, 376.
17. Plutarch, *Lives of the Ancient Greeks and Romans*, 377.
18. Qtd. in Brewster, *Memoirs*, 2:31.
19. Plutarch, *Lives of the Ancient Greeks and Romans*, 378.
20. Godwin, foreword, *Pythagorean Sourcebook*, 11.
21. Newton, "Notes on Ancient Religions," www.newtonproject.sussex.ac.uk/view/translation/TRAN00012.
22. Gjertsen, "Newton's Success," 39–40.
23. Newton, "Original Letter from Isaac Newton to Richard Bentley, dated 10 December 1692," www.newtonproject.sussex.ac.uk/view/texts/normalized/THEM00254.
24. Manuel, *The Religion of Isaac Newton*, 48.
25. Newton, *Optics*, 3, 1, 2nd ed., 543.
26. Dolnick, *The Clockwork Universe*, 307–8.

27. Newton, "A Short Schem of the True Religion," www.newtonproject.sussex.ac.uk/view/texts/normalized/THEM00007.
28. Newton, *Principia*, "General Scholium."
29. Whiston, *New Theory*, book 1, "Lemmata."
30. Newton, *Principia*, 3, 369.
31. The author is indebted to Steven Sittenreich for much of the material in this paragraph.
32. Noam Chomsy, "With Trump Election, We are Now Facing Threats to the Survival of the Species," www.democracynow.org/2017/1/2/noam_chomsky_with_trump_election_we.
33. Dobbs, *Janus Faces of Genius*, 87.
34. Newton, "Untitled Treatise on Revelation," section 1.4, www.newtonproject.sussex.ac.uk/view/texts/normalized/THEM00182.
35. Iliffe, *Newton: A Very Short Introduction*, 97.

INNER TRADITIONS
BEAR & COMPANY

Inner Traditions • Bear & Company
P.O. Box 388
Rochester, VT 05767-0388
U.S.A.

Affix
Postage
Stamp
Here

PLEASE SEND US THIS CARD TO RECEIVE OUR LATEST CATALOG FREE OF CHARGE.

Book in which this card was found _____

☐ Check here to receive our catalog via e-mail.

Company _____

☐ Send me wholesale information

Name _____ Phone _____
Address _____
City _____ State _____ Zip _____ Country _____
E-mail address _____

Please check area(s) of interest to receive related announcements via e-mail:

☐ Health   ☐ Self-help   ☐ Science/Nature   ☐ Shamanism
☐ Ancient Mysteries ☐ New Age/Spirituality ☐ Visionary Plants   ☐ Martial Arts
☐ Spanish Language ☐ Sexuality/Tantra   ☐ Family and Youth   ☐ Religion/Philosophy

Please send a catalog to my friend:

Name _____ Company _____
Address _____ Phone _____
City _____ State _____ Zip _____ Country _____

Order at 1-800-246-8648 • Fax (802) 767-3726

E-mail: customerservice@InnerTraditions.com • Web site: www.InnerTraditions.com

Bibliography

Ackroyd, Peter. *Newton: Ackroyd's Brief Lives*. New York: Nan A. Talese, 2008.
Adams, Mark. "My Quest for Atlantis." *New York Times*, March 20, 2015.
Allen, Richard Hinckley. *Star Names: Their Lore and Meaning*. New York: Dover, 1963.
Amichai, Yehuda. *Poems of Jerusalem*. Tel Aviv: Schocken Publishing, 1987. www.pij.org/details.php?id=1009.
Amihay, Aryeh, and Daniel A. Machiela. "Traditions of the Birth of Noah." In *Noah and his Book(s)*, edited by Michael E. Stone, Aryeh Amihay, and Vered Hillel, 53–69. Atlanta, GA: Society of Biblical Literature, 2010.
Anatolios, Khaled. *Athanasius*. London and New York: Routledge, 2004.
———. *Retrieving Nicaea: The Development and Meaning of Trinitarian Doctrine*. Grand Rapids, MI: Baker Academic, 2011.
Apollonius of Rhodes. *The Voyage of the Argo*. Translated with an introduction by E. V. Rieu. London: Penguin Books, 2006.
Aratus. *Phaenomena*. Translated by Douglas Kidd. Cambridge: Cambridge University Press, 1997.
Aristotle. *Works*. Chicago: Great Books of the Western World/Encyclopedia Britannica, 1952.
Asimov, Isaac. *Asimov on Astronomy*. New York: Bonanza Books, 1979.
———. *Counting the Eons*. New York: Discus/Avon, 1983.
———. *The Double Planet*. New York: Pyramid, 1968
———. *The Solar System and Back*. New York: Discus/Avon, 1970.
Athanasius. *The Life of Antony of Egypt*. Paraphrase by Albert Haase, O.F.M. Downers Grove, IL: IVP Books/InterVarsity Press, 2012.
Augustine. *City of God*. Chicago: Great Books of the Western World/Encyclopedia Britannica, 1952.

Baigent, Michael. *Racing toward Armageddon: The Three Great Religions and the Plot to End the World.* New York: HarperCollins, 2009.

Banville, John. *The Newton Letter.* Boston, MA: David R. Godine, 1999.

Barrett, William. *Irrational Man: A Study in Existential Philosophy.* Garden City, NY: Anchor Books/Doubleday, 1962.

Bauer, Alain. *Isaac Newton's Freemasonry: The Alchemy of Science and Mysticism.* Rochester, VT: Inner Traditions, 2003.

Becker, Allienne R. *Eagle in Flight: The Life of Athanasius, the Apostle of the Trinity.* San Jose, CA: Writers Club Press, 2002.

Ben Uzziel, Jonathan. *The Chaldee Paraphrase on the Prophet Isaiah.* Translated by C. W. H. Pauli. London: London Society's House, 1871.

Bent, J. Theodore. "Did St. John See an Earthquake?" In *Doomsday? How the World Will End—and When,* edited by Martin Ebon, 111–21. New York: New American Library, 1977.

Berlitz, Charles. *The Lost Ship of Noah: In Search of the Ark at Ararat.* New York: G. P. Putnam's, 1987.

Biot, Jean-Baptiste. *Life of Sir Isaac Newton.* Translated by H. Elphinstone, in *Lives of Eminent Persons.* London, 1833. www.newtonproject.sussex.ac.uk/view/texts/normalized/OTHE00089.

Bloom, Harold. *Omens of Millennium: The Gnosis of Angels, Dreams, and Resurrection.* New York: Riverhead/G. P. Putnam's, 1996.

Bodmer, Frederick. *The Loom of Language.* New York: W. W. Norton, 1944.

Bosveld, Jane. "Isaac Newton, World's Most Famous Alchemist." *Discover* (July–August, 2010). http://discovermagazine.com/2010/jul-aug/05-isaac-newton-worlds-most-famous-alchemist.

Bournis, Archimandrite Theodoritos. "St. John on Patmos." In *Doomsday! How the World Will End—and When,* edited by Martin Ebon, 42–49. New York: New American Library, 1977.

[Boyle, Robert.] *Of a Degradation of Gold Made by an Anti-Elixir: A Strange Chymical Narrative.* Published anonymously. London, 1678. www.levity.com/alchemy/boyle.html.

Brewster, David. *Memoirs of the Life, Writings, and Discoveries of Sir Isaac Newton.* Vol. 2. Edinburgh: Edmonston & Douglas, 1855.

Brooke, John. "The God of Isaac Newton." In *Let Newton Be!* edited by John Fauvel, Raymond Flood, Michael Shortland, and Robin Wilson, 171–83. Oxford: Oxford University Press, 1988.

Buchwald, Jed Z., and Mordechai Feingold. *Newton and the Origin of Civilization*. Princeton, NJ: Princeton University Press, 2013.

Burton, Elizabeth. *The Pageant of Stuart England*. New York: Scribner's, 1962.

"Canons from Nicaea 1." http://familybible.org/BeitMidrash/Model/AppendixE.htm.

Card, Orson Scott. *Pastwatch: The Redemption of Christopher Columbus*. New York: Tor Books, 1996.

Casini, P. "Newton: The Classical Scholia." *History of Science* 22, 1–58.

Castillejo, David. *Expanding Force in Newton's Cosmos*. Madrid, Spain: Ediciones de Arte y Bibliofilia, 1981.

Cave, William. *Ecclesiastici: Or the History of the Lives Acts, Death, and Writings of the Most Eminent Fathers of the Church That Flourisht in the Fourth Century*. London, 1683.

Chambers, John. *Victor Hugo's Conversations with the Spirit World: A Literary Genius's Hidden Life*. Rochester, VT: Destiny/Inner Traditions, 2008.

———. "Did Newton Predict the State of Israel?" *Atlantis Rising* 119 (September/October 2016). https://atlantisrisingmagazine.com/article/did-newton-predict-the-state-of-israel.

Champion, Justin A. I. "'Acceptable to Inquisitive Men': Some Simonian Contexts for Newton's Biblical Criticism, 1680–1692." In *Newton and Religion: Context, Nature, and Influence*, edited by James E. Force and Richard H. Popkin, 77–96. Dordrecht, The Netherlands: Kluwer Academic Publishers, 1999.

Chandler, Russell. *Doomsday: The End of the World, A View through Time*. Ann Arbor, MI: Servant Publications, 1993.

Chazelle, Bernard. *On Being* with Krista Tippett. November 13, 2014. American Public Media. www.onbeing.org/program/bernard-chazelle-discovering-the-cosmology-of-bach/7026.

Christianson, Gale E. *Isaac Newton and the Scientific Revolution*. New York: Oxford University Press, 1996.

Clouse, Robert. "The Apocalyptic Interpretation of Thomas Brightman and Joseph Mede." *Bulletin of the Evangelical Theological Society* (Denver) 11, no. 4 (1968): 181–93. www.etsjets.org/files/JETS-PDFs/11/11-4/BETS_11_4_181-193_Clouse.pdf.

"CO_2 at NOAA's Mauna Loa Observatory Reaches New Milestone: Tops 400 ppm." National Oceanic & Atmospheric Administration. www.esrl.noaa.gov/gmd/news/7074.html.

Cochrane, Charles Norris. *Christianity and Classical Culture*. New York: A Galaxy Book/Oxford University Press, 1940/1957.

Conrad, Joseph. *Heart of Darkness*. New York: Norton, 1963.

Copleston, Frederick. *A History of Philosophy*. Vol. 1, *Greece & Rome*, part 2. Garden City, NY: Image Books/Doubleday, 1962.

Cranston, S. L., and Carey Williams, *Reincarnation: A New Horizon in Science, Religion, and Society*. Pasadena, CA: Theosophical University Press, 1984.

Cress, L. "Review." *Newton's Revised History of Ancient Kingdoms: A Complete Chronology*. Amazon.com, July 30, 2009.

Daniel & Revelation: Secrets of Bible Prophecy. Nampa, ID: Amazing Facts Publishing/Pacific Press Publishing Association.

Davies, Paul. "Are We Alone in the Universe?" *New York Times*, November 19, 2013.

Dick, Hugh G., ed. *Francis Bacon: Selected Writings*. New York: The Modern Library, 1955.

Dimont, Max I. *Jews, God and History*. New York: New American Library, 1962.

Diogenes Laertius. *Lives of Eminent Philosophers*. Edited by R. D. Hicks. www.perseus.tufts.edu/hopper/text?doc=Perseus%3Atext%3A1999.01.0258%3Abook%3D8%3Achapter%3D1.

Dobbs, B. J. "Newton's Commentary on the Emerald Tablet of Hermes Trismegistus." In *Hermeticism and the Renaissance*, edited by Ingrid Merkel and Allen G. Debus. Washington: Folger, 1988.

Dobbs, Betty Jo Teeter. *The Janus Faces of Genius: The Role of Alchemy in Newton's Thought*. New York: Cambridge University Press, 1991/2002.

Dolnick, Edward. *The Clockwork Universe: Isaac Newton, the Royal Society and the Birth of the Modern World*. New York: Harper Perennial, 2011.

Dreyer, J. L. E. *History of the Planetary Systems from Thales to Kepler*. Cambrdige: Cambridge University Press, 1906.

Dr. Faustroll, ed. *Pataphysica 4: Pataphysica e Alchimia 2*. New York: iUniverse, 2006.

Drosnin, Michael. *The Bible Code*. New York: Simon & Schuster, 1997.

Dry, Sarah. *The Newton Papers: The Strange & True Odyssey of Isaac Newton's Manuscripts*. New York and Oxford: Oxford University Press, 2014.

Durant, Will, and Ariel Durant. *The Age of Faith*. Vol. 4 of *The Story of Civilization*. New York: Simon and Schuster, 1950.

———. *The Age of Louis XIV*. Vol. 8 of *The Story of Civilization*. New York: Simon and Schuster, 1963.

———. *Caesar and Christ.* Vol. 3 of *The Story of Civilization.* New York: Simon and Schuster, 1944.

———. *The Life of Greece.* Vol. 2 of *The Story of Civilization.* New York: Simon and Schuster, 1939.

Ebon, Martin. "In the Grotto of St. John." In *Doomsday! How the World Will End—and When,* edited by Martin Ebon, 3–15. New York: New American Library, 1977.

Edinger, Edward F. *Ego and Archetype.* Boston & London: Shambhala, 1992.

Ehrman, Bart D. *The Orthodox Corruption of Scripture: The Effect of Early Christological Controversies on the Text of the New Testament.* New York and Oxford: Oxford University Press, 1993.

Ellis, Ralph. *Jesus, King of Edessa.* Kempton, IL: Adventures Unlimited Press, 2013.

Epiphanius. *The Panarion of Epiphanius of Salmis.* Books 1 and 3 (Sects 47-80, De Fide). Leiden: Brill, 1993. http://books.google.com/books/about/The _Panarion_of_Epiphanius_of_Salamis.html?id=brxgNsxJKkUC.

Erasmus, Desiderius. *The Praise of Folly.* With a short life of the author by Hendrik Willem van Loon. Roslyn, NY: Walter J. Black, 1942.

Eusebius Pamphilus. *The Life of the Blessed Emperor Constantine from AD 306 to AD 337.* Merchantville, NJ: Evolution Publishing, 2009.

Fackenheim, Emil. *What Is Judaism? An Interpretation for the Present Age.* Syracuse, NY: Syracuse University Press, 1999.

Fauvel, John, and Raymond Flood, Michael Shortland, and Robin Wilson. *Let Newton Be!* Oxford: Oxford University Press, 1989.

Favaro, John. "Slightly Out of Tune: The Story of Musical Temperament." www.johnfavaro.com.

Ferris, Timothy. "Measuring an Intellect as Limitless as the Universe." Review of James Gleick's *Isaac Newton.* http://articles.latimes.com/2003/jul/20/books/bk-ferris20.

Finney, Gretchen Ludke. "Music: The Breath of Life." *The Centennial Review of Arts & Science* 4, no. 2 (Spring 1960): 179–205.

Flammarion, Camille. *Omega: The Last Days of the World.* Introduction by Robert Silverberg. Lincoln and London: University of Nebraska Press, 1999.

Flaste, Richard, Holcomb Noble, Walter Sullivan, and John Noble Wilford. *New York Times Guide to the Return of Halley's Comet.* New York: Times Books, 1985.

Flaubert, Gustave. "Herodias." In *Three Tales.* Translated by Robert Baldick. London: Penguin, 1961.

Bibliography

———. *The Temptation of Saint Anthony.* Translated by Lafcadio Hearn, introduction by Elizabeth Bisland. New York: Modern Library, 1992.

Fleming, William. *Arts & Ideas.* Forth Worth, TX: Holt, Rinehart, and Winston, 1991.

Fomenko, A. T., et al. *Chronology 1.* Vol. 1 of *History: Fiction or Science?* Bend, OR: Delamere Resources, LLC, 2003–2006.

Force, James E. "Newton, the 'Ancients,' and the 'Moderns.'" In *Newton and Religion: Context, Nature, and Influence,* edited by James E. Force and Richard H. Popkin, 237–57. Dordrecht, The Netherlands: Kluwer Academic Publishers, 1999.

———. "Natural Law, Miracles, and Newtonian Science." In *Newton and Newtonianism: New Studies,* 65–92. Dordrecht, The Netherlands: Kluwer Academic Publishers, 2004.

———. *William Whiston: Honest Newtonian.* Cambridge: Cambridge University Press, 1985.

Force, James E., and Richard H. Popkin, eds. *Newton and Religion: Context, Nature, and Influence.* Dordrecht, The Netherlands: Kluwer Academic Publishers, 1999.

Force, James E., and Sarah Hutton, eds. *Newton and Newtonianism: New Studies.* Dordrecht, The Netherlands: Kluwer Academic Publishers, 2004.

Frye, Northrop. *Anatomy of Criticism.* Princeton, NJ: Princeton University Press, 1971.

———. *The Great Code: The Bible and Literature.* New York and London: Harcourt Brace Jovanovich, 1982.

Gardner, Martin. *Did Adam and Eve Have Navels? Discourses on Reflexology, Numerology, Urine Therapy, and Other Dubious Subjects.* New York: Norton, 2000.

Germanicus. *Les Phénomènes d'Aratos.* Paris: Les Belles Lettres, 2003.

Gibbon, Edward. *The Decline and Fall of the Roman Empire.* Vol. 1. New York: The Modern Library, 1995.

Gilgamesh. Translated by John R. Maier, edited by John Gardner. New York: Knopf, 1984.

Gjertsen, Derek. "Newton's Success." In *Let Newton Be!* edited by John Fauvel, Raymond Flood, Michael Shortland, and Robin Wilson, 23–41. Oxford: Oxford University Press, 1988.

Gladstone, Rick. "26% of Adults Anti-Semitic, Survey Finds." *New York Times,* May 14, 2014.

Glassie, John. *A Man of Misconceptions: The Life of an Eccentric in an Age of Change*. New York: Riverhead, 2012.

Gleick, James. *Isaac Newton*. New York: Pantheon, 2003.

Gleiser, Marcelo. *The Dancing Universe: From Creation Myths to the Big Bang*. New York: Dutton/Penguin Group, 1997.

Goethe. *Selected Verse*. Edited by D. Luke. New York: Penguin, 1981.

Goff, Matthew. "The Millennial Scientist: Isaac Newton Reading Daniel 7." Knowing the Time: Knowing of a Time: 3rd Annual Conference of the Center for Millennial Studies, Boston University, December 6–8, 1998. Conference proceedings.

Goldish, Matt. *Judaism in the Theology of Sir Isaac Newton*. Dordrecht, The Netherlands: Kluwer Academic Publishers, 1998.

Goldsmith, Oliver. *The Vicar of Wakefield*. London: Collins, 1964.

Golinski, Jan. "The Secret Life of an Alchemist." In *Let Newton Be!* edited by John Fauvel, Raymond Flood, Michael Shortland, and Robin Wilson, 146–67. Oxford: Oxford University Press, 1988.

Goonetilleke, D. C. R. A. *Joseph Conrad's "Heart of Darkness": A Routledge Study Guide*. Abingdon, Oxford, UK: Routledge, 2007.

Gouk, Penelope. "The Harmonic Roots of Newtonian Science." In *Let Newton Be!* edited by John Fauvel, Raymond Flood, Michael Shortland, and Robin Wilson, 100–125. Oxford: Oxford University Press, 1988.

———. "Music and the Emergence of Experimental Science in Early Modern Europe." *SoundEffects* 2, no. 1 (2012): 6–21.

Greaves, Richard L., Robert Zaller, and Jennifer Tolbert Roberts. *Civilizations of the West: The Human Adventure*. Vol. 2, *From 1660 to the Present*. New York: HarperCollins, 1992.

Gregory, David. *Elementa astronomiae physicae et geometriae*. https://archive.org/details/elementsofastron00greg.

Grosso, Michael. *The Millennium Myth: Love and Death at the End of Time*. Wheaton, IL: Quest Books, 1995.

Guth, Steven. "The Missing Dark Ages: A Look at the Phantom History Hypothesis." *New Dawn* 7, no. 3 (2013): 26–33.

Guthrie, Kenneth Sylvain, comp. and trans., and David Fideler, ed. *The Pythagorean Sourcebook and Library*. Grand Rapids, MI: Phanes Press, 1988.

Halley, Edmund. "Some Considerations about the Cause of the Universal Deluge, Laid before the Royal Society, on the 12th of December 1694."

R. S. S. Halley, *Philosophical Transactions* (1683–1775). 1753-01-01. 33:118–123. http://archive.org/details/philtrans08240252.

Hammer, Joshua. "What Is Beneath the Temple Mount?" *Smithsonian*, April 2011. www.smithsonianmag.com/history/what-is-beneath-the-temple-mount-920764.

Haycock, David Boyd. *William Stukeley: Science, Religion and Archaeology in Eighteenth-Century England*. Suffolk, UK: Boydell Press, 2002. Newton Project, www.newtonproject.ac.uk.

Heath, Sir Thomas. *Aristarchus of Samos: The Ancient Copernicus*. New York: Dover, 1981. See also https://archive.org/stream/aristarchusofsam00heatuoft#page/n5/mode/2up.

Hecht, Jennifer Michael. *Doubt: A History*. San Francisco: HarperSanFrancisco, 2003.

Herodotus. *The Histories*. Translated by George Rawlinson. www.romanroadsmedia.com/materials/herodotus.pdf.

Hogue, John. *Nostradamus: The Complete Prophecies*. Shaftesbury, Dorset, UK: Element Books, 1997.

Iliffe, Rob. "Digitizing Isaac: The Newton Project and an Electronic Edition of Newton's Papers." In *Newton and Newtonianism: New Studies*, 23–38. Dordrecht, The Netherlands: Kluwer Academic Publishers, 2004.

———. *Newton: A Very Short Introduction*. Oxford: Oxford University Press, 2007.

———. "Prosecuting Athanasius: Protestant Forensics and the Mirrors of Persecution." In *Newton and Newtonianism: New Studies*, 113–54. Dordrecht, The Netherlands: Kluwer Academic Publishers, 2004.

Ingermanson, Randall. *Who Wrote the Bible Code? A Physicist Probes the Current Controversy*. Colorado Springs, CO: WaterBrook Press, 1999.

Isacoff, Stuart. *Temperament: The Idea That Solved Music's Greatest Riddle*. New York: Alfred A. Knopf, 2001.

Jean-Aubry, G. *Joseph Conrad: Life and Letters*. Vol. 1. Garden City, NY: Doubleday, 1927.

Johnson, Paul. *A History of Christianity*. New York: Simon & Schuster, 1976.

———. *A History of the Jews*. New York: HarperCollins, 1988.

Josephus, Flavius. *Complete Works*. Translated by William Whiston. Grand Rapids, MI: Kregel, 1960.

Jung, C. G. *The Basic Writings of C. G. Jung*. New York: The Modern Library, 1959.

———. "The Dark Side of God." In *Doomsday? How the World Will End—and When*, edited by Martin Ebon, 89–100. New York: New American Library, 1977.

———. *Psychology and Alchemy*. Vol. 12, *Collected Works*. Bollingen Series 20. 2nd edition, completely revised. Princeton, NJ: Princeton University Press, 1968.

Kean, Sam. *Newton, the Last Magician*. National Endowment for the Humanities. www.neh.gov/humanities/2011/januaryfebruary/feature/newton-the-last-magician.

Keynes, Geoffrey, ed. *Poetry and Prose of William Blake*. London: Nonesuch Library, 1956.

Keynes, John Maynard. "Newton the Man." In *Essays in Biography*, 310–21. London: Macmillan, 1961.

King, David. *Finding Atlantis: A True Story of Genius, Madness, and an Extraordinary Quest for a Lost World*. New York: Harmony Books, 2005.

Kingsley, Peter. *A Story Waiting to Pierce You: Mongolia, Tibet, and the Destiny of the Western World*. Point Reyes, CA: The Golden Sufi Center, 2011.

Klein, Naomi. *This Changes Everything: Capitalism vs. the Climate*. New York: Simon & Schuster, 2014.

Knoespel, Kenneth J. "Interpretive Strategies in Newton's *Theologiae gentilis origines philosophiae*." In *Newton and Religion: Context, Nature, and Influence*, edited by James E. Force and Richard H. Popkin, 179–202. Dordrecht, The Netherlands: Kluwer Academic Publishers, 1999.

———. "Newton in the School of Time: The *Chronology of Ancient Kingdoms Amended* and the Crisis of Seventeenth-Century Historiography." *The Eighteenth Century: Theory and Interpretation* 30, no. 5 (1989): 19–41.

Koloski-Ostrow, Ann Olga. *The Archaeology of Sanitation in Roman Italy: Toilets, Sewers, and Water Systems*. Studies in the History of Greece and Rome. Chapel Hill: Univeristy of North Carolina Press/Amazon Digital

———. "Raising a Really Big Stink." *Brandeis Magazine* (Spring 2013): 42–43.

Krawcewicz, Wieslaw Z., Gleb V. Nosovskij, and Petr P. Zabreiko. "Investigation of the Correctness of the Historical Dating." World Mysteries, New Tradition Sociological Society, 2002, www.world-mysteries.com/sci_16.htm.

Küng, Hans. *The Catholic Church: A Short History*. New York: The Modern Library, 2001, 2003.

---. *Christianity: Essence, History, and Future.* New York: Continuum, 2003.

Lamartine, Alphonse de. *A Pilgrimage to the Holy Land.* Vol. 2. London: Richard Bentley, 1835.

Lawrence, D. H. *Psychoanalysis and the Unconscious* and *Fantasia of the Unconscious.* Mineola, NY: Dover, 2005.

---. *The Symbolic Meaning: The Uncollected Versions of "Studies in Classic American Literature.* New York: Viking, 1962.

Lequeuvre, Serge. "Le Jardin Secret de Newton." *Historia mensuel,* 668 (August 200): 54–57.

Levitin, Dmitri. "Halley and the Eternity of the World Revisited." *Notes and Records of the Royal Society.* 4 September 2013. http://rsnr.royalsocietypublishing.org/content/67/4/315.

Levitt, Ari. "The Model for Messianic Community." The Family Bible (website). Updated August 3, 2016. http://familybible.org/model/appendix_e.html.

Lovelock, James. *The Ages of Gaia: A Biography of Our Living Brain.* London and New York: W. H. Norton, 1988.

Luce, J. V. *The End of Atlantis: New Light on an Old Legend.* London: Palladin, 1970.

MacCulloch, Diarmaid. *Christianity: The First Three Thousand Years.* New York: Penguin Books, 2009.

Machiela, Daniel A. "The Genesis Apocryphon (1Q20): A Reevaluation of its Text, Interpretive Character, and Relationship to the Book of Jubilees." Ph.D. dissertation, Notre Dame University, 2007. http://etd.nd.edu/ETD-db/theses/available/etd-07022007-205251/unrestricted/MachielaD072007.pdf.

---. "Some Jewish Noah Traditions in Syriac Christian Sources." In *Noah and His Book(s),* edited by Michael E. Stone, Aryeh Amihay, and Vered Hillel, 237–82. Atlanta, GA: Society of Biblical Literature, 2010.

Manuel, Frank E. *Isaac Newton, Historian.* Cambridge, MA: The Belknap Press of Harvard University Press, 1963.

---. *A Portrait of Isaac Newton.* Cambridge, MA: Harvard University Press, 1968.

---. *The Religion of Isaac Newton.* The Fremantle Lectures 1973. Oxford: Clarendon Press, 1974.

Markley, Robert. "Newton, Corruption, and the Tradition of Universal History." In *Newton and Religion: Context, Nature, & Influence,* edited by James E. Force and Richard H. Popkin, 121–43. Dordrecht, The Netherlands: Kluwer Academic Publishers, 1999.

Martin, Malachi. *The Decline and Fall of the Roman Church*. New York: G. P. Putnam's, 1981.

Martin, Thomas R. *Ancient Greece: From Prehistoric to Hellenistic Times*. New Haven & London: Yale University Press, 1996/2000.

Mazza, Ed. "Stephen Hawking Warns That Aggression Could 'Destroy Us All.'" *Huffington Post*, February 23, 2015. www.huffingtonpost.com/2015/02/23/stephen-hawking-aggression_n_6733584.html.

McGuire, J. E., and P. M. Rattansi. "Newton and the 'Pipes of Pan.'" *Notes and Records of the Royal Society of London* 21, no. 2 (December 1966): 108–43.

Mathiesen, Thomas J. *Apollo's Lyre: Greek Music and Musical Theory in Antiquity and the Middle Ages*. Lincoln and London: University of Nebraska Press, 1999.

Mifsud, Anton, Simon Mifsud, Chris Agius Sultana, and Charles Savona Ventura. *Malta: Echoes of Plato's Island*. Malta: Prehistoric Society of Malta, 2001.

Miller, Arthur I. *137: Jung, Pauli, and the Pursuit of a Scientific Obsession*. New York: W. W. Norton, 2009.

Miller, Richard, and Iona Miller. *The Modern Alchemist: A Guide to Personal Transformation*. Grand Rapids, MI: Phanes Press, 1994.

Milton, John. *A Treatise on Christian Doctrine, Compiled from the Holy Scriptures Alone*. Boston: Boston Publishers, 1825. www.newtonproject.sussex.ac.uk/view/texts/normalized/THEM00326.

Moore, Noel Brooke, and Kenneth Bruder. *Philosophy: The Power of Ideas*. 7th edition. New York: McGraw-Hill, 2007.

Moore, Philip. "Newton's Forbidden Works Rescued." In *The End of History Messiah Conspiracy*. www.ramsheadpress.com/messiah/ch11.html#footnote13.

Murrin, Michael. "Newton's Apocalypse." In *Newton and Religion: Context, Nature, and Influence*, edited by James E. Force and Richard H. Popkin, 203–20. Dordrecht, The Netherlands: Kluwer Academic Publishers, 1999.

Needham, Joseph. *Science and Civilization in China*. 7 Vols. Cambridge: Cambridge University Press, 1954–98.

Newman, William R., and Lawrence Principe. *Alchemy Tried in the Fire: Starkey, Boyle and the Fate of Helmontian Chymistry*. Chicago: University of Chicago Press, 2005.

Newton, Sir Isaac. *The Chronology of Ancient Kingdoms Amended*. London: Histories & Mysteries of Man Ltd., 1988.

———. *Correspondence of Isaac Newton.* Vol. 2, *1676–1687.* Edited by H. W. Turnbull. Cambridge: Cambridge University Press, 1960.

———. *Correspondence of Isaac Newton.* Vol. 3, *1688–1694.* Edited by H. W. Turnbull. Cambridge: Cambridge University Press, 1961.

———. "Draft Chapters of a Treatise on the Origin of Religion and its Corruption." Newton Project. www.newtonproject.sussex.ac.uk/view/texts/normalized/THEM00077.

———. "Drafts on the History of the Church." Newton Project. Section 4, www.newtonproject.sussex.ac.uk/view/texts/normalized/THEM00221; Section 7, www.newtonproject.sussex.ac.uk/view/texts/normalized/THEM00237.

———. "Irenicum, or Ecclesiastical Polyty Tending to Peace." Newton Project. www.newtonproject.sussex.ac.uk/view/texts/normalized/THEM00003.

———. *Mathematical Principles of Natural Philosophy* and *Optics.* Chicago: Great Books of the Western World/Encyclopedia Britannica, 1952.

———. "Miscellaneous Draft Portions of 'Theologiæ Gentilis Origines Philosophicæ.'" Newton Project. www.newtonproject.sussex.ac.uk/view/translation/TRAN00010.

———. "Notes on Prophetic Works." Newton Project. www.newtonproject.sussex.ac.uk/view/texts/normalized/THEM00362.

———. *Observations upon the Prophecies of Daniel, and the Apocalypse of St. John.* London: Printed by J. Darby and T. Browne, 1733. Newton Project. www.newtonproject.sussex.ac.uk/view/texts/normalized/THEM00193.

———. "Papers Relating to Chronology and 'Theologiæ Gentilis Origines Philosophicæ.'" Newton Project. www.newtonproject.sussex.ac.uk/view/texts/diplomatic/THEM00098.

———. "Paradoxical Questions concerning the Morals & Actions of Athanasius & His Followers." Newton Project. www.newtonproject.sussex.ac.uk/view/texts/normalized/THEM00117.

———. "A Short Schem of the True Religion." Newton Project. www.newtonproject.sussex.ac.uk/view/texts/normalized/THEM00007.

———. "The Synchronisms of the Three Parts of the Prophetic Interpretation." Newton Project. www.newtonproject.sussex.ac.uk/view/texts/normalized/THEM00049.

———. Theologiae Gentiles Origines Philosophicae. Newton Project. www.newtonproject.sussex.ac.uk/view/texts/normalized/THEM00260.

———. "Treatise on Revelation." Newton Project. www.newtonproject.sussex.ac.uk/view/texts/normalized/THEM00135.

———. "Twenty-Three Quaeries about the Word ὁμοούσιος." Newton Project. www.newtonproject.sussex.ac.uk/view/texts/normalized/THEM00011.

———. *Two Notable Corruptions of Scripture*. Part 1: ff. 1–41. Newton Project. www.newtonproject.sussex.ac.uk/view/texts/normalized/THEM00261.

———. "Untitled Treatise on Revelation." Section 1.4. Newton Project. www.newtonproject.ox.ac.uk/view/texts/normalized/THEM00182.

"Newton's Dark Secrets." *NOVA*, PBS, November 15, 2005. www.pbs.org/wgbh/nova/physics/newton-dark-secrets.html.

Noorbergen, Rene. *Secrets of the Lost Races: New Discoveries of Advanced Technology in Ancient Civilizations*. Indianapolis/New York: Bobbs-Merrill, 1977.

O'Carroll, Eoin. "Judgment Day May 21: When Will the World Actually End?" *Christian Science Monitor*. CSMonitor.com. May 18, 2011.

Pagels, Elaine. *Revelations: Visions, Prophecy, and Politics in the Book of Revelation*. New York: Penguin, 2012.

Pierce, Larry. "Chronology Wars." www.answersingenesis.org/articles/am/v6/n1/chronology-wars.

Plant, Stephen. *Simone Weil: A Brief Introduction*. Maryknoll, NY: Orbis Books, 1996.

Plato. *Critias*. Translated with an introduction by Benjamin Jowett. www.gutenberg.org/files/1571/1571.txt.

———. *The Dialogues of Plato*. Translated by Benjamin Jowett. 2 vols. New York: Random House, 1937.

Plutarch. *The Lives of the Ancient Greeks and Romans*. Translation by John Dryden. New York: Modern Library, 1992.

———. *Plutarch's Lives, vol. 1*. Translated by John Drydeden. New York: Cosimo, 2008 [1859].

Poe, Edgar Allan. *Tales of Edgar Allan Poe*. New York: Random House, 1944.

Pope, Alexander. *The Poetry of Pope: A Selection*. Edited by M. H. Abrams. New York: Appleton-Century-Crofts, 1954.

Popkin, Richard H. "Plans for Publishing Newton's Religious and Alchemical Manuscripts, 1982–1998." In *Newton and Newtonianism: New Studies*, 15–22. Dordrecht, The Netherlands: Kluwer Academic Publishers, 2004.

Principe, Lawrence M. *The Secrets of Alchemy*. Chicago: University of Chicago Press, 2015.

Prisca Sapientia. MathPages. www.mathpages.com/home/kmath066/kmath066.htm.

Pursiful, Darrell. "Mythology and the Table of Nations." October 11, 2007. http://pursiful.com/2007/10/mythology-and-the-table-of-nations.

Rattansi, Piyo. "Newton and the Wisdom of the Ancients." In *Let Newton Be!* edited by John Fauvel, Raymond Flood, Michael Shortland, and Robin Wilson, 185–201. Oxford: Oxford University Press, 1989.

Roberts, J. M. *The Penguin History of the World.* London: Penguin Books, 1992.

Rothstein, Edward. "Finding Archimedes in the Shadows." *New York Times,* October 16, 2011.

Sachar, Abram Leon. *A History of the Jews.* New York: Alfred A. Knopf, 1930.

Santillana, Giorgio de, and Hertha von Dechend. *Hamlet's Mill: An Essay Investigating the Origins of Human Knowledge and Its Transmission through Myth.* Boston: David R. Godine, 1969.

Seibert, Brian. "The Book of Revelation, with Some Teeth in It." *New York Times,* November 13, 2013.

Smith, Alison J. "Everybody's Doomsday." In *Doomsday? How the World Will End—and When,* edited by Martin Ebon, 148–62. New York: New American Library, 1977.

Smoley, Richard. "Apocalypse: The End Times and the Restoration of All Things." *New Dawn,* no. 135 (November–December 2012): 18–19.

———. "2012 and the Annoying Persistence of Time." Reality Sandwich Website. www.realitysandwich.com/2012_and_annoying_persistence_time.

———. "Dreams: Another Reality?" *New Dawn,* no. 135 (November–December 2012): 53–58.

Smolinski, Reiner. "The Logic of Millennial Thought: Sir Isaac Newton Among His Contemporaries." In *Newton and Religion: Context, Nature, and Influence,* edited by James E. Force and Richard H. Popkin, 259–89. Dordrecht, The Netherlands: Kluwer Academic Publishers, 1999.

Smollett, Tobias. *The Expedition of Humphrey Clinker.* New York: Penguin, 1967.

Snobelen, Stephen D. "Isaac Newton, Heretic: The Strategies of a Nicodemite." *British Journal for the History of Sciences* 32 (1999): 381–419.

———. "Isaac Newton (1642–1727): Natural Philosopher, Biblical Scholar and Civil Servant." Encylopedia of the Enlightenment. https://isaacnewtonstheology.files.wordpress.com/2013/06/newton-in-encyclopedia-of-the-enlightenment.pdf.

———. "'The Mystery of This Restitution of All Things': Isaac Newton on the Return of the Jews." In *Millenarianism and Messianism in Early Modern European Culture: The Millenarian Turn,* edited by J. E. Force and

R. H. Popkin, 95–118. Dordrecht, The Netherlands: Kluwer Academic Publishers, 2001.

———. "The Myth of the Clockwork Universe: Newton, Newtonianism, and the Enlightenment." In *The Persistence of the Sacred in Modern Thought*, edited by Chris L. Firestone and Nathan Jacobs, 149–84. Notre Dame, IN: University of Notre Dame Press, 2012.

———. "'Not in the Language of Astronomers': Isaac Newton, Scripture and the Hermeneutics of Accommodation." In *Interpreting Nature and Scripture in the Abrahamic Religions: History of a Dialogue*, vol. 1, edited by Jitse M. van der Meer and Scott H. Mandelbrote, 491–530. Leiden, The Netherlands: Brill, 2008.

———. "Statement on the Date 2060." March 2003. Updated May 2003 and June 2003. http://isaac-newton.org/statement-on-the-date-2060.

———. "'A Time and Times and the Dividing of Time': Isaac Newton, the Apocalypse, and 2060 A.D." *Canadian Journal of History/Annales canadiennes d'histoire* 38 (November–December 2003): 537–51.

Sozomen. *The Ecclesiastical History*. http://biblehub.com/library/sozomen/the_ecclesiastical_history_of_sozomenus/chapter_xxvi_erection_of_a_temple.htm.

Stach, Reiner. *Kafka: The Decisive Years*. Orlando, FL: Harcourt, 2005.

Starr, Chester G. *A History of the Ancient World*. New York/Oxford: Oxford University Press, 1991.

Steiner, George. *Tolstoy or Dostoevsky: An Essay in the Old Criticism*. New York: Dutton, 1971.

Stone, I. F. *The Trial of Socrates*. New York: Random/Anchor Books, 1989.

Stone, Michael E., Aryeh Amihay, and Vered Hillel, eds. *Noah and His Book(s)*. Atlanta, GA: Society of Biblical Literature, 2010.

Stringer, Bruce. "The Mystery of Music, Power of Sound: & the Question of Levitation." *New Dawn*, no. 138 (May–June 2013): 55–62.

Stukeley, William. "Memoir of Newton, Sent to Richard Mead in Four Instalments with Covering Letters, Dated 26 June to 22 July 1727." Newton Project. www.newtonproject.sussex.ac.uk/view/texts/normalized/THEM00158.

Swift, Jonathan. *Gulliver's Travels*. New York: Random House, 1950.

Tacitus, P. Cornelius. *The Annals and the Histories*. Chicago: Great Books of the Western World/Encyclopedia Britannica, 1952.

Theodoret. *Nicene and Post-Nicene Fathers*. Series 2, vol. 3. "The Ecclesiastical

History, Dialogues, and Letters of Theodoret." www.sacred-texts.com/chr/ecf/203/2030048.htm.

Thompson, William Irwin. *At the Edge of History/Passages about Earth*. Aurora, CO: Lindisfarne Press, 1990.

———. *Darkness and Scattered Light: Speculations on the Future*. Garden City, NY: Anchor Books, 1978.

———. *Imaginary Landscape: Making Worlds of Myth and Science*. New York: St. Martin's, 1989.

Tichenor, Henry M. *The Creed of Constantine: Or, the World Needs a New Religion*. Classic reprint ed. London: Forgotten Books, 2015.

Titans and Olympians: Greek & Roman Myth. New York: Time-Life Books, 1997.

Toffler, Alvin. *Future Shock*. New York: Bantam, 1971.

Travels of John Mandeville. www.lib.rochester.edu/camelot/teams/tkfrm.htm.

Tuval, Michal. "The Role of Noah and the Flood in *Judean Antiquities* and *Against Apion* by Flavius Josephus." In *Noah and His Book(s)*, edited by Michael E. Stone, Aryeh Amihay, and Vered Hillel, 167–81. Atlanta, GA: Society of Biblical Literature, 2010.

Twain, Mark. *The Innocents Abroad*. New York: Signet/New American Library, 1966.

Vavra, Robert. *Unicorns I Have Known*. New York: William Morrow, 1983.

Velikovsky, Immanuel. *Worlds in Collision*. New York: Dell, 1967.

Vergano, Dan. "Famed Roman Shipwreck Reveals More Secrets." *USA Today*, January 4, 2013.

Vidal, Gore. *Julian*. New York: Vintage International, 1992.

Vilenkin, Alex. *Many Worlds in One: The Search for Other Universes*. New York: Hill and Wang/Farrar, Strauss and Giroux, 2006.

Voltaire. "Letters on Newton from the Letters on the English or *Lettres Philosophiques*." www.fordham.edu/halsall/mod/1778voltaire-newton.asp.

———. *Philosophical Dictionary*. In *The Portable Voltaire*. New York: Viking, 1948.

———. *Le Siècle de Louis XIV*. 2 vols. Paris: Garnier-Flammarion, 1966.

Von Franz, Marie-Louise. *Alchemy: An Introduction to the Symbolism and the Psychology*. Toronto: Inner City Books, 1980.

Weissman, Judith. *Of Two Minds: Poets Who Hear Voices*. Hanover, NH and London: Wesleyan University Press, 1993.

Wells, H. G. *In the Days of the Comet*. New York: Airmont, 1966.

Westfall, Richard S. *Never at Rest: A Biography of Sir Isaac Newton*. Cambridge: Cambridge University Press, 1980.

Whipple, Fred L. *The Mystery of Comets*. Washington, D.C.: Smithsonian Institution Press, 1985.

Whiston Biography. www-history.mcs.st-andrews.ac.uk/history/Biographies/Whiston.html.

Whiston, William. *A New Theory of the Earth from its Original to the Consummation of all things wherein the Creation of the World in Six Days, the Universal Deluge, and the General Conflagration, are shewn to be perfectly agreeable to Reason and Philosophy*. London, 1737. http://books.google.co.uk/books/about/A_new_Theory_of_the_Earth.html?id=vZI5AAAAcAAJ.

White, Michael. *Isaac Newton: The Last Sorcerer*. Reading, MA: Helix Books/Addison-Wesley, 1997.

White, L. Michael. *From Jesus to Christianity*. New York: HarperOne, 2004.

Willey, Basil. *Nineteenth Century Studies*. London: Chatto & Windus, 1955.

——. *The Seventeenth Century Background*. New York: Doubleday, 1934.

Wills, Garry. *Why I Am a Catholic*. Boston/New York: Houghton Mifflin, 2002.

Yates, Frances A. *The Art of Memory*. Chicago: University of Chicago Press, 1966.

——. *Giordano Bruno and the Hermetic Tradition*. Chicago: University of Chicago Press, 1964.

Yeomans, Donald K. *Comets: A Chronological History of Observation, Science, Myth and Folklore*. New York: John Wiley, 1991.

Zeller, Eduard. *Outlines of the History of Greek Philosophy*. New York: Meridian Books, 1955.

Index

Abraham (biblical figure), 62, 172, 180, 258, 269, 284–85
Adam (biblical figure), 30, 108n, 109n, 195, 201, 231–32, 341
Adams, Mark, 299
alchemy, 310–12, 310n, 313–14
  "astronomical theology," 316–17
  Boyle, 308, 309, 315, 319–20, 321–23
  Cooper, 305–7, 309
  currencies, 273n, 307–8, 308n
  decomposition, 303–4, 313
  Emerald Tablet, 313, 412–13
  Hartlib, 309
  as legal/illegal, 307–10
  Locke, 309, 321–23
  as love/worship of God, 319
  Needham, 311, 314, 320–21
  the Net, 315–17, 316n
  Newman, 315, 316, 317–18, 323–24
  Newton, 273n, 307–10, 312–13, 314–15, 316–19, 321–24, 363, 373
  papacy, 308
  Philosophers' (Angels') Stone, 273n, 308, 312, 315, 317, 319, 322, 326, 327
  Principe, 315–16, 317, 322
  Principia (Newton), 319
  prisca sapientia, 316–17
  as secret of life, 307, 317–18, 324
  Starkey, 318–19
  vegetative spirit, 304–5, 307, 324
Alexander of Alexandria, 69, 70–71, 88, 89, 92, 93
Alexander of Constantinople, 69, 79, 81, 83, 88, 89–90
Alexander the Great, 40, 68, 90, 92, 93, 212, 252–53
Alsted, Johann Heinrich, 178
Amichai, Yehuda, 172
Amihay, Aryeh, 208, 209
Anaxagoras of Clazomene, 337–41, 341n
Anthony (saint)
  and Athanasius, 69–70, 119, 121, 123–25, 126, 127, 129
  death of, 305
  Flaubert on, 120–21
  idolatry/religious relics, 119, 124, 126–27

literacy, 127–28
Newton on, 70, 120, 121–28
supernatural experiences, 122, 124–25, 126
Antichrist, 22, 23, 28, 128, 154, 161, 176, 183
Antigonus Gonatas, 278–80, 281
Antiochus IV Epiphanes, 40–41, 187
anti-Trinitarianism, 47–48
    Arius, 22, 60–61, 69, 71, 75, 78, 79, 83, 88, 89
    corruptions of Scripture, 46, 48–49, 55–56
    Jews, 64, 182
    and love/worship of God, 387–88
    Snobelen, 48, 58
Apion, Peter, 243
Apocalypse, 21, 22, 23, 29, 134–35, 136, 137–38, 144–48, 154–55, 175, 177, 224, 385–87
Apocalypse in 2060, 15–16, 148, 162, 166, 167, 168–70, 169
Aratus, 278, 279, 280–84, 282n
Archimedes, 280, 333, 360, 363, 377, 378–80
Arianism
    Anthanius, 82, 84–85
    Constantius, 82, 114–15
    Melitius, 88
    Newton, 10, 60, 225, 227, 236, 308
    Whiston, 236
Aristarchus of Samos, 330, 331–36, 335n, 360
Aristotle, 8, 220, 302, 338, 339, 347–48, 353, 360

Arius, 60
    anti-Trinitarianism, 22, 60–61, 69, 71, 75, 78, 79, 83, 88
    Council of Nicaea, 74–75, 77, 80–81, 89
    death in boghouse as mythical, 77, 79–81, 83, 113, 114, 119
    excommunication/admittance back into church, 75, 78–79, 98, 109
Ark of the Covenant, 34, 172–73, 176
Armageddon, 154, 162–63, 169, 171, 175, 176
Arsenius, 85, 95, 101–4, 105, 111–12, 113–14, 119
Aspasia, 339, 340, 341, 342
astronomy
    Anaxagoras, 337–38
    Aristarchus, 330
    "astronomical theology," 204, 316–17, 388
    celestial bodies, 229–30, 338, 348, 349–50, 354
    central fire, 344, 347–49
    constellations, 272–73, 278, 279, 283–84, 285–87, 326–27
    day-equals-year formula, 185
    earthquakes, 249, 276
    equinoxes, 276–78
    Flamsteed, 220
    Greenwich Observatory, 220–21
    historical records, 275–76, 278
    music/music of the spheres, 353–54
    orbital velocities, 193, 241, 332, 354
    prophetic hieroglyphs, 317
    star globe, 272, 274, 276, 278, 280, 283–84, 285, 286–87

Athanasius, 70, 92–93
  and Anthony (saint), 69–70, 119, 121, 123–25, 126, 127, 129
  as Antichrist, 22
  as Archbishop of Alexandria, 71, 92–95, 97–106, 109–10, 115–17
  and Arius's mythical death, 81, 83
  and Arsenius's death, 85, 95, 102, 103–4, 105, 111–12, 113–14, 119
  and Constantius, 115–17
  corruptions of Scripture, 50, 69
  Council at Alexandria, 93, 112–13
  Council of Nicaea, 70–71, 74–75, 88, 89, 116
  Council of Tyre, 97–106, 109–10, 113, 119
  exile of, 81–83, 111, 114, 118, 123–24
  forged letters by, 93, 105–6, 111–12, 113–14, 119
  Great Apostasy, 85, 138–39, 151, 156
  idolatry/religious relics, 85, 119, 124, 126–27
  lies, 127, 128, 129
  Melitians versus, 95–96
  monasticism, 122
  severed hand myth, 85, 101–2, 104, 111, 114
  supernatural experiences, 122, 124–25
  Trinitarianism, 22, 55, 60–61, 63, 65, 69, 70–71, 75, 83, 94, 117–18, 138, 156
  Whore of Babylon, 129
Atlantis
  Aristotle, 302
  Chronology (Newton), 289, 293, 294, 295–96, 300–302
  chronology, global, 293
  Egerton Sykes Collection, 289–90
  flood, Ogygian, 289, 294–95
  Luce on, 293–94
  Maltese archipelago, 289, 296–99, 302
  Ogygia/Gozo island, 289, 294–95, 296, 297, 298, 299–300
  Plato, 290, 292, 293, 294, 295–96, 298, 299, 302
  Solon, 290, 291–92, 293–94, 302
atomic bomb, 384–85
Augustine, 127

Bach, Johann Sebastian, 136, 381
Bacon, Francis, 375–76, 377
Bacon, Roger, 305, 318, 327
Barrow, Isaac, 8, 9, 303
Basil of Caesarea, 95
Bedford, Arthur, 178
Belisarius, Flavius, 158
Bentley, Richard, 29, 49, 321
Bible code, 18, 23–26, 153
Blake, William, 223, 232, 331, 344
Bondi, Hermann, 2
Book of Daniel, 38
  day-equals-year formula, 42, 167, 186–87, 188, 190
  divine plan, 35–36, 41
  Great Apostasy, 163–64, 166, 186, 187
  Judgment Day, 179
  prophetic hieroglyphs, 33–37, 144, 163–64
  return of the Jews, 171, 184, 189, 190, 191
  studies of, 28–36, 38–41

synchronistical necessity, 38, 41–42
visions in, 39–40, 375
Book of Revelation, 18–19, 20,
21–22, 23, 129
  Apocalypse, 21, 22, 29, 134–35,
  136, 137–38, 144–47, 148,
  169–70
  channeled texts, 62, 136–38, 143,
  146, 148, 149, 155, 160
  Christianity and Constantine, 76
  corruptions of Scripture, 140
  day-equals-year formula, 42,
  168–69, 184
  divine plan, 32, 35–36, 37, 387
  earthquakes, 22, 149, 150
  flood of fire/*diluvium ignis*, 155,
  158–59
  future history, 42–43, 140–41,
  143, 149–50, 156, 159–60,
  169–71
  Great Apostasy, 139, 139n, 143,
  148, 161, 162–63, 164, 387
  Judgment Day, 152–53
  proof of existence of God, 26, 43
  prophetic hieroglyphs, 33–37,
  143–44, 146, 148, 149–50,
  155–61
  return of the Jews, 162–63
  scholarship, 28–38, 29, 42–43,
  139n
  surrealist images in, 20–21, 33, 147
  synchronistical necessity, 38, 41–42
  Temples, 132–33, 135–36, 140–41
  Whore of Babylon, 21, 29, 129,
  161–63, 186
  woman clothed in the sun, 21, 132,
  161, 162, 168–69

Boyle, Robert, 28, 60, 304, 308, 309,
315, 319–20, 321–23
Brewster, David, 322–23
Bruno, Giordano, 178, 330
Buchwald, Jed Z., 29, 270, 286–87
Buñuel, Luis, 68, 99, 102–3, 104, 116,
117

Cassini, Jean-Dominique, 193
Castillejo, David, 62, 150, 186, 190,
254, 256, 257–58, 263, 270, 314
Cave, William, 71, 110
celibacy, 10, 100, 122–23, 125, 367
Champion, Justin, 49
Chazelle, Bernard, 253, 253n, 254
Chiron, 271–73, 274, 276, 283–86
Chomsky, Noam, 28–29, 385
chronology, global, 251–52, 257–58,
262, 267–69, 272, 287, 292
*Chronology of Ancient Kingdoms
Amended* (Newton), 252–55,
253n, 255n, 256–57, 263, 270,
292–93
  Atlantis, 289, 293, 294, 295–96,
  300–302
  Castillejo, 254, 257–58, 263, 270
  constellations, 272–73
  Cress, 255–56
  Jews, 131, 256, 261, 263, 266,
  267–68, 269
  law of length of reigns, 260,
  261–62, 262n
  law of parsimony, 258–61
  Pierce, 257
  repopulation of world, 263–65
  Saturn (god), 194, 212–13, 214
  Whiston, 236

Church of the Great Martyrium, 108–9, 109n, 110
Church of the Martyrs, 88, 88n, 94
Cleanthes, 305, 334–36, 335n
Clement of Alexandria, 267, 271, 284, 285
comet(s)
   divine plan, 223, 224–27, 226–27n, 229–30, 230–31n
   and Flood, 232–33, 243–44
   Great Comet of 1680, 221–22, 224, 225, 230–31, 230–31n, 232, 233–34, 242, 243, 244–45, 246–49
   Halley's comet, 218–19, 219n, 230, 237, 242–43, 244
   omens, 219–20, 219n
   Principia (Newton), 222–23, 224–25, 244–45
comma Johanneum ("John's phrase"), 49–50, 51, 53, 54, 55, 57–58
Conduitt, Catherine Barton, 11, 246–48
Conduitt, John, 245–46, 248, 249, 249n, 259, 318, 319
Conrad, Joseph, 141–43, 147–48, 149, 151, 159, 161–62, 164, 221, 221n, 374
consciousness, state of, 331, 344, 388
Constantine the Great, 68, 72–73, 82, 109
   and Arius, 78–79, 81, 83, 98, 109
   and Athanasius, 97, 98, 109, 110, 111, 113
   Christianity, 72–73, 76, 88–89, 150, 161

Church of the Great Martyrium, 108–9, 109n, 110
Council of Nicaea, 73, 74, 75, 116
Council of Tyre, 97, 98, 109, 110, 113
idolatry/religious relics, 72, 78, 79, 108–9, 150
statute of, 77–78, 79
Constantius II (emperor), 82, 85, 114–16, 117, 118
Cooper, William, 305–7, 309
Copernicus, Nicholas, 275, 330, 331–32
Coppola, Francis Ford, 142–43
corruptions of Scripture
   anti-Trinitarianism, 49, 55–56
   Athanasius, 69
   Book of Revelation, 140
   Küng, 57–58, 88
   letters from Newton, 44–46, 48–51, 56–58, 63–64, 140, 392–97
   Newton, 69, 76–77, 373
   Trinitarianism, 58
Council at Alexandria, 93, 112–13
Council of Constantinople, 75, 118–19, 139
Council of Nicaea, 63, 68, 73–74, 73n, 88, 149–50
   Arius, 74–75, 77, 80–81, 89
   Athanasius, 70–71, 74–75, 89, 116
   Constantine, 116
   corruptions of Scripture, 50, 58
   Great Apostasy, 139
   Nicene Creed, 89, 139
Cress, L., 255–56

Cromwell, Oliver, 38, 47–48, 178–79, 180
cycles concept, 205–6
Cyprian, 33, 69
Cyril of Alexandria, 91

David (king), 130, 181–82n, 255–56, 267, 269
day-equals-year formula, 42, 167–68, 184, 185, 186–87, 188, 190
decomposition, 303–4, 313
Descartes, René, 8, 361
Dick, Hugh G., 375–76
Dio Chrysostom, 90
Diocletian's persecution of Christians, 70, 74, 84, 86, 98, 105, 148
Dionysius, 20, 99, 100, 101–3, 105, 112
Dobbs, B. J., 61, 223, 224, 225, 226, 324, 387
Dolnick, Edward, 10, 329–30, 382
Dreyer, J. L. E., 332
Drosnin, Michael, 23–24, 25–26
Dry, Sara, 180
Durant, Will, 19, 111
du Temps, Jean, 266

earthquakes, 22, 149, 150, 237, 249, 276
Ehrman, Bart D., 58
Einstein, Albert, 2, 12, 350, 361, 384
Emerald Tablet, 313, 412–13
Empedocles, 185, 358
End Times, 32, 152, 175–76, 177–78, 179, 183–84, 191
Enlightenment, 236, 270, 302, 383
Enoch (patriarch), 128–29, 197–98, 209–10, 326–27
Epicurus, 122, 329, 358

Epiphanius, 87, 88n
Erasmus, Desiderius, 51–55
Eudoxus of Cnidus, 279–80, 281, 282–84, 287
euhemerus, 211–13, 212–13n
Eumalos of Cyrene, 289
Eusebians, 88, 89, 92, 94, 97, 98
Eusebius (archbishop of Nicodemia), 85
Eusebius of Caesarea, 99, 101, 105, 110, 112, 154, 258, 267, 294
Eusebius Pamphilus, 39, 72–73, 73n, 74
Ezekiel (prophet), 131, 167, 177, 181n, 185, 185n, 191–92, 197, 207, 208

Fatio de Duillier, Nicholas, 10–11, 30, 327–28, 366–67, 367n
Feingold, Mordechai, 29, 270, 286–87
Finch, Anne, Viscountess of Conway, 45n, 309, 310n
Flamsteed, John, 10, 220, 221, 222, 222n, 244, 365, 383
Flaubert, Gustave, 120–21, 181n
flood, Ogygian, 289, 294–95
flood of fire (*diluvium ignis*), 17, 152–54, 155, 158–59
Fomenko, A. T., 251–52, 258, 262
Force, James T., 4–5
Four Horsemen of the Apocalypse, 21, 29, 144–48

Galileo Galilei, 8, 195, 330, 364–65
Gibbon, Edward
  on Arius, 80
  on Athanasius, 71, 82, 117–18
  Church of Egypt, 88
  comets and divine plan, 230–31n
  on Constantius, 115

Council of Nicaea, 73n, 74
on Diocletian, 86
on Newton, 115
"Paradoxical Questions" (Newton), 66
repopulation of world, 264–65
on Severus, 147
on Whiston, 227
Gilgamesh epic, 197, 203, 210, 288–89
Gjertsen, Derek, 380–81
Gleick, James, 6
global warming, 17–18, 18n, 22, 152, 152n, 154, 155, 376, 385, 388–89
God, love/worship of
through alchemy, 319
ancestor worship versus, 216
anti-Trinitarianism, 387–88
idolatry/religious relics, 215, 360
John of Patmos era, 26
Manuel, 372
Newton, 59, 60, 63, 139, 161, 162, 205, 372, 376, 381, 386–88
Noah, 204, 216
Numa, 345
Westfall, 211
God, proof of existence of, 6, 26, 31–32, 40, 43, 59–60, 382–83, 383n
God's divine plan
alchemy, 318–19
Book of Daniel, 35–36, 41
celestial bodies, 349–50
comets, 223, 224–27, 226–27n, 230–31n
earthquakes, 237, 249
gravity, 330, 354

Jerusalem, 180–81
mathematics, 352
Moses, 226–27n
Newton, 228, 228n, 238, 248, 381, 382–83, 384
prophetic books, 31–32, 35–36, 37, 41
Revelation, 32, 35–36, 37, 387
Goff, Matt, 32, 41
Goldish, Matt, 134–35
Gozo island, 288, 289, 294–95, 296, 297, 298, 299–300
gravity
and the Flood, 233–34
as God's divine plan, 330, 354
Hooke, 241
Newton, 193, 243, 259, 330, 350, 353–54, 372–73
Plato, 328
Pythagoras, 325, 328, 350
Great Apostasy
Athanasius, 85, 138, 139, 151, 156
Book of Daniel, 163–64, 166, 186, 187
commencement date, 163–65, 167–68, 387–88
Revelation, 139, 139n, 143, 148, 161, 162–63, 163–64, 166, 187, 387
Gregory, David, 249n, 329, 355, 356, 372
Grongnet, Giorgio, 297, 298
Grosso, Michael, 21, 22

Hadrian, 90
Halley, Edmund, 195, 237–40, 240n, 242–43, 244
on the Flood, 195, 233, 239, 243–44

Great Comet of 1680, 242
Halley's comet, 218–19, 219n, 230, 237, 242–43, 244
Principia (Newton), 2, 9, 242, 243, 245, 368, 370–71
theology, 237–39, 243, 244
Hartlib, Samuel, 309
Hawking, Stephen, 2, 28–29, 205
Heath, Thomas, 347
Herodotus, 258, 290–91, 300
Herzl, Theodor, 187–88, 189
Hesiod, 279, 280, 345, 375
Hipparchus, 276, 277, 278, 283, 287
historian, Newton as, 43, 257, 258, 263, 269–70, 276
Hitler, Adolf, 23, 180, 190
Hobbes, Thomas, 343, 365
Homer, 268–69, 278, 279, 281, 288–89, 315–16
Hooke, Robert, 10, 240–41, 329, 383
Huygens, Christiaan, 193–94, 328, 383, 383n
Hypatia, 91

idolatry/religious relics
Anthony (saint), 119, 124, 126–27
Athanasius, 85, 119, 124, 126–27
Constantine, 72, 78, 79, 108–9, 150
love/worship of God, 215, 360
Trinitarianism, 48, 59, 139, 161, 166–67, 182
Iliffe, Rob, 60–61, 110, 113, 151, 156–57, 160, 161, 359
Ingermanson, Randall, 26
Irenaeus, 33, 50, 69
"Irenicum" (Newton), 76, 205

Isacoff, Stuart, 357
Islam, 16, 64, 159–60, 172, 173, 176, 196, 197, 201, 202
Israel, state of, 174, 175–76, 187–89, 190, 192

Jason and the Argonauts, 272–74, 276, 283, 284, 285, 286
Jeffries, George, 368, 369–70, 371, 372, 374
Jerusalem
city of, 22, 28, 108, 109–10, 170, 173, 173n, 174, 175, 180–81, 181n
Temple Mount, 172–77
Temple of, 20, 107, 131, 132–34, 135, 136, 137–38, 140–41, 143, 154, 172, 186–87, 191
Jesus Christ
astronomy, 276
channeled texts, 62, 136–38, 143, 146, 148, 149, 155, 160
divine plan, 225, 226, 227
music of the spheres, 284
as nearly God, 60, 61–62
scholarship, 46–47
Second Coming, 15, 22, 27, 32, 162–63, 175, 182, 183, 188, 189
Jews
Chronology (Newton), 131, 256, 261, 263, 266, 267–68, 269
conversion of, 154, 163, 172, 178, 236
Diaspora, 177, 179, 181, 182, 187, 191, 207
global chronology, 267–69
Judaism, 64, 172, 180, 182, 187, 196, 201, 205

Messiah, 175, 181–82, 181–82n, 181n, 183
prisca sapientia, 267–68, 357
return of the Jews, 162–63, 171, 175, 177, 178–79, 180–81, 181n, 184, 189–90, 191–92, 236
Tabernacle, 130, 131–32, 133–34, 135, 143–44, 145, 150, 236
Torah, 21–22, 24, 25, 26, 32, 35, 76, 134, 140, 150, 180, 187, 208, 253, 261, 268
John (apostle), 19–20
John (bishop), 101–2, 104
John of Patmos, 18–19, 20, 21, 26
Johnson, Paul, 71, 95, 111, 198, 268–69
Josephus, 133–34, 197, 237, 326–27
Jovian (emperor), 118
Jowett, Benjamin, 302
Judgment Day, 122, 152–53, 162–63, 169, 170, 177, 179, 199, 220, 367n
Julian the Apostate, 74, 115, 118
Julian the Apostate (emperor of Rome), 173n
Jung, Carl, 21–22, 361–63
Jupiter (god), 191, 211, 216, 349
Jupiter (planet), 217, 243, 248–49, 249n
Justinian (emperor), 158, 159, 230–31n

Kepler, Johannes, 2, 8, 241, 243, 352, 363
Keynes, John Maynard, 12–14, 312
Kircher, Athanasius, 195, 266, 288
Klein, Naomi, 17–18, 376
Knoespel, Kenneth, 215, 217
Küng, Hans, 57–58, 88
Kurosawa, Akira, 67, 99, 101, 104, 105

Lawrence, D. H., 21, 325–26
Leibniz, Gottfried, 10, 329, 330, 383
Leon, Rabbi Jacob Judah, 131, 132, 132n
Lindsey, Hal, 175–76, 177
literacy, 5, 19–20, 27, 28, 127–28, 268
Locke, John, 44–45, 45n, 330–31
  alchemy, 309, 321–23
  Arianism, 60
  Blake on, 224
  corruptions of Scripture letters, 44–46, 48–51, 56–58, 63–64, 140
  and Newton, 11, 45, 365, 366, 375
  prophetic books of Bible, 28
  on Whiston, 235
London Mint, 10, 235, 245, 367, 375
Lovelock, James, 317
love our neighbors, 63, 161, 204, 372, 386
Lucasian Professor of Mathematics, 8, 9, 303
Luce, J. W., 293–94
Luther, Martin, 22, 28, 156–57, 220

Macarius, 83, 125
MacCulloch, Diarmaid, 64, 71, 72, 80, 95, 177–78
Machiela, Daniel A., 208, 209
Maclaurin, Colin, 355–56
Mahasseh ben Israel, Rabbi, 179, 180
Maltese archipelago, 288, 289, 296–99, 302
Manuel, Frank
  on alchemy and Newton, 317–18
  on Apocalypse and Newton, 169–70
  on the divine plan and Newton, 37, 381
  on Fatio and Newton, 367

on Flamsteed and Newton, 222, 222n
on Newton as historian, 257, 258, 263, 269–70
on Newton and love/worship of God, 372
Newton (father), 6
"Paradoxical Questions" (Newton), 76–77
prisca sapientia and Newton, 359
on prophetic hieroglyphs and Newton, 36
on Solomon's Temple, 130–31
on synchronistical necessity, 38, 41, 42
on Whiston versus Newton, 234–35
Marcellus, Marcus Claudius, 377–78
Marvell, Andrew, 172, 177
mathematics, 2, 9, 13, 37, 45, 185, 193, 350, 351, 352–53, 361, 380
McGuire, J. E., 330
Mede, Joseph, 18, 37–38, 41–43, 156, 167, 168, 184
Melitians, 86, 88, 89, 92, 94, 95–96, 97, 98, 101–2, 105–6, 110
Melitius, 69, 86–88, 88n, 89, 101
Merrill, James, 137
Mersenne, Marin, 354, 355
metaphysics, 75, 76
Methuselah, 209, 305
Mifsud, Anton, 298, 299–300
millenarianism, 27–28, 152, 154, 169, 177, 385
Milton, John, 37, 231–32, 309
More, Henry, 28, 30, 309, 321
Morozov, Nicolai Aleksandrovich, 251, 275n

Moses (biblical figure)
Bible code, 24, 153
corruption of mankind, 368, 370–71
God's divine plan, 226–27n
prisca sapientia, 267–68, 357–58, 359
prophecies, 184
Tabernacle, 130, 131, 133–34, 135
Ten Commandments, 48, 62, 63, 205
Torah, 24, 26
music/music of the spheres, 284–85, 327, 353–57, 388

Needham, Joseph, 311, 314, 320–21
Newman, William, 315–16, 317, 323–24
Newton, Humphrey, 314
Newton, Isaac (father), 5, 6, 46
Newton, Isaac, and biography, 1, 2, 5–11, 30, 155, 364–67, 371–72, 373–75
Newton Code, 18, 26
Newton Project, 4, 14, 16, 60, 359
Noah (biblical figure), 208–9, 209n
altar/eternal flame, 196, 202–4, 202n, 203n
Ark, 194–96, 199–202, 203
children and gods, 211, 213n, 216–17
commandments, 204–5
Flood, 3, 68, 194, 195, 196, 197–99, 198n, 210, 232–33, 233–34, 239, 243–44, 289
immortality, 210–11, 210n, 217, 305
Janus (god), 211, 214, 217
love/worship of God, 204, 216
polytheism, 211, 214, 216
prisca sapientia, 203, 267, 326

prytaneum, religion of, 77, 202, 204–5, 214, 215
"remnant," 171, 194, 204, 208–10, 264, 359–60, 361
repopulation of world, 171, 195–97, 210, 211, 214, 263–65
Nostradamus, 192, 299, 299n
Numa Pompilius, 342–45

Observations . . . (Newton), 18, 23–24, 126, 132–33, 148, 169
"Of Motion" (Newton), 241
Ogygia/Gozo island, 288–89, 296, 297, 298, 299–300
Ogygia island, 289, 294–95
On the Church (Newton), 61–62
Oppenheimer, Robert, 384
optical research, 221, 240, 318, 323–24, 328, 373
Optics (Newton), 2–3, 330, 381–82
Optics/Opticks (Newton), 228n
Origen Adamantius, 50, 100, 100n, 154, 326–27
Oughted, William, 178
ὁμοούσιος ("consubstantial," "of one substance"), 75–77, 398–401

Pagels, Elaine, 20, 128, 129, 139n
Paine, Harrington Spear, 20
Palestine/Palestinians, 12, 28, 34, 87–88, 174–75, 187, 188, 189–90, 191–92
papacy, 22–23, 28, 52–53, 54, 154, 163, 164, 165–67, 187, 219, 219n, 262, 308, 369
"Paradoxical Questions concerning the Morals and Actions of Athanasius and His Followers" (Newton), 12, 63, 65–70, 84, 148, 149–50
  Anthony (saint), 70, 120, 121–28, 129
  Arius's mythical death, 77, 79–81, 83, 113, 114, 119
  celibacy, 123, 125
  corruptions of Scripture, 69, 76–77
  Council of Tyre, 97–106, 109–10, 111–12, 113, 114, 119, 258
  forged letters by Athanasius, 93, 105–6, 111–12, 113–14, 119
  Gibbon on, 115
  idolatry/religious relics, 124, 126–27
  Manuel on, 76–77
  rituals, 125–26
  severed hand myth, 85, 101–2, 104, 111, 114
  supernatural experiences, 122, 124–26
  Westfall on, 66
Parliament, 179–80, 245–46, 307–8, 371–72, 374
Paul (monk), 125
Pauli, Wolfgang, 184, 350, 361–63
Pepys, Samuel, 309, 365–66, 375
Pericles, 336–37, 338–42, 341n
Peter (apostle), 72, 153–54
Peter of Alexandria, 86–88, 88n, 148
Philolaus, 345, 346–48, 349, 352
Philostorgius, 101
Phineus, 31n
physics, 2, 9, 13, 37, 45, 193, 230, 234, 349, 358, 361, 362–63
Pierce, Larry, 257, 292

Plato, 337, 341n
  Atlantis, 290, 292, 293, 294, 295–96, 298, 299, 302
  cycles concept, 205
  Eudoxus on, 280
  gravity, 328
  Newton on, 8
  Philolaus, 346, 347
  science and corruption, 378–79
  writings, 205, 205 292, 290, 292, 299, 302
Plutarch
  altar/eternal flame, 202n
  Archimedes, 378, 378n, 379, 380
  Aspasia, 339
  Cleanthes, 335n
  Marcus Claudius Marcellus, 377
  Numa, 343, 344–45
  Pericles, 340–41, 341n
  Pythagoras, 350
  Solon, 290
Pope, Alexander, 1
Principe, Lawrence, 315–16, 317, 322
Principia Mathematica (Newton), 1–3, 7, 9–10, 44, 45, 71, 136, 366
  alchemy, 319, 363
  atheism, 381
  comets, 222–23, 224–25, 242, 244–45, 246–49
  gravity, 193, 243
  guilt over writing, 370–71, 372, 373, 374, 375, 381
  Halley, 2, 9, 242, 243, 245, 368, 370–71
  prisca sapientia, 328, 329
  proof of existence of God in, 59–60, 382–83

prisca sapientia, 326, 326n, 336
  alchemy, 316–17
  Anaxagoras, 337–41
  ancient science, 358n, 359
  Aristarchus, 330, 331–35
  consciousness, state of, 331, 344
  corruption of mankind, 359–60, 361
  Jews, 267–68
  Moses, 267–68, 357–58, 359
  Newton on, 327–28, 329–31
  Noah, 203, 267, 326
  Numa, 342–45
  Pauli, 362–63
  Philolaus, 345
  Pythagoras, 345, 352, 356, 357, 358–59, 363
prophetic books of Bible, 42–43
prophetic hieroglyphs, 31, 321, 390–91
  astronomy, 317
  Bible, 18, 23, 25, 28–38
  central fire, 349
  Charlemagne, 166–67
  Constantine, 76
  Daniel, Book of, 33–37, 144, 163–64
  Great Comet of 1680, 222
  law of parsimony, 259
  Revelation, 33–37, 143–44, 146, 148, 149–50, 155–61
Protestantism, 22, 28, 156–57, 328, 367, 369
prytaneum, religion of, 77, 202, 204–5, 214, 215–16, 344, 388
Pythagoras/Pythagoreans, 350–52
  Brahmins, 284
  celestial bodies, 348, 354

central fire, 347, 348–49
corruption of mankind, 368, 370–71
Druids, 284
Ezekiel (prophet), 185, 185n
global chronology, 267
gravity, 325, 328, 350, 354
mathematics, 185, 350, 351, 352, 380
music/music of the spheres, 284–85, 327, 353, 355, 388
Newton on, 349, 363, 380
orbital velocities, 354
Philolaus, 346, 347
prisca sapientia, 345, 352, 356, 357, 358–59
silence teachings, 343
Theano, 351, 351n

"Queries Regarding the Word ὁμοούσιος" (Newton), 75–76, 398–401

Rattansi, Piyo, 330, 360–61
Reé, Jonathan, 270
"remnant," 171, 194, 204, 207, 208–10, 233, 264, 296, 359–60, 361
Rips, Eliyahu, 24–25
Royal Society of London, 9, 10, 132n, 233, 236, 239, 240n, 241–42, 243, 245, 246
Rudbeck, Olas, 195–96

Sachar, Abram, 130
Saturn (god), 194, 197, 198n, 211, 212–13, 214, 216, 217, 295–96, 300, 315

Saturn (planet), 193–94, 217, 243, 328, 331
science
ancient science, 358–61, 358n, 410–11
corruption and, 378–79
technological innovations, 12, 24, 215, 255, 273, 300, 372, 375–79, 385–86
Servetus, Miguel, 55–56
Severus, Septimus (emperor), 28, 145, 146, 147
"Short Schem of the True Religion, A" (Newton), 382
Sittenreich, Steven, 185–86n
Snobelen, Stephen, 9, 15–16, 48, 50–51, 58, 167, 169, 181, 192
Socrates, 43, 127, 337, 339, 341n, 342
Solomon (biblical figure), 272, 287
Temple of, 131–33, 134–36, 148, 166, 172–73, 236, 261, 263
Solon, 290–91, 292
Atlantis, 290, 291–92, 293–94, 302
global chronology, 267, 293
Sozomen, 43, 70, 108, 109, 116, 127
Starkey, George, 318–19
Stukeley, William, 7n, 261
Sturm, Johann Christoph, 365
surrealist images, 20–21, 33, 68, 80, 85, 99, 101, 102–3, 104, 147
Swift, Jonathan, 246, 247–48
"Synchronisms . . ." (Newton), 153
System of the World, The (Newton), 329, 410–11

Tabernacle, 130, 131–32, 133–34, 135, 143–44, 145, 150, 236
Tertullian, 28, 33, 50, 69

Theano, 351, 351n
Theodoret, 43, 99–101, 103–4, 110
Theodosius (emperor), 119, 157
"Theologiae Gentiles Origines Philosophicae" (Newton), 216, 263
theological writings of Newton ("History of the corruption of the Soul of Man"), 3–5, 11–13, 12n, 14, 18, 66, 312, 368
Thompson, William Irwin, 317, 364
Thucydides, 275, 340
Timotheus, 100–101
Titans, 47, 193–94, 350
Tolstoy, Leo, 48n, 374
Torah, 21–22, 24, 25, 26, 32, 35, 76, 134, 140, 150, 180, 187, 208, 253, 261, 268
Trinitarianism
  Apocalypse in 2060, 169, 387–88
  Athanasius, 22, 55, 60–61, 63, 65, 69, 70–71, 75, 83, 94, 117–18, 138, 156
  corruptions of Scripture, 58
  councils, 70–71, 74–75, 118–19, 139
  Cromwell, 47–48
  idolatry/religious relics, 48, 59, 139, 161, 166–67, 182
Trump, Donald, 17, 20
"Two Incomplete Treatises on Prophecy" (Newton), 76

"Untitled Treatise on Revelation" (Newton), 138n, 156n, 157, 402

vegetative spirit, 304–5, 307, 324
Velikovsky, Immanuel, 226, 226–27n, 234n

Vidal, Gore, 115
Vilenkin, Alex, 352–53
Vincenzo Galilei, 354–55
Voltaire, 2, 13, 31, 56, 58, 90, 240, 261–62, 266

warfare, 149, 159, 175, 271, 372, 376, 377–79
Westfall, Richard
  on the character of Newton, 371
  Chronology (Newton), 253
  corruptions of Scripture and Newton, 49, 56
  illness and Newton, 365
  London Mint and Newton, 367
  monasticism and Newton, 122–23
  on Newton as historian, 43
  "Paradoxical Questions" (Newton), 66
  patristic literature and Newton, 69
  polytheism and Noah, 211
  Principia (Newton), 242
  prophecy and Newton, 33
  "remnant" and Newton, 207
  supernatural experiences and Newton, 122
  theological writings of Newton, 14
Whiston, William, 227–29, 228, 229, 237
  Arianism, 236
  day-equals-year formula, 185
  divine plan, 228, 229, 230–31, 230–31n, 238
  earthquakes, 237, 249
  Great Comet of 1680, 230, 230–31n, 232, 233, 234n, 235
  Jews, 236

on Newton, 6, 29, 192, 224,
227–29, 235, 238, 379
Revelation, 30–31
as scientist-theologian, 60, 131,
185, 192, 229–30, 229–31, 230n,
232–37, 359
Tabernacle, 131, 236
Temples, 131, 236
writings of, 185, 229–30, 232,
234–35, 236, 237
White, Michael, 324

Whitehead, Alfred North, 360
Wickins, William, 379
Willet, Andrew, 265–66
Wren, Christopher, 240–41

Yahuda, Abraham, 12, 13–14, 180,
256

Zeus (god), 31, 31n, 211, 212, 278,
279, 280, 282, 333–34, 335, 345,
348, 349